Functional Distribution of Anomalous and Nonergodic Diffusion

From Stochastic Processes to PDEs

Functional Distribution of Anomalous and Nonergodic Diffusion

From Stochastic Processes to PDEs

Weihua Deng

Lanzhou University, China

Xudong Wang

Nanjing University of Science and Technology, China

Daxin Nie

Lanzhou University, China

World Scientific

NEW JERSEY · LONDON · SINGAPORE · BEIJING · SHANGHAI · HONG KONG · TAIPEI · CHENNAI · TOKYO

Published by

World Scientific Publishing Co. Pte. Ltd.

5 Toh Tuck Link, Singapore 596224

USA office: 27 Warren Street, Suite 401-402, Hackensack, NJ 07601

UK office: 57 Shelton Street, Covent Garden, London WC2H 9HE

Library of Congress Cataloging-in-Publication Data

Names: Deng, Weihua, author. | Wang, Xudong, author. | Nie, Daxin, 1993–   author.

Title: Functional distribution of anomalous and nonergodic diffusion :
   from stochastic processes to PDEs / Weihua Deng, Lanzhou University, China,
   Xudong Wang, Nanjing University of Science and Technology, China,
   Daxin Nie, Lanzhou University, China.

Description: Hackensack, NJ : World Scientific, [2022] |
   Includes bibliographical references and index.

Identifiers: LCCN 2022007621 | ISBN 9789811250491 (hardcover) |
   ISBN 9789811250507 (ebook for institutions) | ISBN 9789811250514 (ebook for individuals)

Subjects: LCSH: Theory of distributions (Functional analysis)

Classification: LCC QC20.7.T45 D46 2022 | DDC 515/.782--dc23/eng20220503

LC record available at https://lccn.loc.gov/2022007621

British Library Cataloguing-in-Publication Data

A catalogue record for this book is available from the British Library.

For any available supplementary material, please visit
https://www.worldscientific.com/worldscibooks/10.1142/12673#t=suppl

Preface

Diffusion processes are the most fundamental transport processes in the natural world. The archetypal model of diffusion processes is Brownian motion, which is named after Robert Brown, who observed the erratic trajectories of pollen particles suspended in liquid in 1827. Starting from the pioneering work in 1905 by Einstein and Smoluchowski, mathematical treatment carried out by Wiener in 1923, and stochastic calculus developed by Itô in the 1940s, in several decades, Brownian motion plays a supreme role in the kingdom of diffusion processes, and it is even called "normal" diffusion.

With the rapid development of science and technology, nowadays, it is widely acknowledged that non-Brownian motion, commonly perceived as "anomalous" diffusion, is in effect ubiquitous; in other words, "anomalous" should be normal. The cage effect, a particle surrounded by neighboring granular particles, leads to slow dynamics, called subdiffusion. It is observed in glass-forming systems, in motion of lipids on membranes, translocation of polymers, solute transport in porous media, etc. Many cellular processes driven by molecular motor show fast dynamics, called superdiffusion, which are also observed in turbulent flow, optical materials, motion of predators, human travel, etc. They have extensive applications, e.g., it is important to understand human travel, since in some cases human travel is responsible for the spread of infectious diseases, foraging hypothesis predicts that predators adopt Lévy flight strategies where prey is sparse and Brownian motion for abundant prey. The potential applications have greatly pushed forward the developments of anomalous and nonergodic diffusions with more delicate research, e.g., the strong anomalous diffusion, which have multiple modes.

Statistical observables play a central role in studying stochastic processes; they are not only useful physical quantifies with important applications but can help uncover the dynamical mechanisms of the stochastic processes. The statistical observables include first passage time, escape probability, functional, and even the position of the particle, etc., which are usually random variables. It is well known that the probability density function (PDF) of the position of Brownian motion is governed by the classical diffusion equation. All the messages of a random variable are in its PDF. One of the effective ways to get the PDF of a statistical observable is to first derive its governing equation and then find the solution of the equation.

The functionals defined by path integral are a large family of statistical observables, appearing in a wide range of problems across different fields ranging from probability theory, finance, data analysis, disordered systems, etc. In probability theory, the occupation time, i.e., the time spent by a stochastic process in a particular region within a time window of size t, is an important functional of interest; in finance, the integrated stock price up to some target time t is a functional object, the interesting analogy of which is also in a disordered system, where a single overdamped particle moves in a random potential; when describing the stochastic behavior of daily temperature records, the heating degree days measuring the integrated excess temperature up to time t is a functional quantity of interest; the total area under the trajectory of a stochastic process is a functional first studied in the context of economics and later extensively by probabilists and physicists. Here we just name a few examples.

In 1940s, Kac realized that the statistical properties of the Wiener functional can be studied by cleverly using the path integral method devised by Feynman to derive Schrödinger's equation in his unpublished PhD thesis at Princeton, and he presented a unified approach to calculate the distribution function of the Wiener functional. The celebrated Feynman-Kac formula is based on the assumption that the diffusion is Brownian motion. For anomalous cases, the whole machinery of treating Brownian functionals should be modified. The popular microscopic models (stochastic process) to describe anomalous diffusion are continuous time random walks (CTRW) and Langevin picture. Ten years ago, under the CTRW framework, Barkai's group derived the backward and forward Feynman-Kac equations with fractional substantial derivative for the distribution of functionals of the path of a particle undergoing anomalous diffusion; later, Deng's group got the backward and forward fractional Feynman-Kac equations describing the distribution of the functionals of the space- and time-tempered anomalous diffusion, and the governing equations for the functional distributions of aging anomalous diffusion and diffusion with reaction are also obtained. Sometimes the stochastic processes in physics and chemistry are naturally described by Langevin equations. The Langevin picture has the advantage of studying the dynamics with an external force field and analyzing the effect of noise resulting from a fluctuating environment. In the very recent time, Deng's group derived the equations governing the PDFs of the functionals of paths of the Langevin system with both space- and time-dependent force fields and arbitrary multiplicative noise, and the backward version is proposed for a system with arbitrary additive noise or multiplicative Gaussian white noise together with a force field.

The goal of this monograph is to present a pedagogical review of the functional distribution of anomalous and nonergodic diffusion and its numerical simulations, starting from the studied stochastic processes to the deterministic partial differential equations governing the PDF of the functionals. First, we do some preparations for introducing basic knowledge of probability theory and stochastic processes, and introduce the microscopic models characterizing the anomalous and nonergodic diffusion. Then we focus on the central part of the monograph: functional distribution of anomalous and nonergodic diffusion. The importance of the functional, as a statistical observable, is carefully illustrated. We derive in detail the governing equations for the PDFs of the functionals, including the backward and forward versions, the applications of which are also presented.

Once the models have been built, it is natural to turn to their mathematical research. This is not only required in extending the applications of the models, but expected to enrich the knowledge in mathematical field. We discuss the regularity of the backward and forward Feynman-Kac equations, design numerical schemes for the equations, and perform detailed numerical analyses. The main challenge of these works comes from the time-space coupled fractional substantial derivatives.

Hopefully, this monograph can systematically introduce the importance of the functionals in real applications, their mathematical models, the procedure of building the models, and mathematical properties of the models. We do appreciate all the researchers who have ever made contributions to the promising topic, especially our group members for their invaluable scientific contributions to the contents of this book, and their help to the improvement of the presentation.

Contents

Chapter 1

Probability Theory

In the beginning of this book, we will introduce some fundamental concepts of probability theory which is the basis of describing the anomalous and nonergodic processes in the subsequent chapters. We provide the background knowledge and some essential results used later. For more thorough discussions on probability theory, the readers can refer to the books, e.g., [Feller (1971); Øksendal (2005)].

1.1 Random Variables and Probability Distributions

In probability and statistics, a random variable is a variable which takes the value of outcomes of a random phenomenon. For example, when you toss a coin, its outcome is a random variable, which could be head or tail, being also the only two possible values. What will be the outcome is not certain. The two outcomes are named as the events which constitute the sample space $\Omega = \{$head, tail$\}$. Since one of the outcomes must occur, either the event that the coin lands head or tail must have non-zero probability.

The formal mathematical treatment of random variables is a topic in probability theory. In that context, a random variable is understood as a function defined on a sample space whose outputs are numerical values. A random variable $X \colon \Omega \to E$ is a measurable function from a set of possible outcomes Ω to a measurable space E. The technical axiomatic definition requires Ω to be a probability space. Usually X is real-valued (i.e., $E = \mathbb{R}$). When the image of X is finite or countably infinite, it is called a discrete random variable and its distribution is described by a probability mass function which assigns a probability to each value in the image of X. If the image is uncountably infinite, then X is called a continuous random variable, whose distribution can be described by a probability density function (PDF), which assigns probabilities to intervals.

For the example of tossing a coin above, the final outcome is a discrete random variable. We can introduce a real-valued random variable X to model a one dollar payoff for a successful bet on head as:

$$X(\omega) = \begin{cases} 1, \text{ if } \omega = \text{head}, \\ 0, \text{ if } \omega = \text{tail}. \end{cases} \tag{1.1}$$

On the other hand, the coin is assumed to be a fair one. Then the random variable X has a probability mass function f_X given by:

$$f_X(x) = \begin{cases} \frac{1}{2}, \text{ if } x = 1, \\ \frac{1}{2}, \text{ if } x = 0. \end{cases} \tag{1.2}$$

More generally, let the function $P(A)$ be the probability of A and Ω the sample space. Then $P(A)$ satisfies the probability axioms:

(1) $P(A) \geq 0$ for all A;

(2) $P(\Omega) = 1$;

(3) if A_i ($i = 1, 2, 3, \cdots$) is a countable (but possibly infinite) collection of nonoverlapping sets, i.e.,

$$A_i \cap A_j = \varnothing \quad \text{for all} \quad i \neq j, \tag{1.3}$$

then

$$P(\underset{i}{\cup} A) = \sum_i P(A_i). \tag{1.4}$$

Based on these axioms, we also have the two results:

(1) if A^c is the complement of A, then
$$P(A^c) = 1 - P(A); \tag{1.5}$$

(2) $P(\varnothing) = 0$.

1.2   Joint and Conditional Probabilities

1.2.1   *Joint Probability*

In the study of probability, if we are considering two sets of events (e.g., A and B) where $A \cap B$ is nonempty, then the joint probability that the event ω is contained in both A and B is

$$P(A \cap B) = P\{(\omega \in A) \text{ and } (\omega \in B)\}. \tag{1.6}$$

Similarly, for more sets A_i, the joint probability is

$$P\left(\underset{i}{\cap} A\right) = P\{(\omega \in A_1) \text{ and } (\omega \in A_2) \text{ and} \cdots\}. \tag{1.7}$$

1.2.2 *Conditional Probability*

Another important quantity associated with two random variables is conditional probability. For the sets A and B, the conditional probability $P(A|B)$ denotes that $\omega \in A$ (given that we know $\omega \in B$) is given by dividing the probability of joint occurrence by the probability of ($\omega \in B$), i.e.,

$$P(A|B) = P(A \cap B)/P(B). \tag{1.8}$$

In fact, the joint probability and conditional probability are closely related as

$$P(A \cap B) = P(A|B)P(B) = P(B|A)P(A). \tag{1.9}$$

More generally, if we have a collection of sets B_i satisfying

$$B_i \cap B_j = \varnothing \tag{1.10}$$

and

$$\bigcup_i B_i = \Omega, \tag{1.11}$$

then the sets B_i divide the sample space Ω into nonoverlapping subsets. So for any $A \subset \Omega$,

$$\bigcup_i (A \cap B_i) = A \cap (\bigcup_i B_i) = A \cap \Omega = A. \tag{1.12}$$

By using the probability axiom Eq. (1.4), it holds that

$$\sum_i P(A \cap B_i) = P[\bigcup_i (A \cap B_i)] = P(A), \tag{1.13}$$

and thus

$$\sum_i P(A|B_i)P(B_i) = P(A). \tag{1.14}$$

Equations (1.13) and (1.14) imply that summing over all mutually exclusive possibilities of B_i in the joint probability yields a marginal distribution.

1.2.3 *Independence*

Besides the joint and conditional probabilities, one might be also interested in the relation between the two or more random variables, e.g., independence. In a probabilistic way, two sets of events A and B are independent if the specification that a particular event contained in set B has no influence on the probability of that event belonging to set A. Thus, the conditional probability $P(A|B)$ in Eq. (1.8) should be independent of B, i.e.,

$$P(A \cap B) = P(A)P(B). \tag{1.15}$$

Similarly, several events of sets A_i ($i = 1, 2, \cdots, n$) are considered to be independent mutually if

$$P(\bigcap_i A_i) = \prod_i P(A_i). \tag{1.16}$$

1.3   Moments

The common quantities of interest about random variables are given by the moments $\langle X^n \rangle$ $(n = 1, 2, \cdots)$, where the brackets $\langle \cdot \rangle$ denote the ensemble average, i.e., the statistical average over stochastic realizations. In practice, the most important moments are the first and second ones. Now we consider a single random variable X. If X is a discrete random variable, then denote its probability mass function as $P_X(x)$. But if it is a continuous random variable, denote its PDF as $f_X(x)$. Its nth moment is defined as

$$\langle X^n \rangle = \sum_j x_j^n P_X(x_j) \tag{1.17}$$

for discrete random variable X, or

$$\langle X^n \rangle = \int_{-\infty}^{\infty} x^n f_X(x) dx \tag{1.18}$$

for continuous random variable X. Let c be a constant. The first moment satisfies

$$\langle cX \rangle = c\langle X \rangle, \qquad \langle X_1 + X_2 \rangle = \langle X_1 \rangle + \langle X_2 \rangle. \tag{1.19}$$

When $\langle X \rangle$ vanishes, X is said to be a centred random variable [McConnell (1980)].

On the other hand, the nth moment of the random variable X can be obtained from its characteristic function. More precisely, the characteristic function of X is defined as

$$C_X(k) = \langle e^{ikX} \rangle = \int_{-\infty}^{\infty} e^{ikx} f_X(x) dx, \tag{1.20}$$

which yields

$$M_n = \langle X^n \rangle = \frac{1}{i^n} \frac{d^n C_X(k)}{dk^n} \bigg|_{k=0} . \tag{1.21}$$

Conversely, the Taylor expansion of the characteristic function can be expressed as the combination of moments, i.e.,

$$C_X(k) = 1 + \sum_{n=1}^{\infty} (ik)^n M_n / n!, \tag{1.22}$$

which implies that we have the characteristic function if we know all the moments. Since the characteristic function $C_X(k)$ is the Fourier transform of the PDF $f_X(x)$, $f_X(x)$ is the inverse Fourier transform of the characteristic function, i.e.,

$$f_X(x) = \frac{1}{2\pi} \int_{-\infty}^{\infty} C_X(k) e^{-ikx} dk. \tag{1.23}$$

For any function $g(X)$ of a random variable X, its average is

$$\langle g(X) \rangle = \sum_j g(x_j) P_X(x_j) \tag{1.24}$$

for discrete X, or

$$\langle g(X) \rangle = \int_{-\infty}^{\infty} g(x) f_X(x) dx \tag{1.25}$$

for continuous X. But if we are interested in the distribution of the new random variable $Y = g(X)$, we can utilize the technique of δ-function. Take the continuous random variable X as an example. The PDF $f_Y(y)$ of the random variable Y is given by

$$\begin{aligned} f_Y(y) &= \langle \delta(Y - y) \rangle = \langle \delta(g(X) - y) \rangle \\ &= \int_{-\infty}^{\infty} \delta(y - g(x)) f_X(x) dx. \end{aligned} \tag{1.26}$$

If $g_n^{-1}(y)$ is the nth simple root of $g(x) - y = 0$, then the last integral can be evaluated as

$$f_Y(y) = \sum_n f_X(g_n^{-1}(y)) \left[\left| \frac{dg(x)}{dx} \right| \right]^{-1} \Bigg|_{x = g_n^{-1}(y)}. \tag{1.27}$$

The variance $\mathrm{Var}(X)$ of random variable X is defined as

$$\mathrm{Var}(X) = \langle (X - \langle X \rangle)^2 \rangle = \langle X^2 \rangle - \langle X \rangle^2. \tag{1.28}$$

The positive square root of the variance is named as standard deviation. The variance measures the spread or dispersion of the values about the mean. In other words, it is the fluctuation of random variable X. The covariance $\mathrm{Cov}(X, Y)$ of any two random variables X and Y is defined as

$$\mathrm{Cov}(X, Y) = \langle (X - \langle X \rangle)(Y - \langle Y \rangle) \rangle = \langle XY \rangle - \langle X \rangle \langle Y \rangle. \tag{1.29}$$

The covariance $\mathrm{Cov}(X, Y)$ measures how the two random variables X and Y are interrelated. If X and Y are uncorrelated, then

$$\mathrm{Cov}(X, Y) = 0. \tag{1.30}$$

The correlation coefficient, defined as

$$\rho = \frac{\mathrm{Cov}(X, Y)}{\sqrt{\mathrm{Var}(X) \mathrm{Var}(Y)}}, \tag{1.31}$$

is used to measure the correlation between two random variables X and Y.

1.4   Gaussian and Lévy Distributions

1.4.1   *Gaussian Distribution*

One of the most important PDFs is the Gaussian distribution or normal distribution:

$$f_X(x) = \frac{1}{\sqrt{2\pi\sigma^2}} e^{-\frac{(x-\mu)^2}{2\sigma^2}}, \tag{1.32}$$

where $\langle X \rangle = \mu$ and $\text{Var}(X) = \sigma^2$. The Gaussian distribution $f_X(x)$ has the properties that $f_X(x) \to 0$ as $x \to \pm\infty$ and $f_X(x)$ reaches the maximum at $x = \mu$. The distribution in Eq. (1.32) is usually denoted as $N(\mu, \sigma^2)$. The linear transform cX of random variable X is also Gaussian with the distribution $N(c\mu, c^2\sigma^2)$. Likewise, the distribution of

$$Y = \frac{X - \mu}{\sigma} \sim N(0, 1) \tag{1.33}$$

is called the standard normal distribution.

Now we consider n dimensional Gaussian distribution. Let $(X_1, \cdots, X_n)$ be n random variables (not necessarily independent) with means equal to zero. Let $f(x_1, \cdots, x_n)$ be their joint PDF. Then $\mathbf{x} = (X_1, \cdots, X_n)^T$ is normally distributed in n dimensions if

$$f(\mathbf{x}) = \frac{1}{(2\pi)^{n/2}\sqrt{\det\mathbf{M}}} e^{-\mathbf{x}^T\mathbf{M}^{-1}\mathbf{x}/2}, \tag{1.34}$$

where $\mathbf{M}$ is the covariance matrix with the elements

$$[\mathbf{M}]_{ij} = \langle X_i X_j \rangle, \quad i, j = 1, 2, \cdots, n, \tag{1.35}$$

$\mathbf{M}^{-1}$ is the inverse matrix of $\mathbf{M}$, and $\det\mathbf{M}$ is the determinant of matrix $\mathbf{M}$. The marginal distributions $f(x_{i_1}, x_{i_2}, \cdots, x_{i_r})$ for the multidimensional normal distribution can be determined from

$$f(x_{i_1}, \cdots, x_{i_r}) = \int_{-\infty}^{\infty} \cdots \int_{-\infty}^{\infty} f(x_{i_1}, \cdots, x_{i_r}, x_{i_{r+1}}, \cdots, x_{i_n}) dx_{i_{r+1}} \cdots dx_{i_n}. \tag{1.36}$$

If the X_i are independent with each other, then $\langle X_i X_j \rangle = 0$ $(i \neq j)$, $\mathbf{M}$ becomes a diagonal matrix, and

$$f(x_1, \cdots, x_n) = f(x_1)f(x_2) \cdots f(x_n), \tag{1.37}$$

where each $f(x_i)$ is a one-dimensional normal distribution. We have considered the simple case with zero means. If the means of random variable $\mathbf{x}$ is $\mathbf{a}$, then the multidimensional normal distribution is

$$f(\mathbf{x}) = \frac{1}{(2\pi)^{n/2}\sqrt{\det\mathbf{R}}} e^{-(\mathbf{x}-\mathbf{a})^T\mathbf{R}^{-1}(\mathbf{x}-\mathbf{a})/2}, \tag{1.38}$$

where $\mathbf{R}$ is the covariance matrix with elements $\langle X_i X_j \rangle - \langle X_i \rangle \langle X_j \rangle$. The characteristic function of $\mathbf{x}$ with its PDF in Eq. (1.38) is

$$C_{\mathbf{x}}(\mathbf{k}) = \langle e^{i\mathbf{k} \cdot \mathbf{x}} \rangle = e^{i\mathbf{k} \cdot \mathbf{a} - \frac{1}{2}\mathbf{k}^T \mathbf{R}\mathbf{k}}. \tag{1.39}$$

The multidimensional normal distribution is the inverse Fourier transform of the characteristic function Eq. (1.39), i.e.,

$$f(\mathbf{x}) = \frac{1}{(2\pi)^n} \int_{-\infty}^{\infty} e^{-i\mathbf{k} \cdot \mathbf{x}} C_{\mathbf{x}}(\mathbf{k}) d\mathbf{k}. \tag{1.40}$$

1.4.2 *Lévy Distribution*

We begin this section from Lévy stable law. For a positive constant c, we know that the laws of X and cX belong to the same type. The type of law of X is stable means that the laws of X and its addition of independent copies belong to the same type. In detail, if X_1 and X_2 are independent copies of X and c_1, c_2 are positive constants, then there exists a copy X_3 of X and a constant c such that

$$c_1 X_1 + c_2 X_2 \overset{d}{=} cX_3, \tag{1.41}$$

where the symbol $\overset{d}{=}$ represents the equivalence in distribution. We say that the law of X is quasi-stable, if

$$c_1 X_1 + c_2 X_2 \overset{d}{=} cX_3 + d. \tag{1.42}$$

Then we turn to the characteristic function of random variable X obeying the stable law. Denote the characteristic function of X as $C(k)$. Thus, the characteristic functions of $c_1 X_1 + c_2 X_2$ and cX_3 are the product $C(c_1 k)C(c_2 k)$ and $C(ck)$, respectively. Therefore, Eq. (1.41) is equivalent to

$$C(c_1 k)C(c_2 k) = C(ck). \tag{1.43}$$

If the characteristic function has the form

$$C(k) = e^{-\varphi(k)} \tag{1.44}$$

and the real part of $\varphi(k)$ is positive, then the exponent $\varphi(k)$ satisfies

$$\varphi(c_1 k) + \varphi(c_2 k) = \varphi(ck), \tag{1.45}$$

the solution of which is

$$\varphi(k) = c|k|^{\alpha} \tag{1.46}$$

with the real part of c positive.

The classical examples are the Gaussian (normal) law ($\alpha = 2$), the Cauchy law ($\alpha = 1$), and the Lévy-Smirnov law ($\alpha = 1/2$) describing the time when Brownian motion is stopped at a given level [Klafter and Sokolov (2011)]. In these cases the law has an explicit density, while sometimes the law only has the explicit characteristic function. We first consider a Poisson variable X with parameter a. Then

$$\langle e^{ikX} \rangle = \exp\left(a(e^{ik} - 1)\right). \tag{1.47}$$

Therefore, the exponent of the characteristic function of yX is

$$\varphi(k) = a(1 - e^{iky}). \tag{1.48}$$

Similarly, the exponent of the characteristic function of $y(X - \langle X \rangle)$ is

$$\varphi(k) = a(1 - e^{iky} + iky). \tag{1.49}$$

For Eq. (1.48), we consider the function with the form

$$\varphi(k) = \int_0^\infty (1 - e^{iky})\nu(dy), \tag{1.50}$$

where ν is any locally bounded measure on $(0, \infty)$, and both integrals

$$\int_0^1 y\nu(dy) \quad \text{and} \quad \int_1^\infty \nu(dy) \tag{1.51}$$

are finite. Now we choose

$$\nu(dy) = \frac{dy}{y^{1+\alpha}} \quad (0 < \alpha < 1), \tag{1.52}$$

and thus have

$$\varphi(ik) = \int_0^\infty (1 - e^{-ky})\frac{dy}{y^{1+\alpha}} = ck^\alpha \tag{1.53}$$

for some constant $c > 0$ and all $k > 0$. In other words, we obtain the formula

$$\langle e^{-kX} \rangle = e^{-ck^\alpha} \quad (0 < \alpha < 1) \tag{1.54}$$

for the positive stable variable X. Based on Eq. (1.49), similarly, we consider the function with the form

$$\varphi(k) = \int_0^\infty (1 - e^{iky} + iky)\nu(dy), \tag{1.55}$$

and the two integrals

$$\int_0^1 y^2\nu(dy) \quad \text{and} \quad \int_1^\infty y\nu(dy) \tag{1.56}$$

are finite. Choosing

$$\nu(dy) = \frac{dy}{y^{1+\alpha}} \quad (1 < \alpha < 2), \tag{1.57}$$

yields

$$\varphi(ik) = \int_0^\infty (1 - e^{-ky} - ky)\frac{dy}{y^{1+\alpha}} = -ck^\alpha \tag{1.58}$$

for some constant $c > 0$ and all $k > 0$. That is to say, the stable variable X satisfies

$$\langle e^{-kX} \rangle = e^{ck^\alpha} \quad (1 < \alpha < 2), \tag{1.59}$$

and is not positive any more.

The general Lévy distribution $L_{\alpha,\beta,\mu,\sigma}(x)$ is defined with four parameters. Its characteristic function has the form of

$$C_{\alpha,\beta,\mu,\sigma}(k) = \exp(-\sigma^\alpha|k|^\alpha(1 - i\beta\omega(k,\alpha)\text{sign}(k)) + i\mu k) \tag{1.60}$$

with

$$\omega(k,\alpha) \simeq \begin{cases} \tan\frac{\pi\alpha}{2} & \text{for } \alpha \neq 1 \\ -\frac{2}{\pi}\ln|k| & \text{for } \alpha = 1. \end{cases} \tag{1.61}$$

The parameters are the Lévy index $\alpha \in [0,2]$, the skewness (asymmetry) parameter $\beta \in [-1,1]$, the scale parameter (width of the distribution) $\sigma > 0$, and the shift parameter μ. It is stable if μ vanishes. Especially, the parameters μ and σ can be removed by considering the normalized variable $Y = \frac{X-\mu}{\sigma}$. Then

$$L_{\alpha,\beta,\mu,\sigma}(x) = L_{\alpha,\beta,0,1}\left(\frac{x-\mu}{\sigma}\right) =: L_{\alpha,\beta}\left(\frac{x-\mu}{\sigma}\right). \tag{1.62}$$

We omit the scripts of μ and σ for normalized variable for the sake of simplicity. The one-sided distribution with $0 < \alpha < 1$ and $\beta = 1$ is concentrated on the positive half-line, whose characteristic function is

$$C_{\alpha,\beta}(k) = \exp\left(-|k|^\alpha\left(1 - i\tan\frac{\pi\alpha}{2}\text{sign}(k)\right)\right). \tag{1.63}$$

It is consistent with the one in Eq. (1.54) with the constant being

$$c = \frac{1}{\cos(\pi\alpha/2)}. \tag{1.64}$$

Since Lévy distributions are very popular in random walks, we need to generate the random numbers obeying Lévy distribution. The method has been given in [Klafter and Sokolov (2011)]:

(1) Symmetric distributions

- Generate a random variable V distributed homogeneously on $(-\pi/2, \pi/2)$.
- Generate an exponential variable W with mean 1.
- Compute

$$X = \frac{\sin(\alpha V)}{(\cos V)^{1/\alpha}} \left\{ \frac{\cos[(1-\alpha)V]}{W} \right\}^{\frac{1-\alpha}{\alpha}}. \tag{1.65}$$

Then the variable X obeys the symmetric Lévy law $L_{\alpha,0}(x)$.

(2) Asymmetric distributions with $\alpha \neq 1$

- Compute $A = \arctan[\beta \tan(\pi\alpha/2)]$ and the two auxiliary values

$$C = A/\alpha, \qquad D = (\cos A)^{-1/\alpha}. \tag{1.66}$$

- Generate a random variable V distributed homogeneously on $(-\pi/2, \pi/2)$.
- Generate an exponential variable W with mean 1.
- Compute

$$X = D \frac{\sin[\alpha(V+C)]}{(\cos V)^{1/\alpha}} \left\{ \frac{\cos[V - \alpha(V+C)]}{W} \right\}^{\frac{1-\alpha}{\alpha}}. \tag{1.67}$$

Then the variable X obeys the general Lévy law $L_{\alpha,\beta}(x)$.

1.4.3   *Central Limit Theorem*

Central limit theorem is a fundamental statistical theorem. It says that for a sequence of independent random variables $\{X_i\}$ each having a finite second moment, the sum

$$X := \frac{1}{\sqrt{n}} \sum_{i=1}^{n} X_i \tag{1.68}$$

approaches a normally distributed random variable as n tends to infinity. Further, if X_i has the zero mean and variance σ_i^2, then X has the zero mean and variance

$$\sigma^2 = \frac{1}{n} \sum_{i=1}^{n} \sigma_i^2. \tag{1.69}$$

The proof of this theorem can be carried out with the help of characteristic function. Since the mean and variance of X_i are zero and σ_i^2, respectively, its characteristic function can be expanded as

$$\langle e^{ikX_i} \rangle \simeq 1 - \frac{1}{2}\sigma_i^2 k^2 \tag{1.70}$$

for small k. Thus, the characteristic function of the variable X is

$$\langle e^{ikX} \rangle = \left\langle \prod_{i=1}^{n} e^{ikX_i/\sqrt{n}} \right\rangle = \prod_{i=1}^{n} \left\langle e^{i(k/\sqrt{n})X_i} \right\rangle \simeq \prod_{i=1}^{n} \left(1 - \frac{\sigma_i^2}{2n} k^2 \right). \quad (1.71)$$

Taking the logarithm on both sides, we obtain

$$\ln\langle e^{ikX} \rangle \simeq \sum_{i=1}^{n} \ln \left(1 - \frac{\sigma_i^2}{2n} k^2 \right) \simeq \sum_{i=1}^{n} -\frac{\sigma_i^2}{2n} k^2 = -\frac{\sigma^2}{2} k^2, \quad (1.72)$$

which implies that the characteristic function of X is $e^{-\frac{\sigma^2}{2} k^2}$, and the variable X is Gaussian with zero mean and variance σ^2.

We have assumed that the random variables possesses the finite variance in the central limit theorem where the sum of independent random variables tends toward a normal distribution. There are still many variables, however, having divergent second moment, such as the Lévy (stable) distribution with the Lévy index $\alpha < 2$, which plays an important role in anomalous diffusion processes. In this case, we consider a sequence of independent Lévy variables $\{X_i\}$ with their characteristic functions expanded as

$$\langle e^{ikX_i} \rangle \simeq 1 - \sigma_i^\alpha |k|^\alpha \quad (1.73)$$

for small k, which implies the variables $\{X_i\}$ have the different scale parameter σ_i.

Then the characteristic function of

$$X := \frac{1}{n^{1/\alpha}} \sum_{i=1}^{n} X_i \quad (1.74)$$

is

$$\langle e^{ikX} \rangle = \prod_{i=1}^{n} \left\langle e^{i(k/n^{1/\alpha})X_i} \right\rangle \simeq \prod_{i=1}^{n} \left(1 - \frac{\sigma_i^\alpha}{n} |k|^\alpha \right). \quad (1.75)$$

We take the logarithm on both sides and obtain

$$\ln\langle e^{ikX} \rangle \simeq \sum_{i=1}^{n} \ln \left(1 - \frac{\sigma_i^\alpha}{n} |k|^\alpha \right) \simeq \sum_{i=1}^{n} -\frac{\sigma_i^\alpha}{n} |k|^\alpha = -\sigma^\alpha |k|^\alpha, \quad (1.76)$$

which implies that

$$\langle e^{ikX} \rangle \simeq e^{-\sigma^\alpha |k|^\alpha}, \quad (1.77)$$

with

$$\sigma^\alpha = \frac{1}{n} \sum_{i=1}^{n} \sigma_i^\alpha. \quad (1.78)$$

Therefore, with the right way of scaling ($\frac{1}{n^{1/\alpha}}$ in Eq. (1.74)), the variable X obeys the Lévy distribution with parameter σ. This result is named as extended central limit theorem.

1.5   Random Processes

Now we consider a random variable X which depends on time t, i.e., $X = X(t)$. A random process is also known as a stochastic process, which is a family of random variables $\{X(t), t \in [0, T]\}$ with time parameter t. We could extract many instants from the set $[0, T]$, i.e., $t_1 < t_2 < \cdots < t_n < T$ and approximate the random process $X(t)$ by $X(t_1), X(t_2), \cdots, X(t_n)$. For the fixed time t_1, its PDF can be expressed as

$$p_1(x_1, t_1) = \langle \delta(x_1 - X(t_1)) \rangle. \tag{1.79}$$

The probability of finding the random variable $X(t_1)$ in the interval $x_1 \leq X(t_1) \leq x_1 + dx_1$ is given by $p_1(x_1, t_1)dx_1$. Similarly, $p_2(x_2, t_2; x_1, t_1)dx_2dx_1$ is the joint probability of finding $X(t_1)$ in the interval $(x_1, x_1 + dx_1)$ and $X(t_2)$ in the interval $(x_2, x_2 + dx_2)$. By adding more time points, we can get the joint probability with n points as

$$p_n(x_n, t_n; \cdots; x_1, t_1)dx_1 \cdots dx_n, \tag{1.80}$$

where the PDF p_n is given by

$$p_n(x_n, t_n; \cdots; x_1, t_1) = \langle \delta(x_1 - X(t_1)) \cdots \delta(x_n - X(t_n)) \rangle. \tag{1.81}$$

Different from random variable, we need to know all the multi-point joint PDF (i.e., p_n for any n and any $t_i \in [0, T]$) to completely determine the random process $X(t)$ in the interval $[0, T]$. If the PDF in Eq. (1.80) is not changed by replacing t_i by $t_i + t$ for arbitrary t, we call this process stationary.

Then we define the conditional probability density as the PDF of the random variable $X(t)$ at time t_n under the condition that X has the value x_{n-1} at the time $t_{n-1} < t_n$, the value x_{n-2} at the time $t_{n-2} < t_{n-1}, \cdots$, and the value x_1 at the time $t_1 < t_2$:

$$p(x_n, t_n | x_{n-1}, t_{n-1}; \cdots; x_1, t_1)$$
$$= \langle \delta(x_n - X(t_n)) | X(t_{n-1}) = x_{n-1}, \cdots, X(t_1) = x_1 \rangle \tag{1.82}$$

for $t_n > t_{n-1} > \cdots > t_1$, which can be also expressed by

$$p(x_n, t_n | x_{n-1}, t_{n-1}; \cdots; x_1, t_1) = \frac{p_n(x_n, t_n; \cdots; x_1, t_1)}{p_{n-1}(x_{n-1}, t_{n-1}; \cdots; x_1, t_1)}$$
$$= \frac{p_n(x_n, t_n; \cdots; x_1, t_1)}{\int_\infty^\infty p_n(x_n, t_n; \cdots; x_1, t_1)dx_n}. \tag{1.83}$$

1.5.1 *Markov Process*

Now we consider a special kind of random processes named as Markov processes. The conditional probability density of Markov processes only depends on the value $X(t_{n-1}) = x_{n-1}$ at the next latest time but not on the values at time t_{n-2} and so on, i.e.,

$$p(x_n, t_n | x_{n-1}, t_{n-1}; \cdots; x_1, t_1) = p(x_n, t_n | x_{n-1}, t_{n-1}). \qquad (1.84)$$

For the Markov process, Eq. (1.83) can be written as

$$p_n(x_n, t_n; \cdots; x_1, t_1) = p(x_n, t_n | x_{n-1}, t_{n-1}) p_{n-1}(x_{n-1}, t_{n-1}; \cdots; x_1, t_1). \qquad (1.85)$$

Then by using the same argument for p_{n-1} and so on, we find that p_n can be expressed as a product of conditional probabilities and p_1, that is,

$$p_n(x_n, t_n; \cdots; x_1, t_1) = p(x_n, t_n | x_{n-1}, t_{n-1}) p(x_{n-1}, t_{n-1} | x_{n-2}, t_{n-2})$$
$$\cdots p(x_2, t_2 | x_1, t_1) p_1(x_1, t_1). \qquad (1.86)$$

1.5.2 *Chapman-Kolmogorov Equation*

By using Eq. (1.85), the famous Chapman-Kolmogorov equation can be derived. More precisely, the joint PDF p_2 can be expressed as the integral of p_3 over one coordinate, i.e.,

$$p_2(x_3, t_3; x_1, t_1) = \int_{-\infty}^{\infty} p_3(x_3, t_3; x_2, t_2; x_1, t_1) dx_2. \qquad (1.87)$$

Since $X(t)$ is a Markov process, the joint PDF in Eq. (1.87) can be split as the produce of conditional probabilities, i.e.,

$$p(x_3, t_3 | x_1, t_1) p_1(x_1, t_1) = \int_{-\infty}^{\infty} p(x_3, t_3 | x_2, t_2) p(x_2, t_2 | x_1, t_1) p_1(x_1, t_1) dx_2. \qquad (1.88)$$

Since Eq. (1.88) holds for arbitrary $p_1(x_1, t_1)$, we obtain the Chapman-Kolmogorov equation

$$p(x_3, t_3 | x_1, t_1) = \int_{-\infty}^{\infty} p(x_3, t_3 | x_2, t_2) p(x_2, t_2 | x_1, t_1) dx_2. \qquad (1.89)$$

The interpretation of Eq. (1.89) is that the transition probability from x_1 at time t_1 to x_3 at time t_3 is equal to the product of transition probabilities from x_1 at time t_1 to x_2 at time t_2 and from x_2 at time t_2 to x_3 at time t_3 for all possible x_2.

1.5.3 *Lévy Process*

A Lévy process, named after the French mathematician Paul Lévy, is a stochastic process with independent and stationary increments. It can be viewed as the continuous-time analog of a random walk. In strict mathematical definition, a stochastic process $X = \{X(t) : t > 0\}$ is a Lévy process if it satisfies the following properties:

(1) $X(0) = 0$ almost surely;
(2) Independent increments: For any $0 \leq t_1 < t_2 < \cdots < t_n < \infty$, the increments $X(t_2) - X(t_1), X(t_3) - X(t_2), \cdots, X(t_n) - X(t_{n-1})$ are independent.
(3) Stationary increments: For any time $s < t$, the increment $X(t) - X(s)$ is equal in distribution to $X(t - s)$.
(4) Continuity in probability: For any $\epsilon > 0$ and $t \geq 0$, it holds that

$$\lim_{h \to 0} P(|X(t + h) - X(t)| > \epsilon) = 0. \tag{1.90}$$

If $X(t)$ is a Lévy process, then one may construct a version of $X(t)$ to be almost surely right continuous with left limits. Another property of Lévy process $X(t)$ is that it is infinitely divisible for each $t \geq 0$. In detail, for each n, we can write

$$X(t) = Y_1^{(n)}(t) + \cdots + Y_n^{(n)}(t), \tag{1.91}$$

where

$$Y_k^{(n)}(t) = X\left(\frac{kt}{n}\right) - X\left(\frac{(k-1)t}{n}\right). \tag{1.92}$$

The $Y_k^{(n)}(t)$ are independent identically distributed (i.i.d.) random variables for given time t.

Then we show the Lévy-Khintchine formula, first established by Paul Lévy and A. Ya. Khintchine in the 1930s, which gives a characterisation of infinitely divisible random variables through their characteristic funcions. The Lévy-Khintchine formula is quite convenient to describe a Lévy process in high dimensions. Considering an n-dimensional Lévy process $\mathbf{X}(t)$, its characteristic function has a specific form for each $t \geq 0$,

$$\langle e^{i\mathbf{k} \cdot \mathbf{X}(t)} \rangle = e^{t\phi(\mathbf{k})}, \tag{1.93}$$

where

$$\phi(\mathbf{k}) = i\mathbf{k} \cdot \mathbf{b} - \frac{1}{2}\mathbf{k} \cdot \mathbf{a}\mathbf{k} + \int_{\mathbb{R}^n \setminus \{0\}} \left[e^{i\mathbf{k} \cdot \mathbf{Y}} - 1 - i\mathbf{k} \cdot \mathbf{Y}_{\chi\{|\mathbf{Y}|<1\}} \right] \nu(d\mathbf{Y}). \tag{1.94}$$

Here, the vector $\mathbf{b} \in \mathbb{R}^n$, $\mathbf{a}$ is a positive definite symmetric $n \times n$ matrix, χ_I is the indicator function of the set I, and ν is a finite Lévy measure on $\mathbb{R}^n \backslash \{0\}$, implying that $\int_{\mathbb{R}^n \backslash \{0\}} \min\{1, |\mathbf{Y}|^2\} \nu(d\mathbf{Y}) < \infty$. Actually, the $\phi(\mathbf{k})$ is the Lévy symbol of Lévy variable $X(1)$, and $(\mathbf{b}, \mathbf{a}, \nu)$ are called the characteristics of $X(1)$.

The most typical example of Lévy process is Brownian motion (also named Wiener process), which is Gaussian and has continuous sample paths. The Lévy symbol $\phi(\mathbf{k})$ only contains the former two terms in Eq. (1.94), while the last term characterizes the jump feature of Lévy process. For standard n-dimensional Brownian motion $B(t)$, its characteristic function is given by

$$\langle e^{i\mathbf{k} \cdot B(t)} \rangle = e^{-\frac{1}{2} t |\mathbf{k}|^2}. \tag{1.95}$$

The Brownian motion is the only one of the Lévy processes with continuous sample paths. Besides, other common Lévy processes are the Poisson process, compound Poisson process and stable Lévy process. The stable Lévy process at any instant is a stable random variable. Of particular interest is the rotationally invariant case, where the Lévy symbol is

$$\phi(\mathbf{k}) = -\sigma^\alpha |\mathbf{k}|^\alpha, \tag{1.96}$$

where $0 < \alpha \le 2$ is the index of stability and the scale parameter $\sigma > 0$. One noteworthy feature of the stable Lévy process is its self-similarity. In general, a stochastic process $X(t)$ is self-similar with Hurst index $H > 0$ if the two process $X(at)$ and $a^H X(t)$ have the same finite-dimensional distributions for any $a \ge 0$. Within this definition, it can be verified that the rotationally invariant stable Lévy process is self-similar with Hurst index $H = 1/\alpha$. For the special case of Brownian motion ($\alpha = 2$), it is self-similar with Hurst index $H = 1/2$. More self-similar processes can be found in [Embrechts and Maejima (2002)].

1.6   Subordinators

A subordinator is a one-dimensional Lévy process which is non-decreasing (almost sure) [Applebaum (2009)]. Because of the non-decreasing property, a subordinator can be regarded as a random model of time evolution. The strict mathematical definition is: if $X(t)$ is a subordinator, then its Lévy symbol takes the form

$$\phi(k) = ibk + \int_0^\infty (e^{iky} - 1) \nu(dy), \tag{1.97}$$

where $b \geq 0$ and the Lévy measure ν satisfies the additional requirements

$$\nu(-\infty, 0) = 0 \quad \text{and} \quad \int_0^\infty (y \wedge 1)\nu(dy) < \infty. \tag{1.98}$$

Here, the pair (b, ν) is called the characteristics of the subordinator $X(t)$. Since $X(t) \geq 0$, its characteristic function is given by the Laplace transform:

$$\langle e^{-\lambda X(t)} \rangle = e^{-t\Phi(\lambda)}, \tag{1.99}$$

where

$$\Phi(\lambda) = -\phi(i\lambda) = b\lambda + \int_0^\infty (1 - e^{-\lambda y})\nu(dy) \tag{1.100}$$

for any $\lambda > 0$.

1.6.1   *Several Subordinators*

Now we list several common subordinators. The first one is the α-stable subordinator which has the characteristics

$$b = 0 \quad \text{and} \quad \nu(dx) = \frac{\alpha}{\Gamma(1-\alpha)} \frac{dx}{x^{1+\alpha}} \tag{1.101}$$

for $0 < \alpha < 1$. Then the Laplace exponent can be calculated by employing the well-known trick of writing a repeated integral as a double integral and then changing the order of integrations, i.e.,

$$
\begin{aligned}
\Phi(\lambda) &= \frac{\alpha}{\Gamma(1-\alpha)} \int_0^\infty (1 - e^{-\lambda x}) \frac{dx}{x^{1+\alpha}} \\
&= -\frac{\alpha}{\Gamma(1-\alpha)} \int_0^\infty \left(\int_0^x \lambda e^{-\lambda y} dy \right) x^{-1-\alpha} dx \\
&= -\frac{\alpha}{\Gamma(1-\alpha)} \int_0^\infty \left(\int_y^\infty x^{-1-\alpha} dx \right) \lambda e^{-\lambda y} dy \\
&= \frac{\lambda}{\Gamma(1-\alpha)} \int_0^\infty e^{-\lambda y} y^{-\alpha} dy \\
&= \lambda^\alpha.
\end{aligned}
\tag{1.102}
$$

The second one is the Lévy subordinator (i.e., the $\frac{1}{2}$-stable subordinator) has the density given by the Lévy distribution [Revuz and Yor (1990)]

$$f_{X(t)}(s) = \left(\frac{t}{2\sqrt{\pi}} \right) s^{-3/2} e^{-t^2/(4s)} \tag{1.103}$$

for $s \geq 0$. The probabilistic interpretation of the Lévy subordinator is the first hitting time for one-dimensional standard Brownian motion $B(t)$, that is,

$$X(t) = \inf \left\{ s > 0; B(s) = \frac{t}{\sqrt{2}} \right\}. \tag{1.104}$$

1.6.2 *Inverse Subordinator*

Now we consider the inverse subordinator, the inverse process of a subordinator, the definition of which depends on a specific subordinator. The α-stable subordinator $(0 < \alpha < 1)$ is used here and it is denoted as $t = t(s)$ for convenience. Then the corresponding inverse α-stable subordinator $s = t^{-1}(t) = s(t)$ is defined by

$$s(t) = \inf_{s>0}\{s : t(s) > t\}, \tag{1.105}$$

which is the first-passage time of the subordinator $t(s)$. The $t(s)$ and $s(t)$ have the relations

$$\langle \Theta(s - s(t))\rangle = 1 - \langle \Theta(t - t(s))\rangle \tag{1.106}$$

and

$$\begin{aligned}\langle \Theta(s_2 - s(t_2))\Theta(s_1 - s(t_1))\rangle &= 1 - \langle \Theta(t_2 - t(s_2))\rangle \\ &\quad - \langle \Theta(t_1 - t(s_1))\rangle + \langle \Theta(t_2 - t(s_2))\Theta(t_1 - t(s_1))\rangle,\end{aligned} \tag{1.107}$$

where $\Theta(x)$ is the Heaviside step function satisfying

$$\Theta(x) = \begin{cases} 1 & \text{for } x > 0 \\ 0 & \text{for } x < 0 \\ \frac{1}{2} & \text{for } x = 0. \end{cases} \tag{1.108}$$

Since the (two-point) PDF of inverse subordinator $s(t)$ can be expressed as

$$h(s,t) = \langle \delta(s - s(t))\rangle, \tag{1.109}$$

$$h(s_2, t_2; s_1, t_1) = \langle \delta(s_2 - s(t_2))\delta(s_1 - s(t_1))\rangle, \tag{1.110}$$

performing the derivatives with respect to s, s_1, s_2 on Eqs. (1.106) and (1.107) leads to

$$h(s,t) = -\frac{\partial}{\partial s}\langle \Theta(t - t(s))\rangle, \tag{1.111}$$

$$h(s_2, t_2; s_1, t_1) = \frac{\partial}{\partial s_1}\frac{\partial}{\partial s_2}\langle \Theta(t_2 - t(s_2))\Theta(t_1 - t(s_1))\rangle. \tag{1.112}$$

Furthermore, since the initial condition of inverse subordinator is $s(t)|_{t=0} = 0$, the boundary conditions for Eqs. (1.111) and (1.112) are

$$h(s,0) = \delta(s),$$
$$h(s_2, t_2; s_1, 0) = h(s_2, t_2)\delta(s_1), \tag{1.113}$$
$$h(s_2, t_2 \to t_1; s_1, t_1) = \delta(s_2 - s_1)h(s_1, t_1).$$

It can be seen that the right-hand sides of Eqs. (1.111) and (1.112) are the derivatives of the (two-point) PDF of the subordinator $t(s)$, which can be obtained since it is a Lévy process. Denote the (two-point) PDF of the subordinator $t(s)$ as $p(t, s)$ and $p(t_2, s_2; t_1, s_1)$, respectively. Based on the discussions in Sec. 1.4.2, the Laplace transform $(t \to \lambda)$ of $p(t, s)$ is

$$\hat{p}(\lambda, s) = \langle e^{-\lambda t(s)} \rangle = e^{-s\lambda^\alpha}, \tag{1.114}$$

for $0 < \alpha < 1$. The expression in time domain is

$$p(t, s) = \frac{1}{s^{1/\alpha}} L_\alpha \left(\frac{t}{s^{1/\alpha}} \right), \tag{1.115}$$

where $L_\alpha(t)$ denotes the one-sided Lévy-stable distribution whose Laplace transform is $\mathcal{L}\{L_\alpha(t)\} = e^{-\lambda^\alpha t}$. The corresponding two-point PDF can be obtained by using the property of independence and stationarity of increments. More precisely, letting $dt(s)/ds = \tau(s)$, the double Laplace transform $(t_1 \to \lambda_1, t_2 \to \lambda_2)$ of the PDF $p(t_2, s_2; t_1, s_1)$ is

$$\hat{p}(\lambda_2, s_2; \lambda_1, s_1) = \left\langle e^{-\lambda_2 t(s_2) - \lambda_1 t(s_1)} \right\rangle$$

$$= \Theta(s_2 - s_1) \left\langle \exp\left(-\lambda_2 \int_{s_1}^{s_2} ds' \, \tau(s') - (\lambda_1 + \lambda_2) \int_0^{s_1} ds' \, \tau(s') \right) \right\rangle$$

$$+ \Theta(s_1 - s_2) \left\langle \exp\left(-\lambda_1 \int_{s_1}^{s_2} ds' \, \tau(s') - (\lambda_1 + \lambda_2) \int_0^{s_2} ds' \, \tau(s') \right) \right\rangle. \tag{1.116}$$

Then the expectations can be split because of the independence of increments. Thus we have

$$\hat{p}(\lambda_2, s_2; \lambda_1, s_1) = \Theta(s_2 - s_1) e^{-s_1(\lambda_1 + \lambda_2)^\alpha} e^{-(s_2 - s_1)\lambda_2^\alpha}$$

$$+ \Theta(s_1 - s_2) e^{-s_2(\lambda_1 + \lambda_2)^\alpha} e^{-(s_1 - s_2)\lambda_1^\alpha}. \tag{1.117}$$

Combining Eqs. (1.111) and (1.114), we obtain the single-point PDF of inverse subordinator $s(t)$ as

$$\hat{h}(s, \lambda) = -\frac{\partial}{\partial s} \left\langle \frac{1}{\lambda} e^{-\lambda t(s)} \right\rangle = \lambda^{\alpha-1} e^{-s\lambda^\alpha}, \tag{1.118}$$

the inverse Laplace transform of which is known as [Barkai (2001)]

$$h(s, t) = \frac{1}{\alpha} \frac{t}{s^{1+1/\alpha}} L_\alpha \left(\frac{t}{s^{1/\alpha}} \right). \tag{1.119}$$

The normalization of PDF $h(s, t)$ can be obtained by integrating with respect to s in Eq. (1.118), i.e.,

$$\int_0^\infty \hat{h}(s, \lambda) ds = \frac{1}{\lambda}, \tag{1.120}$$

the inverse Laplace transform of which is unit. Especially, the PDF $h(s,t)$ satisfies a fractional evolution equation:

$$\frac{\partial}{\partial t}h(s,t) = -D_t^{1-\alpha}\frac{\partial}{\partial s}h(s,t), \tag{1.121}$$

where the operator $D_t^{1-\alpha}$ denotes the Riemann-Liouville fractional differential operator.

Similarly, combining Eqs. (1.112) and (1.117), we obtain

$$\hat{h}(s_2,\lambda_2;s_1,\lambda_1)$$

$$= \frac{\partial}{\partial s_1}\frac{\partial}{\partial s_2}\frac{1}{\lambda_1\lambda_2}\hat{p}(\lambda_2,s_2;\lambda_1,s_1)$$

$$= \delta(s_2 - s_1)\frac{\lambda_1^\alpha + \lambda_2^\alpha - (\lambda_1 + \lambda_2)^\alpha}{\lambda_1\lambda_2}e^{-s_1(\lambda_1+\lambda_2)^\alpha}$$

$$+ \Theta(s_2 - s_1)\frac{\lambda_2^\alpha[(\lambda_1 + \lambda_2)^\alpha - \lambda_2^\alpha]}{\lambda_1\lambda_2} \tag{1.122}$$

$$\times e^{-s_1(\lambda_1+\lambda_2)^\alpha}e^{-(s_2-s_1)\lambda_2^\alpha}$$

$$+ \Theta(s_1 - s_2)\frac{\lambda_1^\alpha[(\lambda_1 + \lambda_2)^\alpha - \lambda_1^\alpha]}{\lambda_1\lambda_2}$$

$$\times e^{-s_2(\lambda_1+\lambda_2)^\alpha}e^{-(s_1-s_2)\lambda_1^\alpha}.$$

The normalization of the two-point PDF can be verified by integrating with respect to s_1 and s_2 in Eq. (1.122), i.e.,

$$\int_0^\infty \int_0^\infty \hat{h}(s_2,\lambda_2;s_1,\lambda_1)ds_2ds_1 = \frac{1}{\lambda_1\lambda_2}, \tag{1.123}$$

the inverse Laplace transform of which is unit.

1.6.3 *Simulations of Subordinator and Inverse Subordinator*

Here, we present the simulation algorithms for the α-stable subordinator $t(s)$ with $0 < \alpha < 1$ and its corresponding inverse subordinator $s(t)$. Considering the inverse relation between $t(s)$ and $s(t)$, we need to establish two sets of lattices for them, respectively. We approximate the subordinator $t(s)$ on the lattices $\{\tau_j = j\Delta\tau, j = 0, 1, \cdots, M\}$ and the inverse subordinator $s(t)$ on meshes $\{t_i = i\Delta t : i = 0, 1, \cdots, N\}$. It is recommended to choose $\Delta\tau < \Delta t$ for good approximations.

(1) Generate subordinator $t(s)$: Considering the initial condition and the self-similarity with Hurst index $1/\alpha$, the recursive relation

$$t(\tau_0) = 0,$$
$$t(\tau_j) = t(\tau_{j-1}) + \Delta\tau^{1/\alpha}\xi_j \tag{1.124}$$

holds. Here, ξ_j are the i.i.d. totally skewed positive α-stable random variables, which can be generated by the formula in Eq. (1.67). The iterations in Eq. (1.124) end when the subordinator $t(\tau)$ crosses the level T, i.e., when we have $t(\tau_{j_0-1}) \leq T < t(t_{j_0})$ for some $j_0 = M$.

(2) Generate inverse subordinator $s(t)$: The generation is mainly based on the definition of $s(t)$ in Eq. (1.105). For another set of lattices t_i, we find the element τ_j such that $t(\tau_{j-1}) < t_i \leq t(\tau_j)$ is the value of inverse subordinator at time t_i, i.e.,

$$S_{t_i} = \tau_j. \tag{1.125}$$

With these simulations, many compound stochastic processes can be also generated, such as $x(t) := x(s(t))$ being the combination of original process $x(s)$ and inverse subordinator $s(t)$; see more details and other kind of subordinator in [Magdziarz *et al.* (2007); Magdziarz and Weron (2007); Gajda and Magdziarz (2010); Kleinhans and Friedrich (2007); Wang *et al.* (2019b)].

Chapter 2

Anomalous and Nonergodic Diffusion

2.1 Continuous Time Random Walk

The concept of continuous time random walk (CTRW) was originally introduced by Montroll and Weiss in 1965 [Montroll and Weiss (1965)], which extends irregular random walks on lattices to a continuous-time variable. It is characterized by an important distribution related to a stochastic process. This distribution is named as waiting time distribution, permitted to describe both exponential and, what is most significant, power law relaxations as well as normal and anomalous transport and diffusion [Klafter and Zumofen (1994); Zaburdaev et al. (2015)]. Later, the CTRW model was further developed by physicists Scher and Lax in terms of recursion relations [Scher and Lax (1972, 1973a,b)]. This CTRW formalism is more convenient to study as well anomalous transport and diffusion as their anomalous scaling properties [e.g., the nonlinear time growth of mean squared displacement (MSD)]. More quantities can be also evaluated in this CTRW formalism, such as the survival probabilities, first-passage time and the number of distinct sites visited until a given time. All these things are used to characterize the complex systems in the natural world [Paradisi et al. (2015)]. Now, the CTRW has become a foundation of anomalous non-Gaussian transport and diffusion [Scher and Montroll (1975); Pfister and Scher (1978)]. It has been successfully applied in various fields, including the charge carrier transport in amorphous semiconductors [Scher and Montroll (1975)], electron transfer [Nelson (1999)], dispersion in turbulent systems [Solomon et al. (1993)], and so on.

We will show the generalization from discrete-time random walk to the continuous-time one in the following. The CTRW contains two types of random variables: jump lengths and waiting times; more often, they are taken to be i.i.d.. The relationship between jump lengths and waiting times can be uncoupled and coupled. We investigate the subdiffusive CTRW and Lévy flights for the uncoupled case and Lévy walk for the coupled case. Especially, Fokker-Planck equation and fractional Fokker-Planck equation governing the PDFs of the displacement of the particles are derived for different distributions of jump length and waiting time.

2.1.1   *From Discrete to Continuous Random Walk*

2.1.1.1   *Discrete random walk*

Let us start from a simple discrete random walk on a one-dimensional lattice [Klafter and Sokolov (2011)] (see Fig. 2.1). Consider the particle starts at position $x = 0$. It jumps to the neighbouring sites in each step. Different steps are assumed to be independent and share the same law to the right with probability p and to the left with probability $q = 1 - p$. The final position of the particle after n steps is determined by the total number of steps to the right and to the left.

To decide the probability $P_n(j)$ of arriving at site j after n steps, we introduce the expression $pe^{i\theta} + qe^{-i\theta}$ with the coefficients in front of the exponential function being the probability of jumping to the right and left, respectively. For the expression

$$(pe^{i\theta} + qe^{-i\theta})^2 = p^2 e^{2i\theta} + 2pq + q^2 e^{-2i\theta}, \tag{2.1}$$

one can note that the coefficients in Eq. (2.1) are equal to the probability that first two steps were both to the right, one to the right and one to the left, both to the left, respectively. Generalizing this finding to n steps, we see that the coefficient in front of $e^{i\theta j}$ in the expansion of $(pe^{i\theta} + qe^{-i\theta})^n$ is exactly the probability $P_n(j)$. This coefficient can be obtained from Fourier transform:

$$P_n(j) = \frac{1}{2\pi} \int_{-\pi}^{\pi} (pe^{i\theta} + qe^{-i\theta})^n e^{-i\theta j} d\theta. \tag{2.2}$$

Actually, the expression $\phi(\theta) = pe^{i\theta} + qe^{-i\theta}$ is the characteristic function of the distribution of the particle's displacement of each step in this discrete random walk model, i.e.,

$$\phi(\theta) = \int_{-\infty}^{\infty} e^{ix\theta} (p\delta(x - 1) + q\delta(x + 1)) dx. \tag{2.3}$$

For the simple case of a symmetric random walk with $p = q = 1/2$, the probability of arriving at site j after n steps is given by [Klafter and

Sokolov (2011)]

$$P_n(j) = \frac{1 + (-1)^{n+j}}{2^{n+1}} \frac{n!}{\left(\frac{n+j}{2}\right)! \left(\frac{n-j}{2}\right)!}. \tag{2.4}$$

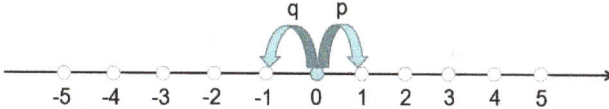

Fig. 2.1: A schematic illustration of the discrete random walk on a one-dimensional lattice. The particle starts at $x = 0$, and jumps to the neighbouring sites with probability p to the right and $q = 1 - p$ to the left.

2.1.1.2 *Continuous random walk*

Instead of the discrete random walk, sometimes, people are more interested in the processes evolving in continuous physical time. Therefore, we introduce the CTRW model as formulated by [Montroll and Weiss (1965)].

In CTRW model, the particle jumps instantaneously from one site to another, following a waiting period on a site whose duration t is drawn from the PDF of waiting time $\psi(t)$. The simplest waiting time distribution is $\psi(t) = \delta(t - \tau)$, costing a positive determined time τ within each step, which leads to a simply way to translate the number of steps n to physical time t. Apart from the simplest one, the waiting time distribution can be various, including the exponential distribution [Debye (1945); Lenk and Gellert (1974); Tschoegl (1989)]

$$\psi(t) = \frac{1}{\tau} e^{-\frac{t}{\tau}}, \tag{2.5}$$

and the power law form [Klafter and Shlesinger (1986); Blumen *et al.* (1986)]

$$\psi(t) = \frac{\alpha \tau_0^\alpha}{(t + \tau_0)^{1+\alpha}}, \tag{2.6}$$

where $\alpha \in (0, 2)$ is the power law exponent and τ_0 is the characteristic time. The difference between the exponential distribution Eq. (2.5) and the power law distribution Eq. (2.6) is significant. The waiting times of different steps almost concentrate on the average τ for exponential distribution while some quite large values of waiting times might be generated for power law distribution. These long waiting times cannot be ignored and they will contribute to abundant anomalous diffusion phenomena.

Since the physical time t is positive at each step, we can take advantage of the Laplace transform to characterize some important quantities. For waiting time distribution $\psi(t)$, its Laplace transform is

$$\hat{\psi}(\lambda) = \int_0^\infty e^{-\lambda t} \psi(t) dt \equiv \langle e^{-\lambda t} \rangle, \tag{2.7}$$

where the angle brackets on the right-hand side represent the ensemble average of the random variable in them (here the random variable is the waiting time t). Since the PDF of the waiting times $\psi(t)$ is normalized, it holds that

$$\hat{\psi}(0) = \int_0^\infty \psi(t) dt = 1. \tag{2.8}$$

The Laplace transform is a powerful tool in probability theory. One of the important properties is that it is the generating function of the corresponding moments. For example, expanding the exponential function $e^{-\lambda t}$ in Eq. (2.7) yields

$$\hat{\psi}(\lambda) = \sum_{n=0}^\infty (-1)^n \frac{\langle t^n \rangle \lambda^n}{n!}, \tag{2.9}$$

which can help us calculate the moments of waiting times from $\hat{\psi}(\lambda)$.

Then we consider the quantity $\psi_n(t)$, the PDF of the occurrence of the nth step at time t, where $t = \sum_{i=1}^n t_i$ and t_i is the waiting time for the ith step of the walker. The Laplace transform of $\psi_n(t)$ can be obtained as

$$\hat{\psi}_n(\lambda) = \langle e^{-\lambda t} \rangle = \left\langle e^{-\lambda \sum_{i=1}^n t_i} \right\rangle = \prod_{i=1}^n \left\langle e^{-\lambda t_i} \right\rangle$$
$$= \left\langle e^{-\lambda t_i} \right\rangle^n = \hat{\psi}^n(\lambda), \tag{2.10}$$

where we write the average over t as a product of averages over different t_i due to the independence of the waiting times of different steps. Another important quantity related to the waiting time distribution $\psi(t)$ is the so-called survival probability $\Psi(t)$, the probability that the waiting time on a site exceeds time t:

$$\Psi(t) = \int_t^\infty \psi(t') dt' = 1 - \int_0^t \psi(t') dt'. \tag{2.11}$$

The Laplace transform of $\Psi(t)$ can be obtained by using the property of Laplace transform of an integral [Abramowitz and Stegun (1972)] and reads

$$\hat{\Psi}(\lambda) = \frac{1 - \hat{\psi}(\lambda)}{\lambda}. \tag{2.12}$$

Based on the two quantities $\psi_n(t)$ and $\Psi(t)$, we obtain the probability of taking exactly n steps up to the time t:

$$\chi_n(t) = \int_0^t \psi_n(\tau)\Psi(t - \tau)d\tau. \tag{2.13}$$

The integrand in Eq. (2.13) represents the joint PDF of occurring nth steps at time τ and not occurring any step with the time interval $[\tau, t]$. Performing Laplace transform on Eq. (2.13) and using the convolution theorem of Laplace transform, we obtain

$$\hat{\chi}_n(\lambda) = \hat{\psi}^n(\lambda)\frac{1 - \hat{\psi}(\lambda)}{\lambda}. \tag{2.14}$$

Now we concentrate on the quantity $p(x, t)$, the PDF of finding the particle at position x at time t, which is given by

$$p(x, t) = \sum_{n=0}^{\infty} p_n(x)\chi_n(t), \tag{2.15}$$

where $p_n(x)$ represents the PDF of finding the particle at position x after n steps. Similar to the discussions of discrete random walk, the Fourier transform of $p_n(x)$ is

$$\tilde{p}_n(k) = \tilde{w}^n(k), \tag{2.16}$$

where $\tilde{w}(k)$ is the characteristic function of the PDF of jump length in each step (i.e., the Fourier transform of the PDF $w(x)$ of jump length). Performing Fourier-Laplace transform on Eq. (2.15), we obtain

$$\begin{aligned}
\tilde{\hat{p}}(k, \lambda) &= \sum_{n=0}^{\infty} \tilde{p}_n(k)\hat{\chi}_n(\lambda) \\
&= \frac{1 - \hat{\psi}(\lambda)}{\lambda} \sum_{n=0}^{\infty} \tilde{w}^n(k)\hat{\psi}^n(\lambda) \\
&= \frac{1 - \hat{\psi}(\lambda)}{\lambda} \frac{1}{1 - \tilde{w}(k)\hat{\psi}(\lambda)}.
\end{aligned} \tag{2.17}$$

Equation (2.17) is the central result of the theory of CTRWs and it is called the Montroll-Weiss equation [Montroll and Weiss (1965)]. This result is also valid for the higher dimensions. The exact expression or the governing equation of $p(x, t)$ can be obtained by taking inverse Fourier-Laplace transform on this equation. In addition, the time-dependent moments of the displacement can be extracted from the Montroll-Weiss equation (2.17), i.e.,

$$M_n(t) = (-i)^n \left. \frac{d^n \tilde{p}(k, t)}{dk^n} \right|_{k=0}. \tag{2.18}$$

In particular, the MSD in Laplace domain is

$$\hat{M}_2(\lambda) = - \left. \frac{d^2 \tilde{\hat{p}}(k, \lambda)}{dk^2} \right|_{k=0} = \frac{\hat{\psi}(\lambda)}{\lambda[1 - \hat{\psi}(\lambda)]} \langle l^2 \rangle, \tag{2.19}$$

where

$$\langle l^2 \rangle = \int_{-\infty}^{\infty} x^2 w(x) dx, \tag{2.20}$$

and we have used the fact $\tilde{\lambda}(0) = 1$ and assumed the PDF of jump length is symmetric, i.e.,

$$\langle l \rangle = \int_{-\infty}^{\infty} x w(x) dx = 0. \tag{2.21}$$

Note that the result in Eq. (2.19) shows the characteristic of space-time independence, since the multiplier of $\langle l^2 \rangle$ is exactly the mean renewal time at time t [Godrèche and Luck (2001)], i.e.,

$$\langle \hat{N}(\lambda) \rangle = \frac{\hat{\psi}(\lambda)}{\lambda[1 - \hat{\psi}(\lambda)]}. \tag{2.22}$$

Let us take some examples by applying different waiting time PDF $\psi(t)$ and jump length PDF $w(x)$ to Eqs. (2.17) and (2.19).

(1) In the first example, we take the waiting time distribution

$$\psi(t) = \frac{1}{\tau} e^{-\frac{t}{\tau}}, \tag{2.23}$$

so its Laplace transform is

$$\hat{\psi}(\lambda) = \frac{1}{1 + \lambda \tau} \simeq 1 - \lambda \tau \tag{2.24}$$

for small λ. The jump length distribution is Gaussian with its Fourier transform

$$\tilde{w}(k) = e^{-\sigma^2 k^2} \simeq 1 - \sigma^2 k^2 \tag{2.25}$$

for small k. Here, the small λ, k limit in Fourier-Laplace space is equivalent to the diffusion limit for long times [Klafter and Silbey (1980); Klafter *et al.* (1987); Bouchaud and Georges (1990); Klafter *et al.* (1990)]. Substituting $\hat{\psi}(\lambda)$ and $\tilde{w}(k)$ into Eq. (2.17), we obtain

$$\tilde{p}(k, \lambda) \simeq \frac{1}{\lambda + Dk^2}, \tag{2.26}$$

where the diffusion coefficient $D = \sigma^2/\tau$. The inverse transform of this expression to the space-time domain yields the Gaussian distribution

$$p(x, t) = \frac{1}{\sqrt{4\pi Dt}} \exp\left(-\frac{x^2}{4Dt}\right). \tag{2.27}$$

The corresponding MSD can be obtained from the propagator $p(x, t)$ in Eq. (2.27) or the formula in Eq. (2.19) as

$$\langle x^2(t) \rangle \simeq 2Dt, \tag{2.28}$$

which displays the normal diffusion. This example implies that any CTRW model will converge to Brownian motion for long times if the mean of waiting times and the second moment of jump length are finite.

(2) Here we extend the jump length distribution to be power law distributed as

$$w(x) \simeq \frac{A_\beta}{\sigma^\beta |x|^{1+\beta}} \tag{2.29}$$

with $0 < \beta < 2$ and its characteristic function

$$\tilde{w}(k) \simeq 1 - \sigma^\beta |k|^\beta. \tag{2.30}$$

The PDF of waiting time is chosen to be exponential, the same as the one in Eq. (2.24). The PDFs of waiting time and jump length result in

$$\tilde{p}(k, \lambda) \simeq \frac{1}{\lambda + K_\beta |k|^\beta} \tag{2.31}$$

with the generalized diffusion coefficient $K_\beta = \sigma^\beta/\tau$. Performing inverse transform yields

$$p(x, t) \simeq \frac{1}{K_\beta^{1/\beta}} L_{\beta,0}\left(\frac{|x|}{(K_\beta t)^{1/\beta}}\right), \tag{2.32}$$

where $L_{\beta,0}(z)$ represents the symmetric stable Lévy distribution with exponent β. The MSD diverges for $0 < \beta < 2$. In this case, the process $x(t)$ is called Lévy flight [Shlesinger *et al.* (1995); Bouchaud and Georges (1990)], which displays superdiffusion.

(3) Now we consider a power law distributed waiting time and a Gaussian distributed jump length. In detail, the PDF of waiting time is

$$\psi(t) \simeq A_\alpha (\tau/t)^{1+\alpha} \tag{2.33}$$

with $0 < \alpha < 1$ and its Laplace transform [Klafter *et al.* (1987); Wolf (1979)]

$$\hat{\psi}(\lambda) \simeq 1 - \tau^\alpha \lambda^\alpha. \tag{2.34}$$

The PDF of jump length is the same as the one in Eq. (2.25). Therefore, the Montroll-Weiss equation (2.17) gives

$$\tilde{p}(k, \lambda) = \frac{\lambda^{\alpha-1}}{\lambda^\alpha + K_\alpha k^2} \tag{2.35}$$

with the generalized diffusion coefficient $K_\alpha = \sigma^2/\tau^\alpha$. The inverse transform of this equation cannot be easily obtained. However, based on the Fox H-function, a closed-form result can be obtained as

$$p(x, t) = \frac{1}{\sqrt{4\pi K_\alpha t^\alpha}} H_{1,2}^{2,0} \left[\frac{x^2}{4K_\alpha t^\alpha} \middle| \begin{array}{l} (1 - \alpha/2, \alpha) \\ (0, 1), (1/2, 1) \end{array} \right]. \tag{2.36}$$

The Fox H-function $H_{p,q}^{m,n}(z)$ is defined as [Srivastava *et al.* (1982); Srivastava and Kashyap (1982); Mathai and Saxena (1978); Mathai *et al.* (2009)]

$$H_{p,q}^{m,n}(z) = H_{p,q}^{m,n} \left[z \middle| \begin{array}{l} (a_p, A_p) \\ (b_q, B_q) \end{array} \right]$$

$$= H_{p,q}^{m,n} \left[z \middle| \begin{array}{l} (a_1, A_1), (a_2, A_2), \cdots, (a_p, A_p) \\ (b_1, B_1), (b_2, B_2), \cdots, (b_q, B_q) \end{array} \right] \tag{2.37}$$

$$= \frac{1}{2\pi i} \int_L \chi(s) z^s ds$$

with

$$\chi(s) = \frac{\prod_{j=1}^m \Gamma(b_j - B_j s) \prod_{j=1}^n \Gamma(1 - a_j + A_j s)}{\prod_{j=m+1}^q \Gamma(1 - b_j + B_j s) \prod_{j=n+1}^p \Gamma(a_j - A_j s)}. \tag{2.38}$$

By using the formula in Eq. (2.19), we find the subdiffusion behavior:

$$\langle x^2(t) \rangle \simeq \frac{2K_\alpha t^\alpha}{\Gamma(1 + \alpha)} \tag{2.39}$$

for $0 < \alpha < 1$.

We show the trajectories of Brownian motion and Lévy flight in Fig. 2.2. As the case (1) shows, the Gaussian shaped PDF of jump length in Eq. (2.25) implies that all step sizes of Brownian motion are almost the same. In contrast, some extremely large steps can be found in Lévy flight due to its power law shaped PDF of jump length in Eq. (2.29) in case (2).

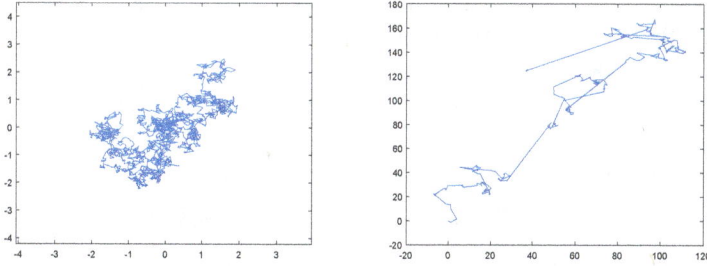

Fig. 2.2: Trajectories of Brownian motion (left) and Lévy flight (right) with $\beta = 1.3$.

2.1.1.3 *Transport equation*

There is an alternative method of deriving the Montroll-Weiss equation in Eq. (2.17), which is named as transport equation [Klafter and Sokolov (2011)]. The PDFs of waiting time $\psi(t)$ and jump length $w(x)$ are still the key elements in this method. The method is based on the transport equation of the flows and valid for both outgoing flow and incoming flow.

Let us first see the transport equation governing the outgoing flows. Let $\gamma(x, t)$ be how many particles leave the point x per unit of time. The equation connects the flux at the current point in space and time to the flux from the neighboring points [Klafter and Silbey (1980)]:

$$\gamma(x, t) = \int_{-\infty}^{\infty} \int_{0}^{t} w(x')\psi(t')\gamma(x - x', t - t')dx'dt' + p_0(x)\psi(t). \quad (2.40)$$

The integration term on the right-hand side denotes the flow $\gamma(x - x', t - t')$ which arrives at the point x at time t through a jump length x' and a waiting time t' within one step. The second term assumes that the particles had an initial distribution $p_0(x)$ and left their initial position according to the waiting time distribution. The next step is to connect the outgoing flow $\gamma(x, t)$ to the current density of particles $p(x, t)$ at a given point in space and time:

$$p(x, t) = \int_{-\infty}^{\infty} \int_{0}^{t} w(x')\Psi(t')\gamma(x - x', t - t')dx'dt' + p_0(x)\Psi(t). \quad (2.41)$$

The integration term denotes the flow $\gamma(x - x', t - t')$ which makes a jump of length x' and stays at position x before time t. The second term on the right-hand side accounts for the particles staying at the initial position until time t. Performing Fourier-Laplace transform $(x \to k,\, t \to \lambda)$ on Eq. (2.40), the flow $\gamma(x, t)$ can be explicitly given in (k, λ) space as

$$\tilde{\hat{\gamma}}(k, \lambda) = \frac{\tilde{p}_0(k)\hat{\psi}(\lambda)}{1 - \tilde{w}(k)\hat{\psi}(\lambda)}. \tag{2.42}$$

Similarly, the Fourier-Laplace transform of Eq. (2.41) gives

$$\tilde{\hat{p}}(k, \lambda) = \tilde{w}(k)\hat{\Psi}(\lambda)\tilde{\hat{\gamma}}(k, \lambda) + \tilde{p}_0(k)\hat{\Psi}(\lambda). \tag{2.43}$$

Combining these two equations, we have

$$\tilde{\hat{p}}(k, \lambda) = \frac{\tilde{p}_0(k)\hat{\Psi}(\lambda)}{1 - \tilde{w}(k)\hat{\psi}(\lambda)}, \tag{2.44}$$

which is equivalent to the Montroll-Weiss equation (2.17) by assuming $p_0(x) = \delta(x)$, i.e., $\tilde{p}_0(k) = 1$.

If we apply the transport equation to the incoming flow $\gamma(x, t)$. The equation connecting the flows from the neighbouring points becomes

$$\gamma(x, t) = \int_{-\infty}^{\infty} \int_{0}^{t} w(x')\psi(t')\gamma(x - x', t - t')dx'dt' + p_0(x)\delta(t). \tag{2.45}$$

Correspondingly, the equation connecting the incoming flow $\gamma(x, t)$ and the current density $p(x, t)$ is

$$p(x, t) = \int_{0}^{t} \Psi(t')\gamma(x, t - t')dt'. \tag{2.46}$$

Still using the technique of Fourier-Laplace transform, we obtain the same result as that in Eq. (2.44).

2.1.2 *Coupled Continuous Time Random Walk and Lévy Walk*

Throughout the discussion on the CTRW model in the previous section, we have assumed that the waiting time and jump length are independent. Some interesting results have been obtained, including the propagators and the diffusion behaviors. However, sometimes the infinite velocity of the model seems to be nonphysical, since the speed of a particle with nonzero mass is finite. The typical example is Lévy flight, whose MSD diverges. In microscopic description of Lévy flight, it is possible for the particles to make a very long jump at a small time scale. Therefore, the coupled CTRW model is developed, which introduces a coupling between the waiting time and jump length. This coupling usually penalizes the long jumps with large waiting time and results in a finite velocity. The typical one of such model is Lévy walk, where the jump length and waiting time are coupled via a constant velocity of the particle.

In Lévy walk model, the particle moves on a straight line at a fixed velocity v_0 for some random time. At the end of the excursion, the particle randomly chooses a new direction of its motion and moves for another random time with the same velocity. The durations of each excursion are independent and drawn from the same distribution $\psi(t)$. Despite the simplicity of this model, it is able to describe various regimes of stochastic transport, from normal diffusion to ballistic superdiffusion. The running time here can be exponential distributed or power law distributed. It is noteworthy that the exponent α of the power law distribution can be larger than one, compared with that of uncoupled CTRW model. Now, the power law distribution is assumed as

$$\psi(t) = \frac{1}{\tau_0} \frac{\alpha}{(1 + t/\tau_0)^{1+\alpha}}. \tag{2.47}$$

Its Laplace transform for small λ is [Rebenshtok *et al.* (2016)]

$$\hat{\psi}(\lambda) \simeq 1 + \sum_{j=1}^{\lfloor \alpha \rfloor} \frac{(-1)^j}{j!} \langle \tau^j \rangle \lambda^j + (-1)^{\lfloor \alpha \rfloor + 1} A \lambda^\alpha, \tag{2.48}$$

where $\lfloor \alpha \rfloor$ denotes the largest integer not larger than α and $\langle \tau^j \rangle$ is the jth moment of flight time, $\langle \tau \rangle = \tau_0/(\alpha - 1)$, $\langle \tau^2 \rangle = 2\tau_0^2/(\alpha - 1)/(\alpha - 2)$ and $A = |\Gamma(1 - \alpha)|\tau_0^\alpha$.

Similar to the derivations in uncoupled CTRW model, we now also present the transport equations for Lévy walk. Let $\gamma(x, t)$ be the frequency of velocity change at position x at time t, which satisfies

$$\gamma(x, t) = \int_{-\infty}^{\infty} \int_0^t \phi(x', t')\gamma(x - x', t - t')dx'dt' + p_0(x)\delta(t), \tag{2.49}$$

where $\phi(x', t')$ is the new introduced coupled transition probability density satisfying

$$\phi(x', t') = \frac{1}{2}\delta(|x'| - v_0 t')\psi(t'). \tag{2.50}$$

It says that the velocity in each duration t' is v_0 or $-v_0$ and thus the displacement x' in this duration should be $v_0 t'$ or $-v_0 t'$. The last term in Eq. (2.49) means that the particles with initial distribution $p_0(x)$ choose velocities at time $t = 0$. On the other hand, the equation connecting $\gamma(x, t)$ and the current density $p(x, t)$ is

$$p(x, t) = \int_{-\infty}^{\infty} \int_0^t \Phi(x', t')\gamma(x - x', t - t')dx'dt', \tag{2.51}$$

where

$$\Phi(x', t') = \frac{1}{2}\delta(|x'| - v_0 t')\Psi(t') \tag{2.52}$$

is the probability density of running a distance x' and remaining in this state. This equation takes care of the fact that a particle is at the point (x, t) if it has started some time t' ago at $x \pm v_0 t'$ and is still in this state of the flight.

Equations (2.49) and (2.51) can be solved by using Fourier-Laplace transform, together with an additional technical complexity due to the coupling between time and space variables [Klafter *et al.* (1987); Zumofen and Klafter (1993)]. We resolve it by using the shift property of Fourier and Laplace transforms and obtain:

$$\tilde{\hat{p}}(k, \lambda) = \frac{[\hat{\Psi}(\lambda + ikv_0) + \hat{\Psi}(\lambda - ikv_0)]\hat{p}_0(k)}{2 - [\hat{\psi}(\lambda + ikv_0) + \psi(\lambda - ikv_0)]}. \tag{2.53}$$

2.1.2.1 *The mean squared displacement of Lévy walk*

It can be seen that $\tilde{\hat{p}}(k, \lambda)$ in Eq. (2.53) is symmetric with respect to spatial variable k, which means that the first moment of displacement is zero. The corresponding MSD can be obtained via the formula

$$\hat{M}_2(\lambda) = - \left. \frac{d^2\tilde{\hat{p}}(k, \lambda)}{dk^2} \right|_{k=0}. \tag{2.54}$$

Similar to the calculations of the uncoupled CTRW model, the asymptotic behavior with $k, \lambda \to 0$ is also considered here. For $1 < \alpha < 2$, two subleading terms in the expansion of $\hat{\psi}(\lambda)$ have to be included in order to capture the effect of the ballistic fronts in Eq. (2.48). By this way, the MSD is obtained for different exponent α:

$$\langle x^2(t) \rangle \propto \begin{cases} t^2 & 0 < \alpha < 1, \\ t^2/\ln t & \alpha = 1, \\ t^{3-\alpha} & 1 < \alpha < 2, \\ t\ln t & \alpha = 2, \\ t & \alpha > 2. \end{cases} \tag{2.55}$$

This equation implies that the diffusion behavior of Lévy walk is between normal and ballistic. A large α is related to a more suppressed diffusion behavior in Eq. (2.55). This phenomenon has an intuitive interpretation from microscopic description. Since the velocity is constant v_0, the farthest distance at measurement time t the particle could arrive at is $v_0 t$ or $-v_0 t$, which implies the fastest diffusion behavior is ballistic. For the very small α, the duration of a unidirection flight can be very long. Many particles undergo the long flights without changing direction, which contributes to the ballistic behavior. With the increase of exponent α, the shape of the PDF $\psi(t)$ of the running time becomes more centralized. It means that the particle is more likely to change the direction of motion and the diffusion behavior gets more suppressed. In contrast to the uncoupled CTRW model, the particle is running all the time without any rest or being trapped in a potential, and the subdiffusion behavior cannot be observed here.

2.1.2.2 *The propagator*

There is no uniform expression for the MSD of Lévy walk in Eq. (2.55) with different α. The characters of Lévy walk with different α are quite different. Besides the MSD, another important quantity is the propagator $p(x,t)$. The propagator $p(x,t)$, the inverse transform of Eq. (2.53), is not easy to be obtained due to the complexity of the coupled spatial and temporal variables. The explicit expression of $p(x,t)$ needs detailed scaling analyses on Eq. (2.53).

For the case of $0 < \alpha < 1$, the mean running time diverges. With the same distribution of running distance (or jump length), the scaling relation of Lévy flight is $x \sim t^{1/\alpha}$, which results in spreading faster than ballistic. For Lévy walk, however, the long time cost makes a truncation of the scaling at $x \sim t$, which results in a ballistic diffusion. For $0 < \alpha < 1$ and the initial distribution $p_0(x) = \delta(x)$, the asymptotic expansion of the propagator in Eq. (2.53) is

$$\tilde{p}(k, \lambda) \simeq \frac{(\lambda + ikv_0)^{\alpha-1} + (\lambda - ikv_0)^{\alpha-1}}{(\lambda + ikv_0)^{\alpha} + (\lambda - ikv_0)^{\alpha}}. \tag{2.56}$$

It can be seen the Fourier and Laplace variables appear in the scaling $\lambda \sim k$, consistent to the scaling $x \sim t$. The inverse Fourier-Laplace transform of Eq. (2.56) has a technical difficulty. However, based on the scaling relation $\lambda \sim k$, this difficulty can be resolved by using the technique in [Godrèche and Luck (2001)]. There is a special case of $\alpha = 1/2$, where the shape of the propagator is

$$p(x, t) = \frac{1}{\pi(v_0^2 t^2 - x^2)^{1/2}}. \tag{2.57}$$

For the case of $1 < \alpha < 2$, Lévy walk displays the subballistic superdiffusion behavior. Now, the situation becomes more complex than the case of $0 < \alpha < 1$. Similar to Lévy flight, the central part of the propagator of Lévy walk is also subjected to the generalized central limit theorem and given by the symmetric Lévy distribution

$$p_{\text{cen}}(x,t) \simeq \frac{1}{(K_\alpha t)^{1/\alpha}} L_{\alpha,0} \left(\frac{x}{(K_\alpha t)^{1/\alpha}} \right). \tag{2.58}$$

While the tail part of the propagator of Lévy walk scales as $x \sim t$ due to the constant velocity v_0. The two kinds of scaling relations coexist in the propagator of Lévy walk. They both play important roles in deciding the moments of Lévy walk and none of them can be ignored. To describe the scaling relation $x \sim t$ at the outmost fronts of the propagator, a scaled ballistic variable $\xi = x/t$ was introduced in [Rebenshtok *et al.* (2014a,b)], where the PDF of this variable is defined as

$$\mathcal{I}(\xi) = \lim_{t \to \infty} t^\alpha p_{\text{tail}}(x/t, t). \tag{2.59}$$

This function is non-normalizable, i.e.,

$$\int_{-\infty}^{\infty} \mathcal{I}(\xi) d\xi = \infty, \tag{2.60}$$

due to the power law singularity in the limit $\xi \to 0$, $\mathcal{I}(\xi) \propto |\xi|^{-1-\alpha}$. Therefore, $\mathcal{I}(\xi)$ is named as infinite density. The infinite density is non-normalizable, the concept of which was thoroughly investigated in mathematics issues [Aaronson (1997); Thaler and Zweimüller (2006)], and has been applied to physics successfully. Here, it is aimed to characterize the ballistic scaling ($x \sim t$) of Lévy walk. In addition, the infinite density is usually discussed together with infinite-ergodic theory, for example, the Brownian motion in a logarithmic potential [Aghion *et al.* (2019)] and the Langevin system with multiplicative noise [Leibovich and Barkai (2019); Wang *et al.* (2019c)]. The infinite density is still valuable in spite of its singularity. In [Rebenshtok *et al.* (2014a,b)], the explicit expression of an infinite density is given as

$$\mathcal{I}(\xi) = B \left[\frac{\alpha \mathcal{Q}_\alpha(|\xi|)}{|\xi|^{1+\alpha}} - \frac{(\alpha-1)\mathcal{Q}_{\alpha-1}(|\xi|)}{|\xi|^\alpha} \right], \tag{2.61}$$

where

$$\mathcal{Q}_\alpha(\xi) = \int_{|\xi|}^{\infty} dv v^\alpha h(v). \tag{2.62}$$

Here, the $h(v)$ is the velocity distribution, reducing to $(\delta(v - v_0) + \delta(v + v_0))/2$ for the case of two-point velocity $\pm v_0$. The two kinds of scaling relations meet in the intermediate region where both functions scale as

$$L_{\alpha,0} \left(\frac{x}{t^{1/\alpha}} \right) \simeq \mathcal{I} \left(\frac{x}{t} \right) \simeq x^{-1-\alpha}. \tag{2.63}$$

Based on the central part and tail part of the propagator $p(x,t)$, the fractional moments can be calculated as

$$\langle |x|^q \rangle \simeq 2 \int_0^{x_c(t)} x^q p_{\text{cen}}(x,t)dx + 2 \int_{x_c(t)}^\infty x^q p_{\text{tail}}(x,t)dx, \quad (2.64)$$

where the two PDFs match nearly perfectly at the point

$$x_c(t) = \left[\frac{c_\alpha}{L_{\alpha,0}(0)} \right]^{1/(1+\alpha)} (K_\alpha t)^{1/\alpha}, \quad (2.65)$$

and the asymptotic forms of $p_{\text{cen}}(x,t)$ and p_{tail} are presented in Eqs. (2.58) and (2.59), respectively. For $q > \alpha$, the second integral plays a leading role in Eq. (2.64), hence we neglect the first integral and obtain

$$\langle |x|^q \rangle \simeq 2t^{q+1-\alpha} \int_0^\infty \mathcal{I}(\xi)\xi^q d\xi. \quad (2.66)$$

While $q < \alpha$, the first integral dominates in the limit $t \to \infty$, and thus we have

$$\langle |x|^q \rangle \simeq 2(K_\alpha t)^{q/\alpha} \int_0^\infty L_{\alpha,0}(y)|y|^q dy. \quad (2.67)$$

The results of MSD in Eqs. (2.66) and (2.67) for subballistic Lévy walk implies that it is a strongly anomalous diffusion process, the definition of which can be illustrated by using a spectrum of fractional moments [Artuso and Cristadoro (2003); Sanders and Larralde (2006); de Anna *et al.* (2013); Rebenshtok *et al.* (2014b)]:

$$\langle |x|^q \rangle = \int_{-\infty}^\infty |x|^q p(x,t)dx \simeq M_q t^{q\nu(q)}. \quad (2.68)$$

Here, M_q is the diffusion coefficient depending on the fractional exponent q. The q-dependent function $\nu(q)$ gives process $x(t)$ a certain quality of strongly anomalous diffusion. If $\nu(q)$ reduces to a constant, such as $\nu(q) \equiv 1/2$ for Brownian motion and $\nu(q) \equiv \alpha/2$ for subdiffusive CTRW, the process is not strongly anomalous. Now for subballistic Lévy walk, the exponent of t is $q\nu(q) = q/\alpha$ for $q < \alpha$ and $q\nu(q) = q+1-\alpha$ for $q > \alpha$. The two equalities meet at the point $q_c = \alpha$.

Based on the analyses of subballistic Lévy walk for the case of $1 < \alpha < 2$, it is not surprising that Lévy walk with $\alpha > 2$ also possesses two kinds of scaling relations. The Gaussian scale $x \sim t^{1/2}$ dominates the central part while the ballistic scale $x \sim t$ makes a truncation at the tail. Therefore, it also displays strongly anomalous diffusion. With some detailed calculations in [Rebenshtok *et al.* (2016)], the fractional moments in Eq. (2.68) are

$$\langle |x|^q \rangle \propto \begin{cases} t^{q/2} & q < 2(\alpha - 1), \\ t^{q+1-\alpha} & q > 2(\alpha - 1). \end{cases} \tag{2.69}$$

Even it displays normal diffusion for the MSD for $\alpha > 2$ [i.e., $q = 2$], the effect of ballistic scale is reflected for higher order moments than $2(\alpha - 1)$.

2.1.2.3   *Extended Lévy walk*

We have discussed the MSD and the propagator of Lévy walk just now. All the quantities are evaluated from the expressions $\tilde{\tilde{p}}(k, \lambda)$ in Fourier-Laplace space. In this section, we will show some generalizations of Lévy walk and the corresponding derivations of the key quantity $\tilde{\tilde{p}}(k, \lambda)$.

The first natural generalization of Lévy walk is to assume that the velocity of the particle is not fixed but is a random variable [Zaburdaev *et al.* (2008)]. Let $h(v)$ be the PDF of velocity. Then the corresponding transport equations of random walks with random velocities are [Zaburdaev *et al.* (2008)]:

$$\gamma(x, t) = \int_{-\infty}^{\infty} \int_0^t \gamma(x - v\tau, t - \tau) h(v) \psi(\tau) d\tau dv + \delta(t) p_0(x) \tag{2.70}$$

and

$$p(x, t) = \int_{-\infty}^{\infty} \int_0^t \gamma(x - v\tau, t - \tau) h(v) \Psi(\tau) d\tau dv. \tag{2.71}$$

These two equations can be solved through Fourier-Laplace transform:

$$\tilde{\tilde{p}}(k, \lambda) = \frac{\int_{-\infty}^{\infty} \hat{\Psi}(\lambda + ikv_0) h(v) dv}{1 - \int_{-\infty}^{\infty} \hat{\Psi}(\lambda + ikv_0) h(v) dv}, \tag{2.72}$$

which recovers Eq. (2.53) for standard Lévy walk when

$$h(v) = \frac{1}{2}[\delta(v - v_0) + \delta(v + v_0)]. \tag{2.73}$$

Another extension leads to the intermittent search process [Bénichou *et al.* (2011)], which switches between local Brownian search phase and Lévy walk relocation phase. The searcher displays a slow reactive motion in the first phase, where the target can be detected. The latter fast phase aims at relocating into unvisited regions to reduce oversampling, during which the searcher is unable to detect the target. The particle might be in any one of the two phases at any time. Therefore, we use the notation "+/−" to represent the "Lévy walk/Brownian" phase for this two-state process. Suppose that the particles are initialized at the origin. We denote the joint PDF of finding the particle at position x and state "$\pm$" at time t as $p_\pm(x,t)$, which is associated with the original propagator by the relation $p(x,t) = p_+(x,t) + p_-(x,t)$. The subscript "$\pm$" will imply an identical meaning for other quantities.

The integral equations for $p_\pm(x,t)$ can be similarly obtained as the transport equations for CTRWs. Besides the sojourn time PDF $\psi_\pm(t)$ and survival probability $\Psi_\pm(t)$, we introduce the notation $G_\pm(x,t)$ to represent the conditional probability that a particle makes a displacement x during sojourn time t at one step in state "$\pm$". Their expressions are given by

$$G_+(x,t) = \delta(|x| - v_0 t)/2 \tag{2.74}$$

and

$$G_-(x,t) = \frac{1}{\sqrt{4\pi Dt}} \exp\left(-\frac{x^2}{4Dt}\right) \tag{2.75}$$

for Lévy walk and Brownian motion, respectively. Then the transport equation governing flux of particles $\gamma_\pm(x,t)$, which defines how many particles leave the position x and change from state "$\mp$" to state "$\pm$" per unit time, can be obtained. This equation connects the flux at the current point to the flux from the neighboring points in the past [Wang *et al.* (2019a)]:

$$\gamma_\pm(x,t) = \int_0^t dt' \int_{-\infty}^\infty dx' \psi_\mp(t') G_\mp(x',t') \gamma_\mp(x - x', t - t') \\ + p_\mp^0 \psi_\mp(t) G_\mp(x,t), \tag{2.76}$$

where we assume that the initial condition is $p_\pm(x, t = 0) = p_\pm^0 \delta(x)$ and the constant $p_\pm^0$ is the initial fraction of two states. The first term on the right-hand side shows that the particles could arrive at position x and state "$\pm$" from another point $x - x'$ after a displacement x' in state "$\mp$". While the second term denotes that the particles initialized in state "$\mp$" turn to state "$\pm$" after a complete step in state "$\mp$". The current density $p_\pm(x,t)$ of particles is connected to the neighboring flux $\gamma_\pm(x,t)$ by

$$p_\pm(x,t) = \int_0^t dt' \int_{-\infty}^\infty dx' \Psi_\pm(t') G_\pm(x',t') \gamma_\pm(x - x', t - t') \\ + p_\pm^0 \Psi_\pm(t) G_\pm(x,t). \tag{2.77}$$

The last term on the right-hand side accounts for the particles initialized in state "$\pm$" staying in this state until the observation time t. The first term sums over all possible neighboring flux $\gamma_\pm(x - x', t - t')$ that could arrive at the point x after displacement x' in state "$\pm$".

By performing Fourier-Laplace transform, we obtain

$$\tilde{\hat{p}}_\pm(k, \lambda) = \frac{p_\pm^0 + p_\mp^0 \tilde{\hat{\phi}}_\mp(k, \lambda)}{1 - \tilde{\hat{\phi}}_+(k, \lambda)\tilde{\hat{\phi}}_-(k, \lambda)} \, \tilde{\hat{\Phi}}_\pm(k, \lambda), \qquad (2.78)$$

where

$$\begin{aligned}
\tilde{\hat{\phi}}_+(k, \lambda) &= [\hat{\psi}_+(\lambda + iv_0k) + \hat{\psi}_+(\lambda - iv_0k)]/2, \\[4pt]
\tilde{\hat{\Phi}}_+(k, \lambda) &= [\hat{\Psi}_+(\lambda + iv_0k) + \hat{\Psi}_+(\lambda - iv_0k)]/2, \\[4pt]
\tilde{\hat{\phi}}_-(k, \lambda) &= \hat{\psi}_-(\lambda + Dk^2), \\[4pt]
\tilde{\hat{\Phi}}_-(k, \lambda) &= \hat{\Psi}_-(\lambda + Dk^2).
\end{aligned} \qquad (2.79)$$

Some information can be extracted from the complex expression of $\tilde{\hat{p}}_\pm(k, \lambda)$ in Eq. (2.78). For example, summing the two states to obtain the propagator $\tilde{\hat{p}}(k, \lambda)$; taking $k = 0$ to get the occupation fraction of two states $p_\pm(t)$ as the marginal density of finding the particles in state "$\pm$" at time t, which is

$$\hat{p}_\pm(\lambda) = \frac{p_\pm^0 + p_\mp^0 \hat{\psi}_\mp(\lambda)}{1 - \hat{\psi}_+(\lambda)\hat{\psi}_-(\lambda)} \frac{1 - \hat{\psi}_\pm(\lambda)}{\lambda}. \qquad (2.80)$$

Besides, by taking the diffusivity $D = 0$, we can obtain an interesting model—Lévy walk interrupted by rest. This process has been observed through the experiments on the flow in a rotating annulus as probed by tracer particles [Solomon *et al.* (1993)] and the frictionless motion of a particle in an "egg-crate" potential in a Hamiltonian system [Klafter and Zumofen (1994)] many years ago. Recently, it is also observed in the transport of the neuronal messenger ribonucleoproteins delivered to their target synapses [Song *et al.* (2018)].

2.1.3 *Fokker-Planck Equation*

Fokker-Planck equation is a partial differential equation which describes the time evolution of the PDF of particle's position. It was introduced in Fokker's thesis and independently obtained by Max Planck. The Fokker-Planck equation describing normal diffusion in an external force field is [de Groot and Mazur (1969); Risken (1989); Coffey *et al.* (2004); van Kampen (1992); Metzler and Klafter (2000b)]

$$\frac{\partial p(x, t)}{\partial t} = \left[\frac{\partial}{\partial x} \frac{V'(x)}{m\eta_1} + K_1 \frac{\partial^2}{\partial x^2} \right] p(x, t), \qquad (2.81)$$

where m is the particle's mass, η_1 represents the friction constant characterising the interaction between the particle and the surroundings, and $V(x)$ is the external potential so that the force is $F(x) = -dV(x)/dx$. The external force field usually exists, such as a constant electrical bias field exerting a force on charge carriers, a harmonic potential describing a bounded particle or a bistable potential in reaction dynamics or molecular switching processes. In the force-free limit, Eq. (2.81) reduces to the classical Fick's second law and the MSD is linear with respect to time t.

Nowadays, anomalous diffusions are attracting more and more interests in the fields of natural science. There have been many literatures focusing on deriving the fractional Fokker-Planck equations for anomalous diffusion processes in an external force field [Metzler and Klafter (2000b)]. For the simple force-free case, the Fokker-Planck equation can be obtained by performing the inverse Fourier-Laplace transform on the Montroll-Weiss equation (2.17). The three different Fokker-Planck equations can be obtained for the three examples given in Sec. 2.1.1.

(1) For normal diffusion with waiting time and jump length PDFs being

$$\hat{\psi}(\lambda) \simeq 1 - \lambda\tau, \qquad \tilde{w}(k) \simeq 1 - \sigma^2 k^2, \qquad (2.82)$$

respectively, the propagator is

$$\hat{\tilde{p}}(k,\lambda) \simeq \frac{1}{\lambda + Dk^2}. \qquad (2.83)$$

Rewriting the expression of propagator as

$$\lambda\hat{\tilde{p}}(k,\lambda) - 1 = -Dk^2\hat{\tilde{p}}(k,\lambda), \qquad (2.84)$$

and performing inverse Fourier-Laplace transform, we obtain

$$\frac{\partial p(x,t)}{\partial t} = D\frac{\partial^2 p(x,t)}{\partial x^2}. \qquad (2.85)$$

(2) For superdiffusion with finite mean waiting time and divergent second moment of jump length, i.e.,

$$\hat{\psi}(\lambda) \simeq 1 - \lambda\tau, \qquad \tilde{w}(k) \simeq 1 - \sigma^\beta|k|^\beta, \qquad (2.86)$$

the propagator satisfies

$$\lambda\hat{\tilde{p}}(k,\lambda) - 1 = -K_\beta|k|^\beta\hat{\tilde{p}}(k,\lambda), \qquad (2.87)$$

the inverse transform of which yields

$$\frac{\partial p(x,t)}{\partial t} = K_\beta\nabla_x^\beta p(x,t) \qquad (2.88)$$

with the notation ∇_x^β being the Riesz space fractional derivative operator [Wu *et al.* (2016); Carmi *et al.* (2010)], defined as

$$\nabla_y^\beta h(y) = -\frac{_{-\infty}D_y^\beta h(y) + _y D_\infty^\beta h(y)}{2\cos(\beta\pi/2)} \qquad (2.89)$$

with Fourier symbol $-|k|^\beta$. Here, for $n-1 < \beta < n$, $_{-\infty}D_y^\beta$ and $_y D_\infty^\beta$ are the Riemann-Liouville fractional derivative, respectively, defined as [Samko *et al.* (1993); Podlubny (1999)]

$$_{-\infty}D_y^\beta h(y) = \frac{1}{\Gamma(n-\beta)}\frac{d^n}{dy^n}\int_{-\infty}^y \frac{h(y')}{(y-y')^{\beta+1-n}}dy', \qquad (2.90)$$

$$_y D_\infty^\beta h(y) = \frac{(-1)^n}{\Gamma(n-\beta)}\frac{d^n}{dy^n}\int_y^\infty \frac{h(y')}{(y'-y)^{\beta+1-n}}dy'. \qquad (2.91)$$

(3) For subdiffusion with divergent mean waiting time and finite second moment of jump length, i.e.,

$$\hat{\psi}(\lambda) \simeq 1 - \tau^\alpha\lambda^\alpha, \qquad \tilde{w}(k) \simeq 1 - \sigma^2 k^2, \qquad (2.92)$$

the propagator solves

$$\lambda\tilde{\hat{p}}(k,\lambda) - 1 = -K_\alpha\lambda^{1-\alpha}k^2\tilde{\hat{p}}(k,\lambda), \qquad (2.93)$$

the inverse Fourier-Laplace transform of which is

$$\frac{\partial p(x,t)}{\partial t} = K_\alpha D_t^{1-\alpha}\frac{\partial^2 p(x,t)}{\partial x^2} \qquad (2.94)$$

with $D_t^{1-\alpha}$ being the temporal Riemann-Liouville fractional derivative operator.

For the case with an external force field, the Fokker-Planck equation cannot be derived from the Montroll-Weiss equation (2.17) directly, especially when both the mean waiting time and the second moment of jump length diverge. In this case, the random walk formulation and the transport equation should be extended by combining the effects of the nonlinear external force $F(x)$ on the system. Let us start from the discrete random walk, where the probabilities to jump right or left, $A(j)$ and $B(j)$, explicitly depend on the position j because of the external force. The corresponding discrete master equation becomes

$$p_j(t+\Delta t) = A_{j-1}p_{j-1}(t) + B_{j+1}p_{j+1}(t). \qquad (2.95)$$

For small space step Δx and time step Δt, the Taylor expansions for the three terms in Eq. (2.95) are

$$p_j(t + \Delta t) \simeq p(x, t) + \Delta t \frac{\partial p(x, t)}{\partial t},$$

$$A_{j-1}p_{j-1}(t) \simeq A(x)p(x, t) - \Delta x \frac{\partial A(x)p(x, t)}{\partial x} + \frac{(\Delta x)^2}{2} \frac{\partial^2 A(x)p(x, t)}{\partial x^2},$$

$$B_{j+1}p_{j+1}(t) \simeq B(x)p(x, t) + \Delta x \frac{\partial B(x)p(x, t)}{\partial x} + \frac{(\Delta x)^2}{2} \frac{\partial^2 B(x)p(x, t)}{\partial x^2},$$

$$(2.96)$$

respectively. Substituting the three expansions into Eq. (2.95) and using the relation $A(x) + B(x) = 1$ lead to the Fokker-Planck equation (2.81), with the appropriate limits

$$\frac{V'(x)}{m\eta_1} \equiv \lim_{\Delta x, \Delta t \to 0} \frac{\Delta x}{\Delta t}[B(x) - A(x)],$$

$$K_1 \equiv \lim_{\Delta x, \Delta t \to 0} \frac{(\Delta x)^2}{2\Delta t}.$$

$$(2.97)$$

The key of this derivation is to assume the time and space steps (Δt and Δx) are small at each step, which enables the Taylor expansions. However, the long rest and big jump might happen and thus result in a nonlocal master equation [Metzler *et al.* (1999b); Metzler and Klafter (2000b); Metzler (2001)]

$$p_j(t + \Delta t) = \sum_{n=1}^{\infty} A_{j,n}p_{j-n}(t) + \sum_{n=1}^{\infty} B_{j,n}p_{j+n}(t), \qquad (2.98)$$

where $A_{j,n}$ and $B_{j,n}$ represent the probabilities of jumping right from position $j - n$ to j and jump left from position $j + n$ to j, respectively. The corresponding continuum limit of this random walk model is

$$p(x, t) = \int_{-\infty}^{\infty} \int_0^t \psi(t - \tau)\Lambda(x, y)p(y, \tau)d\tau dy + \Psi(t)p_0(x), \qquad (2.99)$$

where the transfer kernel is [Metzler *et al.* (1999b); Metzler and Klafter (2000b); Metzler (2001)]

$$\Lambda(x, y) = w(x - y)[A(y)\Theta(x - y) + B(y)\Theta(y - x)], \qquad (2.100)$$

and $\Theta(x - y)$ is the Heaviside step function [$\Theta(x - y) = 1$ for $x > y$ and $\Theta(x - y) = 0$ for $x < y$]. Due to the effect of external force $F(x)$, the jump length PDF $\Lambda(x, y)$ of jumping from position y to x depends on both the departure site y and the arrival site x. The normalisation of Λ is embodied by

$$\int_{-\infty}^{\infty} \Lambda(x, x')dx = 1. \tag{2.101}$$

Through Fourier-Laplace transform on Eq. (2.99), we obtain the generalized master equation

$$\frac{\partial p(x,t)}{\partial t} = \int_{-\infty}^{\infty} \int_{0}^{t} K(x, y; t - \tau)p(y, \tau)d\tau dy, \tag{2.102}$$

where the Laplace transform of the kernel is

$$\hat{K}(x, x'; \lambda) \equiv \lambda\hat{\psi}(\lambda)\frac{\Lambda(x, x') - \delta(x)}{1 - \hat{\psi}(\lambda)}. \tag{2.103}$$

If considering the case with a diverging mean waiting time and a jump length PDF with infinite second moment, one arrives at the fractional Fokker-Planck equation

$$\frac{\partial p(x,t)}{\partial t} = D_t^{1-\alpha}\left[\frac{\partial}{\partial x}\frac{V'(x)}{m\eta_\alpha} + K_\alpha^\beta \nabla_x^\beta\right]p(x,t), \tag{2.104}$$

where the drift and diffusion coefficients are

$$\frac{V'(x)}{m\eta_\alpha} \equiv \frac{2\sigma}{\beta\tau^\alpha}[B(x) - A(x)],$$

$$K_\alpha^\beta \equiv \frac{\sigma^\beta}{\tau^\alpha}, \tag{2.105}$$

respectively. When $\alpha = 1$ and $\beta = 2$, the fractional Fokker-Planck equation (2.105) reduces to the classical one Eq. (2.81).

2.2 Langevin Equation

At the beginning of the twentieth century, many important researches are dedicated to the irregular motion of microscopic particles dispersed in a fluid which is named as Brownian motion. Although the motion is irregular, it reveals some regularity when analyzed statistically, e.g., the linear increase of MSD with respect to time. Einstein made fundamental contribution in this field, whose basic idea was to explain the Brownian motion by going beyond thermodynamics and into kinetic theory. The series of Einstein's articles [Einstein (1956)] together with the work of Smoluchowski [von Smoluchowski (1906)] on diffusions paved the way for the modern theory of Brownian motion. The next step forward was taken by Langevin who wrote down the equation of motion of the Brownian particle according to Newton's laws under the assumptions that the Brownian particles experience two kinds of forces. We will introduce the classical Langevin equation and some extensions on this equation in the following. Based on these equations, many important quantities can be evaluated.

2.2.1 *Classical Langevin Equation*

For a Brownian particle having a mass much larger than the colliding molecules, its motion causes from a great number of successive collisions, which is a condition for the central limit theorem to work and the Gaussian assumption for the Brownian particle is reasonable. The stochastic equation describing this kind of motion of free Brownian particle with mass $m = 1$ in one dimension satisfies Newton's Second Law and can be described as [Langevin (1908); Kubo (1966); Coffey *et al.* (2004)]

$$\ddot{x}(t) = -\gamma\dot{x}(t) + R(t), \tag{2.106}$$

where $x(t)$ and $\dot{x}(t) = v(t)$ is the position and velocity of a Brownian particle respectively, γ is a frictional constant, and the random fluctuation force $R(t)$ is white Gaussian noise with null mean value and correlation function

$$\langle R(t_1)R(t_2)\rangle = 2k_B\mathcal{T}\gamma\delta(t_1 - t_2), \tag{2.107}$$

where k_B is the Boltzmann constant and $\mathcal{T}$ the absolute temperature of the environment. The relation in Eq. (2.107) is known as the fluctuation-dissipation theorem. The effects from the surrounding molecules generally are two folds: one is the random driving force on the Brownian particle and another one is the frictional force for a forced motion. The two effects stem from the same source so that the two kinds of force are closely related through Eq. (2.107). The stochastic equation (2.106) is called underdamped Langevin equation, since the acceleration term $\ddot{x}(t)$ is considered.

It can be seen that the Brownian motion in Eq. (2.106) is a Gaussian process since the random force $R(t)$ is Gaussian. The PDF of a Gaussian process can be fully determined by its first and second moments. By assuming the initial condition $x(0) = 0$ and $v(0) = v_0$, the expression of velocity process $v(t)$ can be obtained from the Laplace transform of Eq. (2.106), i.e.,

$$v(t) = v_0 e^{-\gamma t} + \int_0^t e^{-\gamma(t-t')} R(t')dt', \tag{2.108}$$

which implies the mean value of velocity is

$$\langle v(t) \rangle = v_0 e^{-\gamma t}, \tag{2.109}$$

tending to zero for long times. Therefore, the mean value of displacement $x(t)$ also tends to zero for long times. The velocity correlation function can be obtained by use of the correlation function of noise $R(t)$ in Eq. (2.107), i.e.,

$$\langle v(t_1) v(t_2) \rangle = v_0^2 e^{-\gamma(t_1 + t_2)} + k_B \mathcal{T} \left(e^{-\gamma|t_1 - t_2|} - e^{-\gamma(t_1 + t_2)} \right). \tag{2.110}$$

For large t_1 and t_2, i.e., $\gamma t_1 \gg 1$, $\gamma t_2 \gg 1$, the asymptotic expression of velocity correlation function becomes independent of the initial velocity v_0

$$\langle v(t_1) v(t_2) \rangle \simeq k_B \mathcal{T} e^{-\gamma|t_1 - t_2|}. \tag{2.111}$$

We can see that the velocity correlation function Eq. (2.111) is only dependent on the time difference $|t_1 - t_2|$, which implies that the velocity process is asymptotically stationary. Then the variance of velocity process is $k_B \mathcal{T}$ by taking $t_1 = t_2$ in Eq. (2.111). The average energy of the Brownian particle with $m = 1$ is obtained as

$$\frac{1}{2} \langle v^2(t) \rangle = \frac{1}{2} k_B \mathcal{T}. \tag{2.112}$$

For large time t, the MSD of the stochastic process described by Eq. (2.106) is

$$\langle x^2(t) \rangle = 2 \int_0^t \int_0^{t_1} \langle v(t_1) v(t_2) \rangle dt_2 dt_1 \simeq 2Dt, \tag{2.113}$$

where $D = k_B \mathcal{T}/\gamma$ being the Einstein relation [Kubo (1966)]. That means the normal diffusion of Brownian particle is exhibited.

For the overdamped case, where the friction is very large so that the velocity reaches a steady state in a short time. Therefore, the acceleration can be ignored and only the effect of random force on the displacement is considered in Eq. (2.106) [Langevin (1908)], that is,

$$\dot{x}(t) = \frac{1}{\gamma} R(t). \tag{2.114}$$

The parameters and noise $R(t)$ are the same as the one in Eq. (2.106). So we have the null mean value of displacement and the MSD being

$$\langle x^2(t) \rangle = \frac{1}{\gamma^2} \int_0^t \int_0^t \langle R(t_1) R(t_2) \rangle dt_2 dt_1 = \frac{2k_B \mathcal{T} t}{\gamma}, \tag{2.115}$$

the same as the one in Eq. (2.113) for the underdamped case.

2.2.2 Generalized Langevin Equation

Here we pay attention to the generalized Langevin equation for a free particle with mass $m = 1$ and driven by internal noise. The underdamped generalized Langevin equation is [Coffey *et al.* (2004)]

$$\ddot{x}(t) = -\int_0^t K(t - \tau)\dot{x}(\tau)d\tau + \rho(t), \qquad (2.116)$$

where $x(t)$ denotes the particle displacement, $\dot{x}(t) = v(t)$ is the particle velocity, $K(t)$ is the friction memory kernel, and $\rho(t)$ is a random driving force subject to the conditions $\langle \rho(t) \rangle = 0$ and

$$\langle \rho(t_1)\rho(t_2) \rangle = k_B \mathcal{T} K(t_1 - t_2). \qquad (2.117)$$

Equation (2.117) is the fluctuation-dissipation theorem [Kubo (1966)], which is a generalization of Eq. (2.107). The corresponding overdamped generalized Langevin equation without Newton's acceleration term reads as

$$0 = -\int_0^t K(t - \tau)\dot{x}(\tau)d\tau + \rho(t), \qquad (2.118)$$

where $K(t)$ and $\rho(t)$ are the same as the ones in Eq. (2.116).

The generalizations of Langevin equation include two typical ones: fractional Langevin equation [Lutz (2001); Coffey *et al.* (2004); Kou and Xie (2004); Deng and Barkai (2009)] and tempered fractional Langevin equation [Chen *et al.* (2017); Molina-Garcia *et al.* (2018)], respectively, driven by the fractional Gaussian noise and tempered fractional Gaussian noise as the generalization of white Gaussian noise. The processes described by (tempered) fractional Langevin equation are still Gaussian due to the linearity of the generalized Langevin equation. However, they present anomalous diffusion behavior distinguishing from Brownian motion and classical Langevin equation. More details about these two processes are given as follows.

2.2.2.1 Fractional Brownian-Langevin motion

To introduce the fractional Langevin equation driven by fractional Gaussian noise which is defined through fractional Brownian motion, we have to firstly review the fractional Brownian motion defined as [Mandelbrot and Ness (1968)]

$$B_H(t) := \frac{1}{\Gamma\left(H + \frac{1}{2}\right)} \left(\int_0^t (t - \tau)^{H-1/2} dB(\tau) \right.$$

$$\left. + \int_{-\infty}^0 [(t - \tau)^{H-1/2} - (-\tau)^{H-1/2}] dB(\tau) \right), \qquad (2.119)$$

where the Hurst parameter $0 < H < 1$ and $B(t)$ is ordinary Brownian motion with variance t. Fractional Brownian motion $B_H(t)$ is the extension of Brownian motion $B(t)$ with a memory kernel in power law form. It can be seen that $B_H(t)$ recovers the Brownian motion $B(t)$ when $H = \frac{1}{2}$ in Eq. (2.119). The right-hand side of Eq. (2.119) is a sum of two Gaussian processes. For the first term, it is also named as fractional Brownian motion of Riemann-Liouville type [Lim and Muniandy (2002)]. From the definition of fractional Brownian motion in Eq. (2.119), the correlation function of fractional Brownian motion is [Samorodnitsky and Taqqu (1994); Deng and Barkai (2009)]

$$\langle B_H(t_1)B_H(t_2)\rangle = D_H(t_1^{2H} + t_2^{2H} - |t_1 - t_2|^{2H}), \tag{2.120}$$

where $t_1, t_2 > 0$ and $D_H = [\Gamma(1 - 2H)\cos(H\pi)]/(2H\pi)$. For $t_1 = t_2$, the variance is

$$\langle B_H^2(t)\rangle = 2D_H t^{2H}, \tag{2.121}$$

which means the subdiffusion phenomenon for $0 < H < 1/2$ and superdiffusion for $1/2 < H < 1$. Fractional Brownian motion is the Gaussian self-similar process with stationary nonindependent increment [Mandelbrot and Ness (1968)].

The fractional Gaussian noise, the random driving force of the fractional Langevin equation, is defined as the nonindependent increment process of fractional Brownian motion: [Mandelbrot and Ness (1968); Deng and Barkai (2009)]

$$\gamma(t) = \frac{B_H(t + h) - B_H(t)}{h} \tag{2.122}$$

for $h \ll t$, which is also formally denoted as $\gamma(t) = dB_H(t)/dt$. It is a stationary Gaussian process with the mean $\langle \gamma(t)\rangle = 0$ and the correlation function

$$\langle \gamma(t_1)\gamma(t_2)\rangle = 2D_H H(2H - 1)|t_1 - t_2|^{2H-2}, \tag{2.123}$$

by taking the derivatives with respect to t_1 and t_2 in the correlation function of fractional Brownian motion defined in Eq. (2.120). For $0 < H < 1$, the power law exponent in Eq. (2.123) is negative, which means that the correlation decreases with the increase of time difference.

Based on the correlation function of fractional Gaussian noise in Eq. (2.123) and fluctation-dissipation theorem, the fractional Langevin equation is given by

$$\ddot{x}(t) = -\int_0^t (t - \tau)^{2H-2}\dot{x}(\tau)d\tau + \varrho\gamma(t), \tag{2.124}$$

where the coefficient $\varrho = [k_B\mathcal{T}/(2D_H H(2H-1))]^{1/2}$. Considering the integrability of the memory kernel in Eq. (2.124), fractional Langevin equation is meaningful for $1/2 < H < 1$. By assuming the initial condition of position and velocity that $x(0) = 0$ and $v(0) = v_0$, the sample path $x(t)$ can be obtained analytically from the fractional Langevin equation (2.124) through the technique of Laplace transform:

$$x(t) = v_0 t E_{2H,2}(-\Gamma(2H-1)t^{2H})$$
$$+ \varrho \int_0^t (t-\tau) E_{2H,2}(-\Gamma(2H-1)(t-\tau)^{2H})\gamma(\tau)d\tau,$$

(2.125)

where the generalized Mittag-Leffler function is defined as

$$E_{\alpha,\beta}(t) = \sum_{n=0}^{\infty} \frac{t^n}{\Gamma(\alpha n + \beta)}.$$

(2.126)

By virtue of the asymptotic expression

$$E_{\alpha,\beta}(-t) \simeq \frac{t^{-1}}{\Gamma(\beta - \alpha)}$$

(2.127)

and the correlation function of noise $\gamma(t)$, the mean value of the sample path $x(t)$ is [Porrà *et al.* (1996); Deng and Barkai (2009); Chen *et al.* (2017)]

$$\langle x(t) \rangle = v_0 t E_{2H,2}(-\Gamma(2H-1)t^{2H})$$
$$\simeq \frac{v_0}{\Gamma(2H-1)\Gamma(2-2H)} t^{1-2H},$$

(2.128)

tending to zero for long times, and the MSD (denoted as $\langle (\Delta x(t))^2 \rangle$ due to the nonzero mean) is

$$\langle (\Delta x(t))^2 \rangle = 2k_B\mathcal{T}t^2 E_{2H,3}(-\Gamma(2H-1)t^{2H})$$
$$\simeq \frac{2k_B\mathcal{T}}{\Gamma(2H-1)\Gamma(3-2H)} t^{2-2H},$$

(2.129)

where the thermal initial condition $v_0^2 = k_B\mathcal{T}/m$ with $m = 1$ is assumed. Since $1/2 < H < 1$, the Langevin system described by Eq. (2.124) undergoes subdiffusion, which can model the dynamics of a single protein molecule [Kou and Xie (2004)].

The overdamped fractional Langevin equation [Min *et al.* (2005); Deng and Barkai (2009)] is

$$0 = -\int_0^t (t-\tau)^{2H-2}\dot{x}(\tau)d\tau + \varrho\gamma(t),$$

(2.130)

by neglecting the Newton's acceleration term. With similar procedures to the underdamped case, we obtain the position correlation function

$$\langle x(t_1)x(t_2)\rangle = D_F(t_1^{2-2H} + t_2^{2-2H} - |t_1 - t_2|^{2-2H}), \qquad (2.131)$$

where

$$D_F = \frac{k_B T \pi \csc[\pi(2 - 2H)]}{(2 - 2H)\Gamma^2(2H - 1)\Gamma^2(2 - 2H)}. \qquad (2.132)$$

The MSD of the overdamped Langevin equation is $\langle x^2(t)\rangle \simeq 2D_F t^{2-2H}$ obtained from Eq. (2.131), which displays the same subdiffusive behavior as Eq. (2.129) of underdamped case.

2.2.2.2   *Tempered fractional Brownian-Langevin motion*

Brownian motion can be also extended to tempered fractional Brownian motion, introduced by [Meerschaert and Sabzikar (2013)] with its definition

$$B_{H,\mu}(t) := \int_0^t e^{-\mu(t-x)}(t - x)^{H-1/2}dB(x)$$

$$+ \int_{-\infty}^0 [e^{-\mu(t-x)}(t - x)^{H-1/2} - e^{-\mu(-x)}(-x)^{H-1/2}]dB(x), \qquad (2.133)$$

where $\mu > 0$, the Hurst index $H > 0$. Tempered fractional Brownian motion modifies the power law kernel in the moving average representation of a fractional Brownian motion Eq. (2.119) by adding an exponential tempering which aims at limiting the memory since the memory in the natural world might not be infinitely long. The basic theory about tempered fractional Brownian motion was developed with application to modeling wind speed [Meerschaert and Sabzikar (2013)]. Tempered fractional Brownian motion is still Gaussian but not self-similar any more. However, it can be called generalized self-similar in the sense that for any $c > 0$

$$\{B_{H,\mu}(ct)\}_{t\in\mathbb{R}} \overset{d}{=} \{c^H B_{H,c\mu}(t)\}_{t\in\mathbb{R}}. \qquad (2.134)$$

The correlation function of tempered fractional Brownian motion is

$$\langle B_{H,\mu}(t_1)B_{H,\mu}(t_2)\rangle = \frac{1}{2}\left[C_{t_1}^2|t_1|^{2H} + C_{t_2}^2|t_2|^{2H} - C_{t_1-t_2}^2|t_1 - t_2|^{2H}\right], \qquad (2.135)$$

where

$$C_t^2 = \frac{2\Gamma(2H)}{(2\mu|t|)^{2H}} - \frac{2\Gamma(H + \frac{1}{2})K_H(\mu|t|)}{\sqrt{\pi}(2\mu|t|)^H} \qquad (2.136)$$

with $t \neq 0$ and $K_H(t)$ being the modified Bessel function of second kind. It is obvious that $\langle B_{H,\mu}(t) \rangle = 0$ and the variance of tempered fractional Brownian motion is

$$\langle B_{H,\mu}^2(t) \rangle = C_t^2 |t|^{2H} \simeq 2\Gamma(2H)(2\mu)^{-2H} \tag{2.137}$$

as $t \to \infty$ on account of $K_H(t) \simeq \sqrt{\pi}(2t)^{-\frac{1}{2}}e^{-t}$, which means that tempered fractional Brownian motion is a localization diffusion process; see Fig. 2.3.

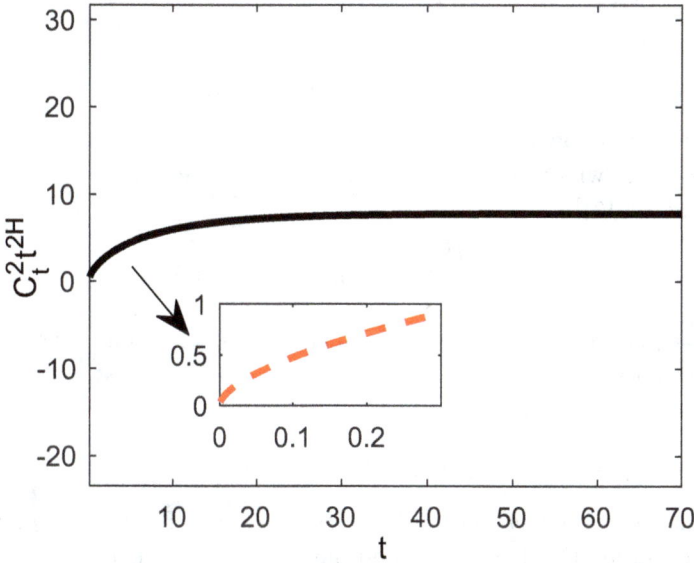

Fig. 2.3: Evolution of the asymptotic $C_t^2 |t|^{2H}$. The parameters $H = 0.3$ and $\mu = 0.1$. It tends to a constant for long times, which means the tempered fractional Brownian motion is localized.

Given a tempered fractional Brownian motion Eq. (2.133), the tempered fractional Gaussian noise can be defined as

$$\gamma(t) = \frac{B_{H,\mu}(t+h) - B_{H,\mu}(t)}{h}, \tag{2.138}$$

being similar to the definition of fractional Gaussian noise [Mandelbrot and Ness (1968)], where h is small and $h \ll t$. The mean of tempered fractional

Gaussian noise is $\langle\gamma(t)\rangle = 0$ and its correlation function is

$$\langle\gamma(t_1)\gamma(t_2)\rangle = \frac{1}{2h^2}(C_{t_1-t_2+h}^2|t_1-t_2+h|^{2H} + C_{t_1-t_2-h}^2|t_1-t_2-h|^{2H}$$
$$- 2C_{t_1-t_2}^2|t_1-t_2|^{2H}),$$

(2.139)

which means that the tempered fractional Gaussian noise is a stationary Gaussian process. For a fixed μ, the asymptotic behavior of the correlation function is

$$\langle\gamma(0)\gamma(t)\rangle \simeq -\Gamma(H+1/2)\mu^2(2\mu)^{-H-1/2}t^{H-1/2}e^{-\mu t}$$

(2.140)

for large t and

$$\langle\gamma(0)\gamma(t)\rangle \simeq 2D_H\Gamma^2(H+1/2)H(2H-1)t^{2H-2}.$$

(2.141)

Tempered fractional Langevin equation is driven by tempered fractional Gaussian noise, which is also a Gaussian process and can be written as [Chen *et al.* (2017)]

$$\ddot{x}(t) = -\int_0^t K(t-\tau)\dot{x}(\tau)d\tau + \varrho\gamma(t),$$

(2.142)

where $\varrho = \sqrt{2k_B\mathcal{T}}$. According to the fluctuation-dissipation theorem, the friction memory kernel is

$$K(t) = 2\langle\gamma(0)\gamma(t)\rangle$$
$$= h^{-2}(C_{t+h}^2|t+h|^{2H} + C_{t-h}^2|t-h|^{2H} - 2C_t^2|t|^{2H})$$

(2.143)

for a sufficiently small h, with $1/2 < H < 1$. For fixed small μ, with the time evolution the first and second moments of particle trajectory $x(t)$ behave like

$$\langle x(t)\rangle: \quad \sqrt{k_B\mathcal{T}}t \to Bt^{1-2H} \to At$$

(2.144)

and

$$\langle x^2(t)\rangle: \quad Dt^{2+2H} + k_B\mathcal{T}t^2 \to Ct^{2-2H} \to \sqrt{k_B\mathcal{T}}At^2.$$

(2.145)

Here

$$A = \sqrt{k_B\mathcal{T}}/[1 + 2\Gamma(2H)(2\mu)^{-2H}],$$
$$B = \sqrt{k_B\mathcal{T}}/[2D_H\Gamma^2(H+1/2)\Gamma(2H+1)\Gamma(2-2H)],$$
$$C = k_B\mathcal{T}/[D_H\Gamma^2(H+1/2)\Gamma(2H+1)\Gamma(3-2H)],$$
$$D = 4D_H\Gamma^2(H+1/2)k_B\mathcal{T}/(H+1).$$

(2.146)

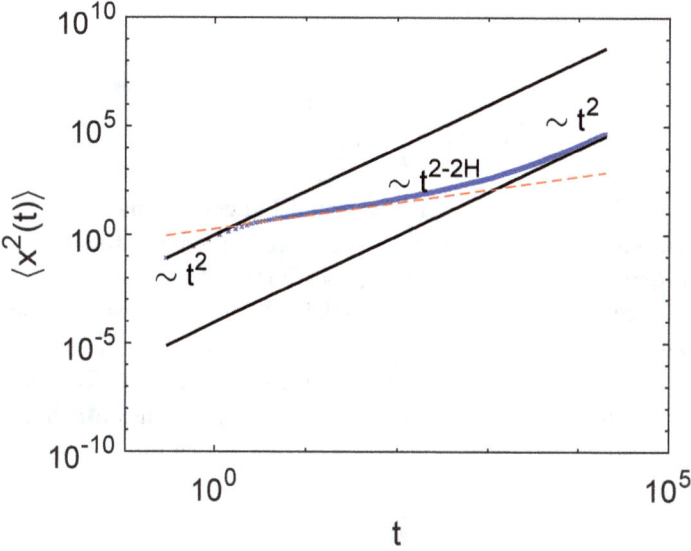

Fig. 2.4: Evolution of second moment in Eq. (2.145). The two parallel black solid lines are the theoretical results t^2 for short and long times, while the red dash line denotes the theoretical result t^{2-2H}.

In particular, for short times, from Eq. (2.145), it can be seen that $k_B \mathcal{T} t^2$ dominates MSD. While for long times, the particle displays ballistic diffusion, a special superdiffusion; see Fig. 2.4. For the overdamped case, the transition of the diffusion type is from subdiffusion to ballistic diffusion (from t^{2-2H} to t^2).

Another model displaying ballistic diffusion is the celebrated Lévy walk with power law exponent of waiting time less than 1 [Zaburdaev *et al.* (2015)], which has been introduced in Sec. 2.1.2. However, they are significantly different since the process $x(t)$ described by tempered fractional Langevin equation (2.142) is Gaussian while Lévy walk is not. The connection between Lévy walk model and the corresponding coupled Langevin system with α-stable subordinator is presented in [Eule *et al.* (2012)].

Another kind of tempered fractional Langevin equation is established with a truncated Mittag-Leffler memory kernel [Liemert *et al.* (2017)], which reads as

$$K(t) = \frac{\gamma}{\tau^{\alpha\delta}} e^{-bt} t^{\beta-1} E_{\alpha,\beta}^{\delta}\left(-\frac{t^\alpha}{\tau^\alpha}\right), \qquad (2.147)$$

where the parameters $\alpha, \beta, \gamma > 0$, $b, \delta \geq 0$, $\tau > 0$ is the characteristic time scale. The three-parameter Mittag-Leffler function is defined as [Podlubny (1999)]

$$E_{\alpha,\beta}^{\delta}(z) = \sum_{k=0}^{\infty} \frac{(\delta)_k}{\Gamma(\alpha k + \beta)} \frac{z^k}{k!}, \qquad (2.148)$$

where $(\delta)_k = \Gamma(\delta + k)/\Gamma(\delta)$. The fractional Langevin equation with truncated three-parameter Mittag-Leffler function is unrelated to the one driving by tempered fractional Gaussian noise. For the former case, it is shown that in presence of truncation, the particle changes from subdiffusive behavior in short times to normal diffusion in long times, distinguished from the localization of tempered fractional Langevin equation. For the overdamped case, the subdiffusion behavior and normal diffusion behavior are also found in short and long times, respectively, which are the same as the underdamped case [Liemert *et al.* (2017)].

2.2.3   *Langevin Equation Coupled with a Subordinator*

For some data in real life, such as biology, financial time series, ecology, and physics, the time-changed stochastic process is needed, where the deterministic time variable is replaced by a positive non-decreasing random process and thus a combination of two independent random processes is produced. One of the processes is called external process (or the original process), and another one is called internal process (or a subordinator). The idea of subordination was put forward by [Bochner (1949)] in 1949. Here we consider some time-changed Langevin systems, subordinated by inverse α-dependent process.

We have introduced the α-stable subordinator and the inverse α-stable subordinator in Sec. 1.6. Now we extend the subordinator $t(s)$ to be α-dependent, which has the characteristic function $(t \to \lambda)$:

$$\hat{z}(\lambda, s) = \langle e^{-\lambda t(s)} \rangle = e^{-s\Phi(\lambda)}. \qquad (2.149)$$

The Laplace symbol $\Phi(\lambda)$ can be [Baule and Friedrich (2005); Wang *et al.* (2019b)]

$$\Phi(\lambda) = \begin{cases} \lambda^{\alpha}, & 0 < \alpha < 1, \\ \mu_1 \lambda - \mu_\alpha \lambda^{\alpha}, & 1 < \alpha < 2. \end{cases} \qquad (2.150)$$

With the similar procedure in Sec. 1.6, the two-point PDF $z(t_1, s_1; t_2, s_2)$ of the subordinator $t(s)$ in Laplace space $(t_1 \to \lambda_1, t_2 \to \lambda_2)$ can be found by virtue of the statistical independence of the increments of subordinator $t(s)$ [Baule and Friedrich (2005)]

$$\hat{z}(\lambda_1, s_1; \lambda_2, s_2) = \langle e^{-\lambda_1 t(s_1)} e^{-\lambda_2 t(s_2)} \rangle$$

$$= \Theta(s_2 - s_1) e^{-s_1 \Phi(\lambda_1 + \lambda_2)} e^{-(s_2 - s_1)\Phi(\lambda_2)} \tag{2.151}$$

$$+ \Theta(s_1 - s_2) e^{-s_2 \Phi(\lambda_1 + \lambda_2)} e^{-(s_1 - s_2)\Phi(\lambda_1)}.$$

The PDF of the inverse α-dependent subordinator $s(t)$ in Laplace space $(t \to \lambda)$ is [Baule and Friedrich (2005)]

$$\hat{h}(s, \lambda) = \frac{\Phi(\lambda)}{\lambda} e^{-s\Phi(\lambda)}. \tag{2.152}$$

The two-point PDF $h(s_1, t_1; s_2, t_2)$ of the inverse subordinator $s(t)$ has the Laplace transform [Baule and Friedrich (2005); Wang *et al.* (2019b)]

$$\hat{h}(s_1, \lambda_1; s_2, \lambda_2)$$

$$= \frac{\partial}{\partial s_1} \frac{\partial}{\partial s_2} \frac{1}{\lambda_1 \lambda_2} z(\lambda_1, s_1; \lambda_2, s_2)$$

$$= \delta(s_2 - s_1) \frac{\Phi(\lambda_1) + \Phi(\lambda_2) - \Phi(\lambda_1 + \lambda_2)}{\lambda_1 \lambda_2} e^{-s_1 \Phi(\lambda_1 + \lambda_2)}$$

$$+ \Theta(s_2 - s_1) \frac{\Phi(\lambda_2)(\Phi(\lambda_1 + \lambda_2) - \Phi(\lambda_2))}{\lambda_1 \lambda_2} \tag{2.153}$$

$$\times e^{-s_1 \Phi(\lambda_1 + \lambda_2)} e^{-(s_2 - s_1)\Phi(\lambda_2)}$$

$$+ \Theta(s_1 - s_2) \frac{\Phi(\lambda_1)(\Phi(\lambda_1 + \lambda_2) - \Phi(\lambda_1))}{\lambda_1 \lambda_2}$$

$$\times e^{-s_2 \Phi(\lambda_1 + \lambda_2)} e^{-(s_1 - s_2)\Phi(\lambda_1)}.$$

Based on the generalized results of the α-dependent subordinator and inverse subordinator, the PDF $p(y, t)$ of the subordinated process $y(t) := y(s(t))$ can be written through the PDF $p_0(y, s)$ of the original process $y(s)$ [Baule and Friedrich (2005); Barkai (2001)]

$$p(y, t) = \int_0^\infty p_0(y, s) h(s, t) ds. \tag{2.154}$$

Then the moments of the subordinated process $y(t)$ in Laplace space can be obtained as

$$\mathcal{L}[\langle y^n(t) \rangle] = \int_0^\infty \langle y^n(s) \rangle \mathcal{L}[h(s, t)] ds, \tag{2.155}$$

which could be gotten by multiplying y^n on both sides of Eq. (2.154) and integrating about y. Similarly, the two-point PDF $p(y_1, t_1; y_2, t_2)$ of $y(t)$

can be connected with the two-point PDF $p_0(y_1, s_1; y_2, s_2)$ of the original stochastic process $y(s)$

$$p(y_1, t_1; y_2, t_2) = \int_0^\infty \int_0^\infty p_0(y_1, s_1; y_2, s_2) h(s_1, t_1; s_2, t_2) ds_1 ds_2. \quad (2.156)$$

The correlation function of $y(t)$ in Laplace space is

$$\langle \hat{y}(\lambda_1) \hat{y}(\lambda_2) \rangle = \int_0^\infty \int_0^\infty \langle y(s_1) y(s_2) \rangle \hat{h}(s_1, \lambda_1; s_2, \lambda_2) ds_1 ds_2. \quad (2.157)$$

2.2.3.1 *Subordinated Langevin equation*

The subordinated Langevin equation can be studied by using the technique of (inverse) subordinator. The most basic Gaussian process, describing normal diffusion, is Brownian motion with its corresponding Langevin equation [Coffey *et al.* (2004)]

$$\dot{x}(t) = \sqrt{2D}\xi(t), \quad (2.158)$$

where $D = k_B T / \gamma$, the random fluctuation force $\xi(t)$ is Gaussian white noise with null mean value and correlation function $\langle \xi(t_1) \xi(t_2) \rangle = \delta(t_1 - t_2)$. Modify the physical time t as the operational time s and consider the coupled Langevin equation

$$\dot{x}(s) = \sqrt{2D}\xi(s), \quad \dot{t}(s) = \eta(s), \quad (2.159)$$

where the Gaussian white noise $\xi(s)$ and the fully skewed α-stable Lévy noise $\eta(s)$ [Schertzer *et al.* (2001)] are independent noise sources. The random time transformation function $t(s)$ is an α-stable subordinator with $0 < \alpha < 1$ as usual. With the inverse α-stable subordinator $s(t)$, the combined process in physical time t is $x(t) := x(s(t))$. The coupled Langevin system in Eq. (2.159) describing subdiffusion is the continuous realization of the CTRW models with power law distributed waiting time and normal distributed jump length [Metzler and Klafter (2000b)], which has been proposed by Fogedby in [Fogedby (1994)]. Compared to Eq. (2.158), the subordinator $t(s)$ in Eq. (2.159) essentially changes the distribution of waiting time and thus eventually slows down the diffusion, i.e., turning normal diffusion into subdiffusion with MSD

$$\langle x^2(t) \rangle = \frac{2D}{\Gamma(1 + \alpha)} t^\alpha. \quad (2.160)$$

Different from the subordinated overdamped Langevin equation (2.159) which is equivalent to the subdiffusive process in CTRW framework, we consider another kind of subordinated Langevin equation—underdamped

Langevin system coupled with a subordinator [Wang *et al.* (2019b)]:

$$\dot{x}(t) = v(t), \qquad \dot{v}(s) = -\gamma v(s) + \sqrt{2D}\xi(s), \qquad \dot{t}(s) = \eta(s), \qquad (2.161)$$

which could describe the Lévy-walk-like process saying that a particle moves ballistically for a random time and then randomly changes direction but keeps the same magnitude of velocity [Zaburdaev *et al.* (2015)]. In the case of $\gamma \neq 0$ in Eq. (3.206), the velocity $v(s)$ in Eq. (3.206) is not a Brownian motion, but an Ornstein-Uhlenbeck process [Risken (1989)], which ensures a steady state of the diffusivity dynamics with respect to velocity for long times. The velocity process $v(s)$ can be analytically expressed as Eq. (2.108) after substituting the operational time s for physical time t. Then the correlation function of velocity process is Eq. (2.110) (substituting the operational time s for physical time t). By virtue of the subordination method Eq. (2.157), the correlation function of velocity process in Laplace space ($t_1 \to \lambda_1, t_2 \to \lambda_2$) with small λ_1 and λ_2 is

$$\langle \hat{v}(\lambda_1)\hat{v}(\lambda_2) \rangle \simeq \frac{D}{\gamma \lambda_1 \lambda_2} \cdot \frac{\Phi(\lambda_1) + \Phi(\lambda_2) - \Phi(\lambda_1 + \lambda_2)}{\Phi(\lambda_1 + \lambda_2)}. \qquad (2.162)$$

By performing the inverse Laplace transform, for large times t_1, t_2 and $t_1 < t_2$, the velocity correlation function can be gotten as [Wang *et al.* (2019b)]

$$\langle v(t_1)v(t_2) \rangle \simeq \begin{cases} \frac{D}{\gamma}\frac{\sin(\pi\alpha)}{\pi} B\left(\frac{t_1}{t_2}; \alpha, 1-\alpha\right), & 0 < \alpha < 1 \\ \frac{D}{\gamma}\left[(t_2 - t_1)^{1-\alpha} - t_2^{1-\alpha}\right], & 1 < \alpha < 2. \end{cases} \qquad (2.163)$$

Then the MSD $\langle x^2(t) \rangle = \int_0^t \int_0^t \langle v(t_1')v(t_2') \rangle dt_1' dt_2'$ of the stochastic process described by Eq. (2.161) is

$$\langle x^2(t) \rangle \simeq \begin{cases} \frac{D}{\gamma}(1-\alpha)t^2, & 0 < \alpha < 1, \\ \frac{D}{\gamma}\frac{2(\alpha-1)}{(2-\alpha)(3-\alpha)}t^{3-\alpha}, & 1 < \alpha < 2, \end{cases} \qquad (2.164)$$

which is the same as the one in Lévy walk model [Froemberg and Barkai (2013c)].

As the discussions above, Fogedby proposed the coupled Langevin equation (2.159) to describe the process in CTRWs [Fogedby (1994)], where $\xi(s)$ and $\eta(s)$ are independent with each other, characterizing the jump lengths and waiting times, respectively. But for Lévy walk, its corresponding Langevin picture should be presented like Eq. (3.206), where the derivative of position x with respect to physical time t is velocity v and the subordinator $t(s)$ characterizes the distribution of duration of each flight.

The second equation in Eq. (3.206) gives the distribution of velocity v. One special case that $\eta(s)$ has α-stable distribution ($0 < \alpha < 1$) and

$$v(s) = \gamma^{-1}\xi_d(s)$$

has been pointed out in [Eule *et al.* (2012)], where $\xi_d(s)$ is a dichotomous noise source, i.e., a random sequence of the values -1 and 1. It is just a one-to-one correspondence to the standard Lévy walk with the exponent of waiting time distribution less than 1. In general, the distribution of velocity v could be various, such as, Gaussian distribution, exponential distribution, and uniform distribution [Rebenshtok *et al.* (2014b,a)]. In more general cases, velocity v may be fluctuant due to a random force [Karatsas and Shreve (1977)] and thus its distribution becomes time-dependent. All in all, velocity v can be described by a Langevin equation, i.e., the second equation of Eq. (2.161). The nonzero constant γ makes sure a steady state of velocity v could be reached for long times, analogously to the finite moments of v in Lévy walk. In a word, the overdamped Langevin equation with a subordinator Eq. (2.159) corresponds to CTRWs, while the underdamped Langevin equation coupled with a subordinator Eq. (2.161) corresponds to Lévy walks.

We simulate one trajectory of the Langevin equation (2.161) in Fig. 2.5, including the inverse subordinator $s(t)$, the velocity process $v(t)$, and position process $x(t)$. In either the left one or the right one, it can be seen that the constant time periods of $s(t)$ represent the trapping events, where $v(t)$ is keeping its current state and $x(t)$ is experiencing the flight events, like Lévy walk. For the comparison of the two cases $\gamma = 0$ and $\gamma \neq 0$, we take one specific realization $\eta(s)$ in Eq. (2.161). So the trajectories of $s(t)$ are the same in the two panels of Fig. 2.5. The fluctuation of velocity $v(t)$ in case $\gamma = 0$ is more obvious than that for $\gamma = 1$, which implies the ensemble averaged MSD for $\gamma = 0$ is much larger than that for $\gamma \neq 0$.

2.2.3.2 *Subordinated generalized Langevin equation*

If the solution $x(t)$ of fractional Langevin equation (2.124) is subordinated by an inverse α-stable subordinator $s(t)$ with $0 < \alpha < 1$, then the subordinated stochastic process could be described by the following coupled fractional Langevin equation

$$\ddot{x}(s) = -\int_0^s (s-\tau)^{2H-2}\dot{x}(\tau)d\tau + \varrho\gamma(s), \qquad \dot{t}(s) = \eta(s). \qquad (2.165)$$

By virtue of the relation in Eq. (2.155) between the moments of the subordinated process and the one of original process, the MSD of subordinated

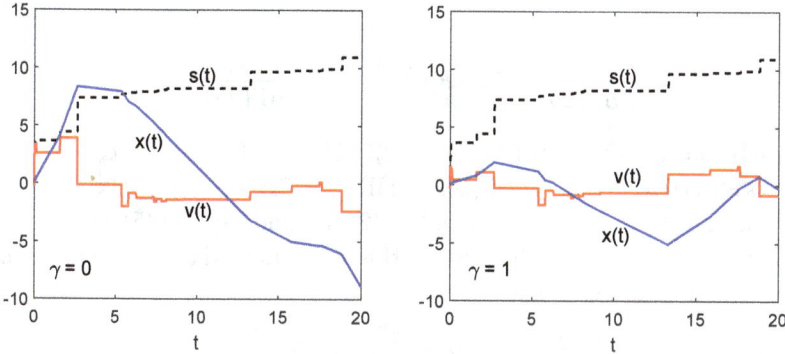

Fig. 2.5: Sample trajectories of inverse subordinator $s(t)$ (black dash line), velocity process $v(t)$ (red solid line), and position process $x(t)$ (blue solid line) of Langevin system in Eq. (2.161) for $T = 20$, $D_v = 1$, $\tau_0 = 1$, $\alpha = 1.8$, $\gamma = 0$ (left) and $\gamma = 1$ (right). We take one specific realization of $\eta(s)$ in Eq. (2.161) for the comparison of two cases $\gamma = 0$ and $\gamma \neq 0$, implying that the samples of inverse subordinator $s(t)$ (black dash line) used in the two cases are the same.

process $x(t)$ for large physical time t is

$$\langle (\Delta x(t))^2 \rangle \simeq \frac{2k_B T}{\Gamma(2H - 1)\Gamma((2 - 2H)\beta + 1)} t^{(2-2H)\alpha} \qquad (2.166)$$

with $0 < \alpha < 1$. It also undergoes subdiffusion and becomes slower than original process in Eq. (2.129). It can be seen that after performing the α-stable subordination on the fractional Langevin equation, the corresponding MSD can be easily obtained by replacing the parameter H in the MSD of fractional Langevin equation with $1 - (1 - H)\alpha$.

Now, we turn to the time-changed tempered fractional Langevin equation coupled with α-stable subordinator [Chen *et al.* (2018)]

$$\ddot{x}(s) = -\int_0^s K(s - \tau)\dot{x}(\tau)d\tau + \varrho\gamma(s), \quad \dot{t}(s) = \eta(s), \qquad (2.167)$$

where the first equation has been shown in Eq. (2.142). According to Eq. (2.155), the first and second moments of the subordinated process $x(t) := x(s(t))$ behave as

$$\langle x(t) \rangle : \frac{\sqrt{k_B T}}{\alpha \Gamma(\alpha)} t^\alpha \to E\, t^{(1-2H)\alpha} \to \frac{A}{\alpha \Gamma(\alpha)} t^\alpha, \qquad (2.168)$$

and

$$\langle x^2(t)\rangle : \quad \frac{k_B T}{\alpha\Gamma(2\alpha)}\, t^{2\alpha} \rightarrow F\,t^{(2-2H)\alpha} \rightarrow \frac{\sqrt{k_B T}\,A}{\alpha\Gamma(2\alpha)}\, t^{2\alpha}, \qquad (2.169)$$

where $E = \sqrt{k_B T}/[2D_H\Gamma^2(H+1/2)\Gamma(2H+1)\Gamma((1-2H)\alpha+1)]$ and $F = k_B T/[D_H\Gamma^2(H+1/2)\Gamma(2H+1)\Gamma((2-2H)\alpha+1)]$. These asymptotic behaviors are consistent with Eq. (2.144) and Eq. (2.145) when $\alpha = 1$. With the time evolution, the MSD of this subordinated tempered fractional Langevin equation goes like

$$\langle(\Delta x(t))^2\rangle : \left(\frac{k_B T}{\alpha\Gamma(2\alpha)} - \frac{k_B T}{\alpha^2\Gamma^2(\alpha)}\right) t^{2\alpha} \rightarrow F t^{(2-2H)\alpha} - E^2 t^{2(1-2H)\alpha}$$

$$\rightarrow \left(\frac{\sqrt{k_B T}\,A}{\alpha\Gamma(2\alpha)} - \frac{A^2}{(\alpha\Gamma(\alpha))^2}\right) t^{2\alpha}.$$

$$(2.170)$$

The simulation results of MSD for different α are given in Fig. 2.6, displaying the transition procedure with the time evolution. To observe the middle stage clearly, we take a moderately small $\mu = 0.001$. In Fig. 2.6, it can be found that the simulation results of MSD are consistent with the theoretical ones in Eq. (2.170) through the whole procedure. Comparing the measurement times in Fig. 2.6 (a) and (b), we find that for bigger α, the diffusion rate becomes faster, then the time reaching the final state becomes shorter. Especially, for large times, the diffusion of particle described by the subordinated tempered fractional Langevin equation (2.167) is slower than the original process exhibiting ballistic diffusion, and could be subdiffusion when $0 < \alpha < 1/2$, superdiffusion when $1/2 < \alpha < 1$, and even normal diffusion as $\alpha = 1/2$.

2.2.4   *Langevin Equation with External Force*

In the natural world, it is hard to find the real free particles; actually almost all the time, they are in some kinds of external potentials. The motion of particles in complex disordered systems generally is no longer Brownian, exhibiting anomalous diffusion behavior. Compared with CTRW, the Langevin equation is a convenient model for describing the diffusion trajectory of particles under the influence of external forces. In this subsection, we present the subdiffusion particle in different kinds of external force field. Different external forces and different modes of action of the same external force on particles affect the final motion of the particles.

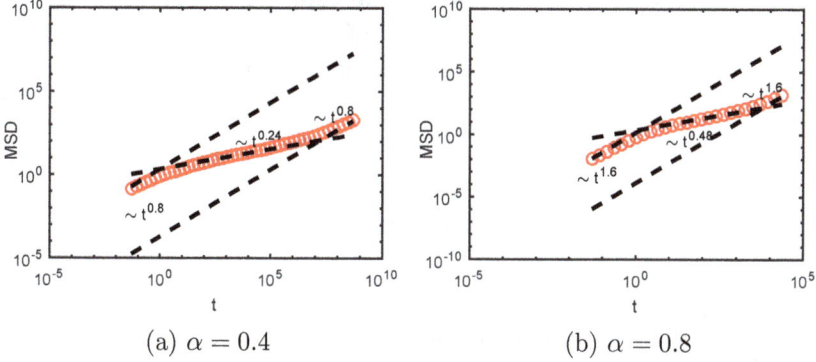

(a) $\alpha = 0.4$ (b) $\alpha = 0.8$

Fig. 2.6: Evolution of the MSD of the subordinated tempered fractional Langevin equation for different α. Dashed lines are the analytical results in Eq. (2.170) and the circle-markers are the computer simulation results. The measurement times are 4.9×10^8 in (a) and 2.2×10^4 in (b) respectively. Parameter values: $H = 0.7$, $\mu = 0.001$, $k_B \mathcal{T} = 1$, $\alpha = 0.4$ in (a) and $\alpha = 0.8$ in (b).

2.2.4.1 *Subordinated Langevin equation with external force*

As we have known, the phenomena of subdiffusion are widely observed in physical and biological systems. To investigate the effects of external potentials, the subordinated Langevin equation is a powerful tool. Next, we will introduce the subordinated Langevin equation with some external forces (e.g., harmonic potential, linear potential, and time dependent force), and show some statistics describing diffusion properties and the corresponding Fokker-Planck equations.

Consider the Langevin system with a harmonic potential on the original process $x(s)$ [Eule *et al.* (2007); Burov *et al.* (2010)]

$$\dot{x}(s) = -\gamma x(s) + \sqrt{2D}\xi(s), \qquad \dot{t}(s) = \eta(s), \qquad (2.171)$$

where γ is a positive constant. The harmonic potential $V(x) = \gamma x^2/2$ leads to a friction-like force $F(x) = -dV(x)/dx = -\gamma x$ in the first equation of Eq. (2.171). Based on Eq. (2.171), a new single Langevin equation in physical time t of the subordinated process $y(t) = x(s(t))$ can be obtained as [Chen *et al.* (2019)]

$$\dot{y}(t) = -\gamma y(t)\dot{s}(t) + \sqrt{2D}\,\overline{\xi}(t), \qquad (2.172)$$

with $\overline{\xi}(t) = \int_0^\infty \xi(\tau)\delta[t - t(\tau)]d\tau$, or equivalently, $\overline{\xi}(t) = \xi(s(t))\dot{s}(t)$. The noise $\overline{\xi}(t)$ here can be regarded as the formal derivative of the time-changed

Brownian motion $B(s(t))$. The external force in Eq. (2.171) only changes the motion of the particles at the instant of jumps; in fact, this mechanism can be easily found from the equivalent Langevin equation in physical time in Eq. (2.172), i.e., when a particle suffers a trapping event before next jump, the internal time process $s(t)$ remains a constant and the external force becomes zero due to $\dot{s}(t) = 0$ in Eq. (2.172).

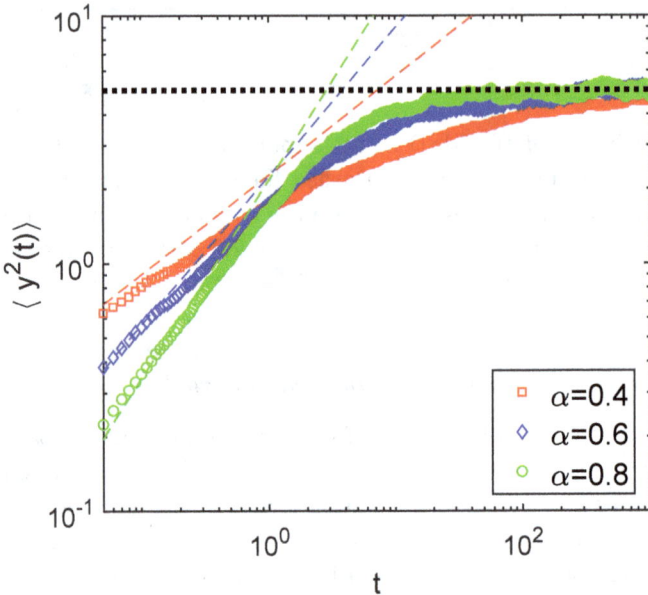

Fig. 2.7: Simulation results of the MSD of stochastic process in Eq. (2.171) for different α. Color markers represent the simulation results of MSD with parameters $D = 1$ and $\gamma = 0.2$ averaging over 2000 trajectories. Dashed lines (red for $\alpha = 0.4$, blue for $\alpha = 0.6$ and green for $\alpha = 0.8$) represent the asymptotic theoretical values of MSD for short times; Dotted black line represents the asymptotic saturation plateau value.

Using the subordinated method shown in Sec. 2.2.3, it can be gotten that the first moment of the stochastic process $y(t)$ is zero due to symmetry and the MSD is

$$\langle y^2(t) \rangle = \frac{D}{\gamma} - \frac{D}{\gamma} E_\alpha(-2\gamma t^\alpha), \qquad (2.173)$$

by utilizing $\langle x^2(s) \rangle = \frac{D}{\gamma}(1 - e^{-2\gamma s})$ with the initial position $x_0 = 0$, where

$E_\alpha(z)$ is the Mittag-Leffler function [Erdelyi (1981)]. For long times $t \gg (2\gamma)^{-\frac{1}{\alpha}}$, one has

$$\langle y^2(t) \rangle \simeq \frac{D}{\gamma} - \frac{D}{2\gamma^2\Gamma(1-\alpha)} t^{-\alpha}, \qquad (2.174)$$

because of the asymptotic expression

$$E_\alpha(-2\gamma t^\alpha) \simeq \frac{t^{-\alpha}}{2\gamma\Gamma(1-\alpha)} \quad \text{for large } t. \qquad (2.175)$$

The saturation plateau value, denoted as $\langle y^2 \rangle_{\text{th}} = \frac{D}{\gamma}$, is approached at the power law rate. Compared with the original process $x(s)$, the MSD of which relaxes to the value $\frac{D}{\gamma}$ exponentially, the subordinator $s(t)$ in this model only changes the convergence rate but keeps the same saturation plateau value. The simulation results for different α are shown in Fig. 2.7. It can be seen that the MSD with a smaller α tends to the saturation plateau value more slowly, being an expected dynamical behavior within a confined harmonic potential due to smaller α corresponding to longer waiting time. This process behaves as a localized diffusion for long times. Compared with the original process $x(s)$, the MSD of which relaxes to the value $\frac{\sigma}{\gamma}$ exponentially, the subordinator $s(t)$ in this model only changes the convergence rate but keeps the same saturation plateau value.

The Fokker-Planck equation is the equation governing the PDF of finding the particle at position x at time t. The Fokker-Planck equation corresponding to the Langevin equation (2.171) is [Metzler *et al.* (1999a); Metzler and Klafter (2000b); Gajda and Magdziarz (2010)],

$$\frac{\partial p(y,t)}{\partial t} = \mathcal{L}_{\text{FP}} D_t^{1-\alpha} p(y,t) \qquad (2.176)$$

with the Fokker-Planck operator $\mathcal{L}_{\text{FP}} = -\frac{\partial}{\partial y}F(y) + D\frac{\partial^2}{\partial y^2}$ (here $F(y) = -\gamma y$), which can be derived through the relation between the PDF of subordinated process and original process [Chen *et al.* (2018)].

Then we pay attention to the case that the harmonic potential acts on subordinated process $y(t)$ in physical times [Cairoli and Baule (2015b)]:

$$\dot{y}(t) = -\gamma y(t) + \sqrt{2D}\,\bar{\xi}(t), \qquad (2.177)$$

where $\bar{\xi}(t)$ is the same noise as the one in Eq. (2.172). Here the force keep acting on the system all the time, even in the constant period of inverse subordinator $s(t)$. Actually, the Langevin equation (2.177) can be rewritten as a coupled Langevin system with a subordinator as

$$\dot{x}(s) = -\gamma x(s)\eta(s) + \sqrt{2D}\xi(s), \qquad \dot{t}(s) = \eta(s). \qquad (2.178)$$

More precisely, the solution of Eq. (2.177) is

$$y(t) = \sqrt{2D} \int_0^t e^{-\gamma(t-\tau)} \overline{\xi}(\tau) d\tau \qquad (2.179)$$

with initial condition $y_0 = 0$, which is equivalent to

$$x(s) = \sqrt{2D} \int_0^s e^{-\gamma(t(s)-t(\tau))} dB(\tau), \qquad (2.180)$$

by replacing s with $s(t)$. Compared with Eq. (2.171), the friction term $-\gamma x(s)$ is multiplied by the Lévy noise $\eta(s)$, which acts as a multiplicative noise in the first equation in Eq. (2.178).

From Eq. (2.179), it can be calculated that the mean of $y(t)$ is zero and the MSD is

$$\langle y^2(t) \rangle = \frac{2D}{\Gamma(1+\alpha)} e^{-2\gamma t} t^\alpha \, {}_1F_1(\alpha, 1+\alpha; 2\gamma t), \qquad (2.181)$$

where the confluent hypergeometric function is [Abramowitz and Stegun (1972)]

$$ {}_1F_1(a, b; z) = \frac{\Gamma(b)}{\Gamma(a)\Gamma(b-a)} \int_0^1 e^{zu} u^{a-1}(1-u)^{b-a-1} du. \qquad (2.182)$$

For long times $t \gg (2\gamma)^{-1}$, using the asymptotic expansion

$$ {}_1F_1(a, b; z) \simeq \Gamma(b) \left(e^z z^{a-b}/\Gamma(a) + (-z)^{-a}/\Gamma(b-a) \right) \qquad (2.183)$$

for large z [Abramowitz and Stegun (1972)], we get

$$\langle y^2(t) \rangle \simeq \frac{D}{\gamma\Gamma(\alpha)} t^{\alpha-1}, \qquad (2.184)$$

which tends to zero at the power law rate. The consistency between simulation and the theoretical results about the MSD of model in Eq. (2.177) can be found in Fig. 2.8. There is an interesting physical interpretation of the model in Eq. (2.177) that it is equivalent to the problem of a sub-diffusive CTRW particle moving in an exponentially contracting medium by comparing the Langevin equation (2.177) with Eqs. (9) and (32) of Ref. [Yuste *et al.* (2016)] and Eq. (47) of Ref. [Le Vot *et al.* (2017)] replacing H by $-\gamma$. In addition, the MSDs in Eqs. (2.181) and (2.184) here are equal to Eqs. (79) and (83) of Ref. [Le Vot *et al.* (2017)].

Different from Eq. (2.171), the subordinator in this model changes not only the convergence rate but also the stationary value of MSD for long times. The external force in this model damps the oscillation of the particle in harmonic potential and drags it towards zero for all times; while the

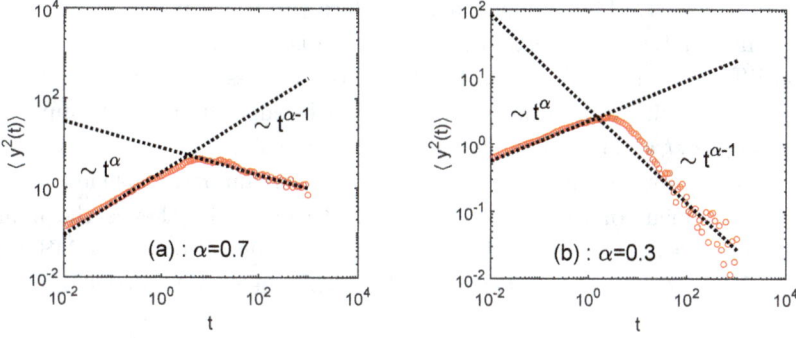

Fig. 2.8: Simulation results of the MSD of stochastic process described by the Langevin equation (2.177) for different α. The parameters are, respectively, taken as $D = 1$, $\gamma = 0.1$, $\alpha = 0.7$ (a) or $\alpha = 0.3$ (b), and the initial position $y_0 = 0$. Dotted black lines represent the asymptotic theoretical values in Eq. (2.184). Red circle-markers are the simulation results.

subordinated process described by Eq. (2.171) does not get dragged to zero position during waiting times since the external force is zero during these time periods.

In the following, we focus on the subdiffusive dynamics in linear potential, which could be described by the subordinated Langevin equation with constant external force [Cairoli and Baule (2015a); Dieterich *et al.* (2015)]

$$\dot{x}(s) = F + \sqrt{2D}\xi(s), \qquad \dot{t}(s) = \eta(s). \qquad (2.185)$$

The corresponding single Langevin equation of $y(t) = x(s(t))$ in physical time t is

$$\dot{y}(t) = F\dot{s}(t) + \sqrt{2D}\,\bar{\xi}(t), \qquad (2.186)$$

which evidently shows that the external force only acts at the moments of jump and does not affect the particle during waiting times (trapping events). In addition, the Fokker-Planck equation with respect to this Langevin system is also Eq. (2.176) by replacing $F(y)$ with F.

The first moment and MSD of the subordinated process $y(t) = x(s(t))$ are obtained as [Compte *et al.* (1997); Metzler and Klafter (2000b); Dieterich *et al.* (2015); Chen *et al.* (2018)]

$$\langle y(t) \rangle = \frac{F}{\Gamma(1+\alpha)} t^\alpha$$

$$\langle (\Delta y(t))^2 \rangle = \left(\frac{F^2}{\alpha\Gamma(2\alpha)} - \frac{F^2}{\alpha^2\Gamma^2(\alpha)} \right) t^{2\alpha} \qquad (2.187)$$

with $0 < \alpha < 1$. The simulation results are presented in Fig. 2.9. This subordinated Langevin system shows subdiffusion when $0 < \alpha < \frac{1}{2}$ and superdiffusion when $\frac{1}{2} < \alpha < 1$. It is more or less interesting that the waiting time with infinite mean value produces superdiffusion. Compte *et al.* [Compte *et al.* (1997)] explained this phenomenon that some stagnated particles are not continuously dragged by the stream and thus slow down the advancement of the center of mass of the particles, hence the main dispersion mechanism should be convection. The deviation of the MSD of Langevin systems with constant force from the one of free particle implies this external force is a biasing force.

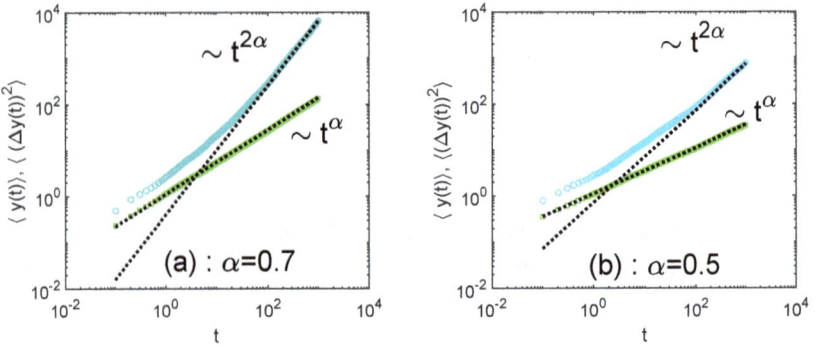

Fig. 2.9: Simulation results of the mean value and MSD of stochastic process described by the Langevin equation (2.185) for different α. The parameters are, respectively, taken as $D = 1$, $F = 1$, $\alpha = 0.7$ (a) or $\alpha = 0.5$ (b), and the initial position $y_0 = 0$. Dotted black lines represent the theoretical results. While the cyan circle-markers and green square-markers represent the simulation results of MSD and first moment, respectively.

What about the differences if the external constant force F acts directly on the subordinated process $y(t)$ and continues to affect the stochastic process all the time. In this case, the Langevin equation is [Cairoli and Baule (2015b); Eule and Friedrich (2009)]

$$\dot{y}(t) = F + \sqrt{2D}\,\bar{\xi}(t) \tag{2.188}$$

with the equivalent coupled Langevin equation [Chen *et al.* (2019)]

$$\dot{x}(s) = F\eta(s) + \sqrt{2D}\xi(s), \qquad \dot{t}(s) = \eta(s). \tag{2.189}$$

Using the solution of the exact trajectory, $y(t) = Ft + \sqrt{2D}\int_0^t \bar{\xi}(\tau)d\tau$, the

MSD of stochastic process $y(t)$ is

$$\langle (\Delta y(t))^2 \rangle = \frac{2D}{\Gamma(1+\alpha)} t^\alpha, \tag{2.190}$$

which coincides with the MSD of free particle in Eq. (2.160); see Fig. 2.10 for the simulation results. It implies that the external force does not change the subdiffusion behavior and behaves as a decoupled force. Hence, the subdiffusion model described by Eq. (2.188) is Galilean invariant [Metzler and Klafter (2000b)].

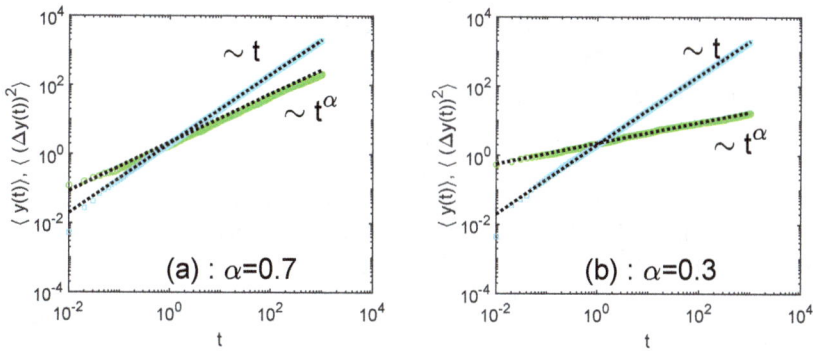

Fig. 2.10: First moment and MSD of stochastic process described by the Langevin equation (2.188) for different α. The parameters are, respectively, taken as $D = 1$, $F = 2$, $\alpha = 0.7$ in (a) or $\alpha = 0.3$ in (b). Dotted black lines represent the theoretical results of first moment Ft and MSD in Eq. (2.190). While the green circle-markers and cyan square-markers represent the simulation results of MSD and first moment, respectively.

The corresponding Fokker-Planck equation of model described by Eq. (2.188) is [Cairoli *et al.* (2018)]

$$\frac{\partial p(y,t)}{\partial t} = -F \frac{\partial p(y,t)}{\partial y} + D \frac{\partial^2}{\partial y^2} \mathcal{D}_t^{1-\alpha} p(y,t), \tag{2.191}$$

where

$$\mathcal{D}_t^{1-\alpha} p(y,t) = \frac{1}{\Gamma(\alpha)} \left[\frac{\partial}{\partial t} + F \frac{\partial}{\partial y} \right] \int_0^t \frac{p(y - F(t-\tau), \tau)}{(t-\tau)^{1-\alpha}} d\tau \tag{2.192}$$

is fractional substantial derivative [Friedrich *et al.* (2006b)] with the Fourier-Laplace transform

$$\mathcal{F}_{y \to k} [\mathcal{L}_{t \to \lambda} [\mathcal{D}_t^{1-\alpha} p(y,t)]] = (\lambda - ikF)^{1-\alpha} p(k, \lambda). \tag{2.193}$$

When $F = 0$, $\mathcal{D}_t^{1-\alpha}$ recovers the Riemann-Liouville fractional derivative $D_t^{1-\alpha}$ and the Fokker-Planck equation goes back to the free subdiffusion case.

Then we consider the time-dependent force. The Langevin system with the time-dependent force acting on the original process $x(s)$ is expressed as [Magdziarz *et al.* (2008); Weron *et al.* (2008)]

$$\dot{x}(s) = F(t(s)) + \sqrt{2D}\xi(s), \qquad \dot{t}(s) = \eta(s). \tag{2.194}$$

Noting the subordination of the original process $x(s)$, the force term $F(t(s(t))) = F(t)$ has reasonable physical meaning, since a physical force should act on a system at physical time t not internal time s [Magdziarz *et al.* (2008)]. Its corresponding single Langevin equation describing the subordinated process $y(t) = x(s(t))$ in physical time is

$$\dot{y}(t) = F(t)\dot{s}(t) + \sqrt{2D}\,\bar{\xi}(s). \tag{2.195}$$

It is obvious that the time-dependent force $F(t)$ acts on the system only at the moments of jump, and the corresponding Fokker-Planck equation is [Sokolov and Klafter (2006); Magdziarz *et al.* (2008); Eule and Friedrich (2009); Cairoli and Baule (2017); Sokolov and Klafter (2006); Magdziarz (2009)]

$$\frac{\partial p(y,t)}{\partial t} = \left[-\frac{\partial}{\partial y}F(t) + D\frac{\partial^2}{\partial y^2}\right]D_t^{1-\alpha}p(y,t), \tag{2.196}$$

where the Riemann-Liouville fractional derivative $D_t^{1-\alpha}$ cannot be interchanged with $-\frac{\partial}{\partial y}F(t) + D\frac{\partial^2}{\partial y^2}$.

Using the Fokker-Planck equation (2.196) derived from CTRW model, Sokolov *et al.* obtained the recursive relation of the moments $r_n(t) := \langle y^n(t)\rangle$ [Sokolov and Klafter (2006)]

$$\frac{dr_n(t)}{dt} = nF(t)D_t^{1-\alpha}r_{n-1}(t) + \frac{n(n-1)}{2}D_t^{1-\alpha}r_{n-2}(t) \tag{2.197}$$

with $r_0(t) = 1, r_{-1}(t) = 0$, and $n \in \mathbb{N}$. Two years later, Magdziarz *et al.* derived the same recursive relation of the moments through the Langevin equation (2.195) [Magdziarz *et al.* (2008)]. Hence, the correspondence between the Fokker-Planck equation and the Langevin equation with a time-dependent force is established.

The case in which the time-dependent external force acting on the system all the time is

$$\dot{y}(t) = F(t) + \sqrt{2D}\,\bar{\xi}(t). \tag{2.198}$$

The corresponding coupled Langevin equation is [Eule and Friedrich (2009)]

$$\dot{x}(s) = F(s)\eta(s) + \sqrt{2D}\xi(s), \qquad \dot{t}(s) = \eta(s). \tag{2.199}$$

The MSD of this model can be easily obtained as

$$\langle (\Delta y(t))^2 \rangle = \frac{2D}{\Gamma(1+\alpha)} t^\alpha, \tag{2.200}$$

which is identical with the case of free particle in Eq. (2.160). It means that the time-dependent external force field here acts as a decoupled force, independent of the diffusion behavior. It is Galilean invariant model while another model Eq. (2.194) breaks Galilean invariance. The Fokker-Planck equation corresponding to the Langevin equation (2.198) is [Chen *et al.* (2019)]

$$\frac{\partial p(y,t)}{\partial t} = -\frac{\partial}{\partial y} F(t)p(y,t) + D\frac{\partial^2}{\partial y^2}\mathcal{A}_t^{1-\alpha}p(y,t) \tag{2.201}$$

with the operator in Fourier space

$$\mathcal{F}_{y\to k}[\mathcal{A}_t^{1-\alpha}p(y,t)] = e^{ik\int_0^t F(t')dt'} D_t^{1-\alpha} e^{-ik\int_0^t F(t')dt'} p(k,t). \tag{2.202}$$

Taking the constant force $F(t) = F$, the operator $\mathcal{A}_t^{1-\alpha}$ reduces to the fractional substantial derivative $\mathcal{D}_t^{1-\alpha}$ and the Fokker-Planck equation goes back to Eq. (2.191).

2.2.4.2 *Generalized Langevin equation with external force*

Generalized Langevin equation effected by external potential $U(x)$ reads as [Burov and Barkai (2008b)]

$$\ddot{x}(t) = -\int_0^t K(t-\tau)\dot{x}(\tau)d\tau - U'(x) + \rho(t), \tag{2.203}$$

where $-U'(x)$ is an external force, $\rho(t)$ is the random force. If the external potential is a harmonic one $U(x) = \frac{1}{2}\omega^2 x^2(t)$, where ω is the frequency of the oscillator, then we have

$$\ddot{x}(t) = -\int_0^t K(t-\tau)\dot{x}(\tau)d\tau - \omega^2 x(t) + \rho(t). \tag{2.204}$$

In the following, we consider the affects of harmonic potential on the fractional Langevin equation and tempered fractional Langevin equation by presenting the normalized displacement correlation function, which is defined as

$$C_x(t) = \frac{\langle x(t)x(0) \rangle}{\langle x^2(0) \rangle},$$

under the thermal initial conditions $\langle F(t)x(0)\rangle = 0$, $\langle x^2(0)\rangle = \frac{k_B T}{m\omega^2}$, and $\langle x(0)v(0)\rangle = 0$. The normalized displacement correlation function describes the relevance between particle position at time t and zero. Performing Laplace transform $(t \to \lambda)$ on Eq. (2.204) leads to

$$\hat{x}(\lambda) = \frac{(\lambda + \hat{K}(\lambda))x(0) + \hat{\rho}(\lambda) + v(0)}{\lambda^2 + \lambda\hat{K}(\lambda) + \omega^2}. \tag{2.205}$$

Then through the thermal initial conditions, one has

$$\hat{C}_x(\lambda) = \frac{\lambda + \hat{K}(\lambda)}{\lambda^2 + \lambda\hat{K}(\lambda) + \omega^2}, \tag{2.206}$$

which results in

$$C_x(t) = 1 - m\omega^2 I(t) \tag{2.207}$$

with $I(t)$ being the inverse Laplace transform of

$$\hat{I}(\lambda) = \frac{\lambda^{-1}}{\lambda^2 + \lambda\hat{K}(\lambda) + \omega^2}. \tag{2.208}$$

For fractional Langevin equation shown in Eq. (2.124) with harmonic potential, the asymptotic behavior of the normalized displacement correlation function for $t \to \infty$ is [Burov and Barkai (2008a)]

$$C_x(t) \sim \frac{1}{\omega^2}t^{2H-2}, \tag{2.209}$$

which decays monotonically for long times. While for the tempered fractional Langevin equation driven by tempered fractional Gaussian noise in Eq. (2.142) effected by harmonic potential, the normalized displacement correlation function with $H = 0.5$ and small ω is [Chen *et al.* (2017)]

$$C_x(t) = 2\mu\sqrt{\frac{a}{4a\mu^2 - \omega^2}}e^{-\frac{\omega^2}{2a\mu}t}\sin\left(\frac{\omega\sqrt{4a\mu^2 - \omega^2}}{2a\mu}t + \theta\right), \tag{2.210}$$

where $\theta = \arctan(-\frac{\sqrt{4a\mu^2-\omega^2}}{\omega})$. The phase, amplitude, and period are θ, $2\mu\sqrt{\frac{a}{4a\mu^2-\omega^2}}e^{-\frac{\omega^2}{2a\mu}t}$, and $\frac{4\pi a\mu}{\omega\sqrt{4a\mu^2-\omega^2}}$, respectively.

For the tempered fractional Langevin equation with a truncated Mittag-Leffler memory kernel shown in Eq. (2.147) effected by a harmonic potential, if we take a special case $\beta = \delta = 1$ and $\tau \to 0$, the tempered power-law memory kernel $K(t) = e^{-bt}\frac{t^{-\alpha}}{\Gamma(1-\alpha)}$, $0 < \alpha < 1$ is obtained. In this case, the normalized displacement correlation function is represented as [Liemert *et al.* (2017)]

$$C_x(t) = 1 - \omega^2\sum_{n=0}^{\infty}(-\gamma)^n \varepsilon_{0+;2,n+3}^{-\omega^2;n+1}\left(e^{-bt}\frac{t^{(1-\alpha)n-1}}{\Gamma(n(1-\alpha))}\right), \tag{2.211}$$

where

$$\left(\varepsilon_{0+;\alpha,\beta}^{\omega;\delta}f\right)(t) = \int_0^t (t-t')^{\beta-1} E_{\alpha,\beta}^{\delta}(\omega(t-t')^{\alpha}) f(t') dt', \qquad (2.212)$$

and $\beta, \delta, \omega \in \mathbb{C}, \mathfrak{R}(\alpha) > 0$, is the Prabhakar integral operator [Prabhakar (1971)]. The normalized displacement correlation function shows different behaviors as the fractional Langevin equation case: monotonic decay, non-monotonic decay without zero crossings, critical behavior which distinguishes the cases with and without zero crossings, and oscillation like behavior with zero crossings [Liemert *et al.* (2017)]. These behaviors are based on the cage effects of the environment as shown by [Burov and Barkai (2008a,b)]. This means that, depending on the values of the friction memory kernel parameters, the friction caused by the complex environment may force either diffusion or oscillations.

2.3 Ergodic and Nonergodic Behavior

As we all know, ensemble averaged MSD is a powerful statistical quality to reflect the diffusion phenomenon of a stochastic process. However, it is the statistical average over many stochastic realizations and is not easy to measure in experiments. In 1914, Nordlund determined the diffusion coefficients of the traced droplets from separate analysis of single trajectory. After that, the recorded time series is evaluated in terms of the time averaged MSD for a free particle [Metzler *et al.* (2014)]

$$\overline{\delta^2(\Delta)} = \frac{1}{T-\Delta} \int_0^{T-\Delta} [x(t+\Delta) - x(t)]^2 dt, \qquad (2.213)$$

where Δ is the lag time, separates the displacement between trajectory points and is much shorter than measurement time T. The time averaged MSD is considered in the limit $\Delta \ll T$ to obtain good statistics. Especially, for the particle affected by the external force, the first moment of the stochastic process cannot be ignored in the time averaged MSD [Burov *et al.* (2011); Akimoto *et al.* (2018)]:

$$\overline{\delta^2(\Delta)} = \frac{1}{T-\Delta} \int_0^{T-\Delta} [x(t+\Delta) - x(t) - \langle x(t+\Delta) - x(t)\rangle]^2 dt. \qquad (2.214)$$

The single particle tracking techniques have been widely employed to study diffusion of particles in living cell [Golding and Cox (2006); Weber *et al.* (2010); Bronstein *et al.* (2009)]. We call the process is ergodic if the ensemble averaged MSD and time averaged MSD are equivalent in the limit of long measurement times.

In the following three subsections, we show the ergodicity of some stochastic processes in CTRW or Langevin framework. One can see that the Brownian motion is ergodic while for most of the anomalous diffusion processes with non-stationary increments, the mean of time averaged MSD is not equal to the corresponding ensemble averaged MSD in the limit of long measurement times, i.e.,

$$\langle x^2(\Delta) \rangle \neq \lim_{T \to \infty} \langle \overline{\delta^2(\Delta)} \rangle, \tag{2.215}$$

which is named as ergodicity breaking phenomenon.

2.3.1 *Ergodicity Property in Continuous Time Random Walk Framework*

Let us pay attention to the Brownian motion firstly, which exhibits normal diffusion and can be described by CTRW model with finite first moment of waiting time $\langle \tau \rangle$ and second moment of jump length $\langle l^2 \rangle$. The diffusion constant of Brownian motion is defined as $D = \langle l^2 \rangle/(2\langle \tau \rangle)$ in the limiting sense of a random walk [Metzler and Klafter (2000b); Metzler *et al.* (2014)]. Note that the waiting times are i.i.d. random variables, as well as jump lengths, and that as long as the lag time Δ is much longer than the characteristic time $\langle \tau \rangle$ for a single jump, the average number of jumps during this time span equals $\Delta/\langle \tau \rangle$. Thus, the kernel $[x(t + \Delta) - x(t)]^2$ in Eq. (2.213) on average is given by $\langle l^2 \rangle \Delta/\langle \tau \rangle$. The result is

$$\lim_{T \to \infty} \overline{\delta^2(\Delta)} = 2D\Delta. \tag{2.216}$$

Identifying the lag time Δ with the regular time t in the ensemble averaged MSD, we indeed find the equivalence

$$\langle x^2(\Delta) \rangle = \lim_{T \to \infty} \overline{\delta^2(\Delta)} \tag{2.217}$$

between the ensemble averaged MSD and time averaged MSD, which means the ergodicity of the Brownian motion.

For the subdiffusive process described by CTRW with the divergent first moment of waiting time whose distribution is

$$\psi(t) \simeq \frac{At^{-1-\alpha}}{|\Gamma(-\alpha)|}, \tag{2.218}$$

we have

$$x(t + \Delta) - x(t) = \sum_{i=1}^{n_{t,t+\Delta}} l_i, \tag{2.219}$$

where l_i are independent random jump lengths with zero mean value, and $n_{t,t+\Delta}$ is the number of jumps in the time interval $(t, t + \Delta)$. Using the independence and the mean value of the jump length, we have

$$\langle [x(t + \Delta) - x(t)]^2 \rangle = \langle l^2 \rangle \langle n_{t,t+\Delta} \rangle. \qquad (2.220)$$

The difference is the value of the number of jumps $n_{t,t+\Delta}$ in the time interval $(t, t + \Delta)$. Due to long traps in subdiffusive CTRW, the number of jump is less than normal diffusion case. The average number of jumps in time interval $(t, t + \Delta)$ is $\langle n_{t,t+\Delta} \rangle = \langle n_{0,t+\Delta} \rangle - \langle n_{0,t} \rangle$. Using the method in [Godrèche and Luck (2001)], it holds that

$$\langle n_{0,t} \rangle \simeq \frac{t^\alpha}{A\Gamma(1 + \alpha)} \qquad (2.221)$$

with $0 < \alpha < 1$, and one obtains [He *et al.* (2008)]

$$\overline{\langle \delta^2(\Delta) \rangle} = \frac{\langle l^2 \rangle}{A\Gamma(1 + \alpha)} \frac{T^{1+\alpha} - \Delta^{1+\alpha} - (T - \Delta)^{1+\alpha}}{(1 + \alpha)(T - \Delta)}. \qquad (2.222)$$

Finally, in the limit $\Delta \ll T$, one has

$$\overline{\langle \delta^2(\Delta) \rangle} \simeq \frac{2D_\alpha}{\Gamma(1 + \alpha)} \frac{\Delta}{T^{1-\alpha}}, \qquad (2.223)$$

where we use the generalized Einstein relation $D_\alpha = \langle l^2 \rangle/(2A)$ [Barkai *et al.* (2000)]. The inconsistency between the time averaged MSD and ensemble averaged MSD means the non-ergodicity of the subdiffusion process. Furthermore, the time averages remain random even for very long averaging times, and time averaged physical observables are thus irreproducible [Burov *et al.* (2010)]. The randomness of the time averaged MSD is quantified in terms of the dimensionless variable

$$\xi = \frac{\overline{\delta^2(\Delta)}}{\langle \overline{\delta^2(\Delta)} \rangle}. \qquad (2.224)$$

In the following, we briefly introduce the derivation of the distribution $\phi(\xi)$ of ξ of the subdiffusion process; see [He *et al.* (2008); Sokolov *et al.* (2009)] for more detailed derivation.

Let the PDF $p(n, t)$ be the probability of making n steps in time t. Then in Laplace domain, it reads

$$\hat{p}(n, \lambda) = \frac{1 - \hat{\psi}(\lambda)}{\lambda} [\hat{\psi}(\lambda)]^n. \qquad (2.225)$$

For large time t, i.e., $\lambda \to 0$, substituting the asymptotic expression $\hat{\psi}(\lambda) \simeq 1 - A\lambda^\alpha$ of waiting time into Eq. (2.225) leads to

$$\hat{p}(n, \lambda) \simeq A\lambda^{\alpha-1} e^{-nA\lambda^\alpha} \qquad (2.226)$$

in the limit of $n \to \infty$. Inverting to the time domain, we have

$$p(n,t) \simeq \frac{t}{\alpha A^{1/\alpha} n^{1+1/\alpha}} L_\alpha \left(\frac{t}{A^{1/\alpha} n^{1/\alpha}} \right), \qquad (2.227)$$

where $L_\alpha(t)$ is the one-sided Lévy stable PDF, whose Laplace pair is $e^{-\lambda^\alpha}$. We note that $\langle \overline{\delta^2} \rangle = C\langle n \rangle / t$ with $\langle n \rangle \simeq t^\alpha / A\Gamma(1+\alpha)$. By change of variables we obtain that the PDF of the dimensionless random variable

$$\xi = \frac{\overline{\delta^2}}{\langle \overline{\delta^2} \rangle} = \frac{n}{\langle n \rangle} \qquad (2.228)$$

is

$$\lim_{T \to \infty} \phi(\xi) = \frac{\Gamma^{1/\alpha}(1+\alpha)}{\alpha \xi^{1+1/\alpha}} L_\alpha \left[\frac{\Gamma^{1/\alpha}(1+\alpha)}{\xi^{1/\alpha}} \right]. \qquad (2.229)$$

2.3.2   *Ergodicity Property in Langevin Picture*

As we all know, one of the advantages of the Langevin equation is that the velocity can be solved analytically by using Laplace transform and the velocity correlation function can be obtained directly. Then the ensemble average of the integrand in the time averaged MSD can be gotten easily by use of the velocity correlation function

$$\langle [x(t+\Delta) - x(t)]^2 \rangle = \int_t^{t+\Delta} \int_t^{t+\Delta} \langle v(t_1)v(t_2) \rangle dt_1 dt_2. \qquad (2.230)$$

Another method to obtain the ensemble averaged MSD and time averaged MSD for a free stochastic process is the generalized Green-Kubo formula [Dechant *et al.* (2014); Meyer *et al.* (2017)], which says that: if the velocity correlation function of the concerned stochastic process has the scaling form

$$\langle v(t)v(t+\tau) \rangle = \mathcal{C}_{EA} t^{\nu-2} \phi_{EA} \left(\frac{\tau}{t} \right) \qquad (2.231)$$

with $\phi_{EA}(q) \simeq cq^{-\delta}$ for small q, and the variance of velocity is asymptotically as $\langle v^2(t) \rangle \simeq a\mathcal{C}_{EA} t^\beta$, then the ensemble averaged MSD and time averaged MSD for a free stochastic process are

$$\langle x^2(t) \rangle \simeq 2D_{EA} t^\nu \qquad (2.232)$$

with

$$D_{EA} = \frac{\mathcal{C}_{EA}}{\nu} \int_0^\infty (1+q)^{-\nu} \phi_{EA}(q) dq, \qquad (2.233)$$

and

$$\langle \overline{\delta^2(\Delta)} \rangle \simeq D_{TA} T^\beta \Delta^{\nu-\beta} \qquad (2.234)$$

with

$$D_{TA} = \frac{2c\,\mathcal{C}_{EA}}{(\beta+1)(\nu-\beta-1)(\nu-\beta)},$$ (2.235)

respectively.

Noting that for a stationary process, its velocity correlation function only depends on the time difference τ. The explicit dependence on time t in Eq. (2.231) implies that the aging effects can be important in this system. Generally, any system may show aging for short enough times, but for a wide class of systems, relaxation into the stationary state is exponentially fast. They can thus be fully described by their stationary behavior on time scales larger than the characteristic relaxation time. A non-negligible aging effect is usually together with the ergodicity breaking of a diffusion process. The aging phenomena in anomalous diffusion processes have been discussed in CTRW [Barkai and Cheng (2003)], renewal theory [Schulz *et al.* (2014)], ballistic Lévy walks [Magdziarz and Zorawik (2017)], and two-state alternating process [Wang *et al.* (2019a)].

To be more self-contained, we briefly present the derivations of Eqs. (2.232)-(2.235) here. Based on the scaling form of the velocity correlation function in Eq. (2.231), the relative MSD is given by

$$\langle(x(t_0+t)-x(t_0))^2\rangle = \int_{t_0}^{t_0+t}\int_{t_0}^{t_0+t}\langle v(t_1)v(t_2)\rangle dt_1 dt_2$$

$$\simeq 2\mathcal{C}_{EA}\int_0^t\int_0^{t_2}(t_1+t_0)^{\nu-2}\phi_{EA}\left(\frac{t_2-t_1}{t_1+t_0}\right)dt_1 dt_2.$$ (2.236)

With the variable substitution $s = (t_2-t_1)/(t_1+t_0)$ and $z = t_2/t$, the MSD can be written as

$$\langle(x(t_0+t)-x(t_0))^2\rangle \simeq 2D_\nu^{t/t_0}t^\nu$$ (2.237)

with

$$D_\nu^{t/t_0} = \mathcal{C}_{EA}\int_0^1 z^{\nu-1}\left(1+\frac{1}{z\frac{t}{t_0}}\right)^{\nu-1}\int_0^{z\frac{t}{t_0}}(s+1)^{-\nu}\phi_{EA}(s)dsdz.$$ (2.238)

The anomalous diffusion coefficient D_ν^{t/t_0} depends on the ratio t/t_0. In the limit $t \gg t_0$, we can neglect the second term in parentheses and take the upper bound of the second integral to infinity to obtain the expression given by the scaling Green-Kubo relation independent of t_0

$$D_\nu^{t/t_0} \simeq D_\nu^\infty = \frac{\mathcal{C}_{EA}}{\nu}\int_0^\infty (s+1)^{-\nu}\phi_{EA}(s)ds.$$ (2.239)

In the opposite limit $t \ll t_0$, the main term resulting from the small s expansion of the scaling function $\phi_{EA}(s)$ is

$$D_\nu^{t/t_0} \simeq \mathcal{C}_{EA} \left(\frac{t_0}{t} \right)^{\nu-1} \int_0^1 \int_0^{z\frac{t}{t_0}} \phi_{EA}(s) ds dz. \qquad (2.240)$$

Furthermore, the asymptotic expression of velocity variance is assumed as $\langle v^2(t) \rangle \simeq a\mathcal{C}_{EA} t^\beta$ with some positive constant a and $0 \leq \beta < \nu - 1$, which could be a constant or increases with respect to time. By using the asymptotic expression of $\phi_{EA}(s) \simeq cs^{-\delta}$ with $\delta = 2 - \nu + \beta$, the anomalous diffusion coefficient D_ν^{t/t_0} in this case can be obtained as

$$D_\nu^{t/t_0} \simeq \frac{c\,\mathcal{C}_{EA}}{(\nu - \beta - 1)(\nu - \beta)} \left(\frac{t_0}{t} \right)^\beta. \qquad (2.241)$$

Thus the MSD $\langle (x(t_0 + t) - x(t_0))^2 \rangle$ is

$$\langle (x(t_0 + t) - x(t_0))^2 \rangle \simeq \frac{2c\,\mathcal{C}_{EA}}{(\nu - \beta - 1)(\nu - \beta)} t_0^\beta t^{\nu-\beta}. \qquad (2.242)$$

Then by using the definition of time averaged MSD, we get Eq. (2.234).

For the Langevin equation (2.158) with Gaussian white noise $\xi(t)$ showing normal diffusion phenomenon, by use of the definition of time averaged MSD in Eq. (2.213), it can be seen that $\overline{\delta^2(\Delta)} = 2D\Delta$ as long as the measurement time is sufficiently long. The equivalence of ensemble averaged MSD and time averaged MSD $\overline{\delta^2(\Delta)} = \langle x^2(\Delta) \rangle$ means the ergodicity of the Brownian motion.

Different from the ergodic Brownian motion, the anomalous diffusion processes with non-stationary increments always exhibit weak ergodicity breaking phenomenon. Here we present the ergodicity or nonergodicity of several representative anomalous diffusion processes, such as fractional Brownian motion, fractional Langevin equation and Lévy walk.

2.3.2.1 *The ergodicity of fractional Brownian-Langevin motion*

For the fractional Brownian motion, using the correlation function and the variance shown in Eq. (2.120) and Eq. (2.121), we can obtain the mean of time averaged MSD of fractional Brownian motion

$$\langle \overline{\delta^2(\Delta)} \rangle = \frac{1}{T - \Delta} \int_0^{T-\Delta} \langle [x(t + \Delta) - x(t)]^2 \rangle dt = 2D_H \Delta^{2H}, \qquad (2.243)$$

which is the same as the ensemble averaged MSD $\langle x^2(\Delta) \rangle$ for all times. It can be also obtained by the property of stationary increments of the fractional Brownian motion so that the integrand in Eq. (2.243). In addition, the variance of $\overline{\delta^2}$ is $\mathrm{Var}[\overline{\delta^2}] = \langle[\overline{\delta^2}]^2\rangle - \langle\overline{\delta^2}\rangle^2$, with

$$\langle[\overline{\delta^2}]^2\rangle = \frac{\int_0^{T-\Delta}\int_0^{T-\Delta}\langle[x(t_1+\Delta)-x(t_1)]^2[x(t_2+\Delta)-x(t_2)]^2\rangle dt_1 dt_2}{(T-\Delta)^2}.$$

(2.244)

A dimensionless measure of ergodicity breaking is the parameter

$$E_B = \frac{\mathrm{Var}[\overline{\delta^2}]}{\langle\overline{\delta^2}\rangle^2},$$

(2.245)

which is zero in the limit $T \to \infty$ if the process is ergodic. To obtain the ergodicity breaking parameter of the fractional Brownian motion, the formula for Gaussian process with mean zero can be used:

$$\langle x(t_1)x(t_2)x(t_3)x(t_4)\rangle = \langle x(t_1)x(t_2)\rangle\langle x(t_3)x(t_4)\rangle$$
$$+ \langle x(t_1)x(t_3)\rangle\langle x(t_2)x(t_4)\rangle \qquad (2.246)$$
$$+ \langle x(t_1)x(t_4)\rangle\langle x(t_2)x(t_3)\rangle.$$

Then we obtain

$$\langle[x(t_1+\Delta)-x(t_1)]^2[x(t_2+\Delta)-x(t_2)]^2\rangle$$
$$= 4D_H^2\Delta^{4H} + 2D_H^2[|t_1+\Delta-t_2|^{2H} + |t_2+\Delta-t_1|^{2H} - 2|t_1-t_2|^{2H}]^2.$$

(2.247)

Finally, for large measurement time T, the ergodicity breaking parameter of the fractional Brownian motion is [Deng and Barkai (2009)]

$$E_B \simeq \begin{cases} k(H)\frac{\Delta}{T}, & 0 < H < \frac{3}{4} \\ k(H)\frac{\Delta}{T}\ln T, & H = \frac{3}{4} \\ k(H)\left(\frac{\Delta}{T}\right)^{4-4H}, & \frac{3}{4} < H < 1, \end{cases}$$

(2.248)

where

$$k(H) = \begin{cases} \int_0^\infty[(\tau+1)^{2H} + |\tau-1|^{2H} - 2\tau^{2H}]^2 d\tau, & 0 < H < \frac{3}{4} \\ 4H^2(2H-1)^2 = \frac{9}{16}, & H = \frac{3}{4} \\ \left(\frac{4}{4H-3} - \frac{4}{4H-2}\right)H^2(2H-1)^2, & \frac{3}{4} < H < 1. \end{cases}$$

(2.249)

The most remarkable result is that $k(H)$ diverges when $H \to 3/4$, which marks a nonsmooth transition in the properties of fractional Brownian motion. The asymptotic convergence when $0 < H < \frac{3}{4}$ is faster than the case $H > \frac{3}{4}$. Especially when $H \to 0$, it holds that $k(H) \to 0$ implying a fastest convergence. By contrast, when $H \to 1$, we have $E_B \to 2$ indicating ergodicity breaking.

For underdamped fractional Langevin equation shown in Eq. (2.124), the correlation function of $x(t)$ is given by [Deng and Barkai (2009)]

$$\langle x(t_1)x(t_2)\rangle = v_0^2 t_1 t_2 E_{2H,2}(-\Gamma(2H-1)t_1^{2H})E_{2H,2}(-\Gamma(2H-1)t_2^{2H})$$

$$+ k_B T \int_0^{t_2}\int_0^{t_1} d\tau ds(t_1-\tau)E_{2H,2}[-\Gamma(2H-1)(t_1-\tau)^{2H}]$$

$$\times (t_2-s)E_{2H,2}[-\Gamma(2H-1)(t_2-s)^{2H}]|\tau-s|^{2H-2}.$$

$$(2.250)$$

When t_1 and t_2 tend to infinity, the asymptotic expression of correlation function of fractional Langevin equation is

$$\langle x(t_1)x(t_2)\rangle \simeq \frac{k_B T}{\Gamma^2(2H-1)\Gamma^2(2-2H)}\int_0^{t_2}\int_0^{t_1} d\tau ds$$

$$\times (t_1-\tau)^{1-2H}(t_2-s)^{1-2H}|\tau-s|^{2H-2}.$$

$$(2.251)$$

The correlation function of $x(t)$ approximates to the ones of overdamped Langevin equation, where the correlation function is

$$\langle y(t_1)y(t_2)\rangle = D_F(t_1^{2-2H}+t_2^{2-2H}-|t_1-t_2|^{2-2H}) \qquad (2.252)$$

with

$$D_F = \frac{k_B T \pi \csc[\pi(2-2H)]}{(2-2H)\Gamma^2(2H-1)\Gamma^2(2-2H)}. \qquad (2.253)$$

So we can expect in the long time limit

$$\langle \overline{\delta^2(x)}\rangle \simeq \langle \overline{\delta^2(y)}\rangle \simeq \langle \overline{\delta^2(B_H)}\rangle \qquad (2.254)$$

and

$$E_B(x) \simeq E_B(y) \simeq E_B(B_H), \qquad (2.255)$$

where x, y, B_H represent the underdamped fractional Langevin equation, overdamped fractional Langevin equation and fractional Brownian motion, respectively. From the above discussion, we can see that the fractional Langevin equation and fractional Brownian motion are all ergodic although they are anomalous diffusion processes.

Similar to above derivations, the tempered fractional Brownian motion in Eq. (2.133), which has stationary increments, is also ergodic that the ensemble averaged MSD equals the ensemble averaged time averaged MSD for all times

$$\langle x^2(\Delta)\rangle = \langle \overline{\delta^2(\Delta)}\rangle = C_\Delta^2 \Delta^{2H}. \qquad (2.256)$$

For long times and moderate μ, the variance of $\overline{\delta^2}$ is [Chen *et al.* (2017)]

$$\text{Var}[\overline{\delta^2}] \simeq k(H)\frac{4\Gamma^2(H+\frac{1}{2})}{(2\mu)^{2H+1}}\Delta^{2H}t^{-1}, \tag{2.257}$$

where

$$k(H) = \int_0^\infty [(1+\tau)^{H-\frac{1}{2}}e^{-\mu\Delta(1+\tau)} + |\tau-1|^{H-\frac{1}{2}}e^{-\mu|\Delta(\tau-1)|}$$
$$- 2\tau^{H-\frac{1}{2}}e^{-\mu\Delta\tau}]^2 d\tau. \tag{2.258}$$

Then the ergodicity breaking parameter for the tempered fractional Brownian motion is

$$E_B \simeq \frac{4k(H)\Gamma^2(H+1/2)}{(2\mu)^{2H+1}C_\Delta^4\Delta^{2H}}T^{-1}, \tag{2.259}$$

which tends to zero at the rate T^{-1} for $H \in (0,1)$ as $T \to \infty$.

2.3.2.2 *The nonergodicity of Lévy walk*

Except for the fractional Brownian motion and fractional Langevin equation which are both ergodic, most of the anomalous diffusion processes display the nonergodic behavior. The typical one is Lévy walk which exhibits super-diffusion phenomenon for the power law exponent of waiting times less than 2. As discussed above, Lévy walk can be described by the coupled CTRW in Sec. 2.1.2 and the subordinated Langevin equation in Sec. 2.2.3.

Considering a particle alternating its velocity between v_0 and $-v_0$ at random times. The time difference τ between successive turning events are i.i.d. random variables with a common distribution in Eq. (2.218). Then the velocity correlation function can be expressed by the probability $p_0(t_1,t_2)$ for no renewal happening in the time interval (t_1,t_2):

$$\langle v(t_1)v(t_2)\rangle = v_0^2 p_0(t_1,t_2), \tag{2.260}$$

where the terms $p_n(t_1,t_2)$ with $n \neq 0$ make no contribution to the velocity correlation function due to symmetry. The expression of $p_0(t_1,t_2)$ can be obtained from the renewal theory [Godrèche and Luck (2001)]. Here we present a brief deduction of the expression of $p_0(t_1,t_2)$. Denoting the number of events occurring between 0 and t as N. The time of occurrence of the last event before t is $t_N = \tau_1 + \cdots + \tau_N$, where τ_i are the i.i.d. renewal times. Let E_t denote the forward recurrence time, i.e., the time interval between t and the next event, $E_t = t_{N+1} - t$. Then the distributions of E_t can be obtained as

$$f_{E,N}(t;y,n) = \langle\delta(y-t_{n+1}+t)I(t_n < t < t_{n+1})\rangle, \tag{2.261}$$

where $I(\cdot)$ is the indicator function. In Laplace space $(t \to \lambda, y \to u)$, one has

$$\hat{f}_{E,N}(\lambda; u, n) = \left\langle \int_{t_n}^{t_{n+1}} e^{-\lambda t} e^{-u(t_{n+1}-t)} \right\rangle$$

$$= \frac{\hat{\psi}(\lambda) - \hat{\psi}(u)}{u - \lambda} \hat{\psi}^n(\lambda) \tag{2.262}$$

for $n \geq 0$ and $\hat{\psi}(\lambda)$ is the Laplace transform of the PDF of renewal time. Summing over n, finally, the distribution $f_E(t; y)$ of E_t in Laplace space is

$$\hat{f}_E(\lambda; u) = \frac{\hat{\psi}(\lambda) - \hat{\psi}(u)}{u - \lambda} \frac{1}{1 - \hat{\psi}(\lambda)}. \tag{2.263}$$

The probability $p_0(t, t + t')$ that no renewal happens between t and $t + t'$ equals the survival probability of E_t, that is

$$p_0(t, t + t') = P(E_t > t') = \int_{t'}^{\infty} f_E(t; y) dy. \tag{2.264}$$

Then the Laplace transform $(t \to \lambda, t' \to u)$ of $p_0(t, t + t')$ is

$$\hat{p}_0(\lambda, u) = \frac{1 - \lambda \hat{f}_E(\lambda; u)}{\lambda u}$$

$$= \frac{\lambda - u + u\hat{\psi}(\lambda) - \lambda\hat{\psi}(u)}{\lambda(\lambda - u)(1 - \hat{\psi}(\lambda))u}, \tag{2.265}$$

which is related to the Laplace transform $(t \to \lambda, y \to u)$ of $f_E(t; y)$. But it seems not easy to perform the inverse Laplace transform. Instead, we can obtain the expression of $p_0(t_1, t_2)$ in Laplace space $(t_1 \to \lambda_1, t_2 \to \lambda_2)$ by substituting variables [Wang *et al.* (2019a)]. For convenience of marking for $p_0(t_1, t_2)$ and $p_0(t, t + t')$, we denote $p_0(t, t + t') = f(t, t')$ and obtain

$$\hat{p}_0(\lambda_1, \lambda_2) = \int_0^{\infty} dt_1 \int_0^{\infty} dt_2 \, e^{-\lambda_1 t_1} e^{-\lambda_2 t_2} p_0(t_1, t_2)$$

$$= \int_0^{\infty} dt_1 \int_{t_1}^{\infty} dt_2 e^{-\lambda_1 t_1} e^{-\lambda_2 t_2} p_0(t_1, t_2) \tag{2.266}$$

$$+ \int_0^{\infty} dt_2 \int_{t_2}^{\infty} dt_1 e^{-\lambda_1 t_1} e^{-\lambda_2 t_2} p_0(t_1, t_2).$$

The first term can be rewritten through substitution $t_2 = t_1 + t'$ as

$$\int_0^{\infty} dt_1 \int_0^{\infty} dt' e^{-\lambda_1 t_1} e^{-\lambda_2(t_1+t')} p_0(t_1, t_1 + t')$$

$$= \int_0^{\infty} dt_1 \int_0^{\infty} dt' e^{-(\lambda_1+\lambda_2)t_1} e^{-\lambda_2 t'} f(t_1, t') = \hat{f}(\lambda_1 + \lambda_2, \lambda_2). \tag{2.267}$$

Similarly, the second term is equal to $\hat{f}(\lambda_1 + \lambda_2, \lambda_1)$ due to the symmetry of λ_1 and λ_2. Hence, we have

$$\hat{p}_0(\lambda_1, \lambda_2) = \hat{f}(\lambda_1 + \lambda_2, \lambda_1) + \hat{f}(\lambda_1 + \lambda_2, \lambda_2). \tag{2.268}$$

Finally, after the inverse Laplace transform, we get the expression of $p_0(t_1, t_2)$ for large t_1, t_2 and $t_1 < t_2$

$$p_0(t_1, t_2) \simeq \begin{cases} \frac{\sin(\pi\alpha)}{\pi} B\left(\frac{t_1}{t_2}; \alpha, 1 - \alpha\right), & 0 < \alpha < 1, \\ (t_2 - t_1)^{1-\alpha} - t_2^{1-\alpha}, & 1 < \alpha < 2. \end{cases} \tag{2.269}$$

The function $B(t; a, b) = \int_0^t x^{a-1}(1-x)^{b-1}dx$ is the incomplete Beta function.

Another method to derive the velocity correlation function of Lévy walk is from the Langevin description of Lévy walk presented in Eq. (3.206)

$$\dot{x}(t) = v(t), \quad \dot{v}(s) = -\gamma v(s) + \sqrt{2D}\xi(s), \quad \dot{t}(s) = \eta(s). \tag{2.270}$$

Using the subordination method, the velocity correlation function is the same as above

$$\langle v(t_1)v(t_2)\rangle \simeq \begin{cases} \frac{D}{\gamma} \frac{\sin(\pi\alpha)}{\pi} B\left(\frac{t_1}{t_2}; \alpha, 1 - \alpha\right), & 0 < \alpha < 1 \\ \frac{D}{\gamma}\left[(t_2 - t_1)^{1-\alpha} - t_2^{1-\alpha}\right], & 1 < \alpha < 2 \end{cases} \tag{2.271}$$

with $v_0^2 = D/\gamma$.

According to the generalized Green-Kubo formula [Dechant *et al.* (2014); Meyer *et al.* (2017)], one should first rewrite the velocity correlation function in Eq. (2.271) as the scaling form, i.e.,

$$\langle v(t)v(t + \tau)\rangle = \mathcal{C}_{EA}t^{\nu-2}\phi_{EA}\left(\frac{\tau}{t}\right). \tag{2.272}$$

And we assume that $\phi_{EA}(q) \to cq^{-\delta}$ as $q \to 0$.

For the case $0 < \alpha < 1$, the velocity correlation function can be written as

$$\langle v(t)v(t + \tau)\rangle = \frac{D}{\gamma} \frac{1}{\Gamma(1-\alpha)\Gamma(\alpha)}\phi_{EA}\left(\frac{\tau}{t}\right), \tag{2.273}$$

where

$$\phi_{EA}(q) = B\left(\frac{1}{1+q}; \alpha, 1 - \alpha\right). \tag{2.274}$$

When $q \to 0$, $\phi_{EA}(q)$ converges to the constant $c = \Gamma(\alpha)\Gamma(1 - \alpha)$. Then we obtain the parameters:

$$\mathcal{C}_{EA} = \frac{D}{\gamma} \frac{1}{\Gamma(1-\alpha)\Gamma(\alpha)}, \quad \nu = 2. \tag{2.275}$$

Substituting these parameters into the generalized Green-Kubo formula, the second moment of Lévy walk can be obtained as

$$\langle x^2(t) \rangle = \frac{D(1-\alpha)}{\gamma} t^2. \tag{2.276}$$

As for the ensemble averaged time averaged MSD of the Lévy walk, the variance of the velocity is a constant for any α, so that $\beta = 0$ in $\langle v^2(t) \rangle \propto t^\beta$. According to the generalized Green-Kubo formula for the ensemble averaged time averaged MSD in Eq. (2.234) and the parameters we need, we can obtain the ensemble averaged time averaged MSD of Lévy walk:

$$\langle \overline{\delta^2(\Delta)} \rangle = \frac{D}{\gamma} \Delta^2, \tag{2.277}$$

which is different from the ensemble averaged MSD in Eq. (2.276) by a constant multiplier.

For the case $1 < \alpha < 2$, the velocity correlation function is

$$\langle v(t)v(t+\tau) \rangle = \frac{D}{\gamma} t^{1-\alpha} \phi_{EA} \left(\frac{\tau}{t} \right), \tag{2.278}$$

where

$$\phi_{EA}(q) = q^{1-\alpha} - (1+q)^{1-\alpha}. \tag{2.279}$$

When $q \to 0$, we have the asymptotic expression $\phi_{EA}(q) \simeq q^{1-\alpha}$. Then we obtain the parameters we need: $\mathcal{C}_{EA} = \frac{D}{\gamma}$, $\nu = 3 - \alpha$, $c = 1$, $\beta = 0$, $D_{EA} = \frac{D(\alpha-1)}{\gamma(2-\alpha)(3-\alpha)}$. Finally, the ensemble averaged MSD is

$$\langle x^2(t) \rangle = \frac{2D(\alpha-1)}{\gamma(2-\alpha)(3-\alpha)} t^{3-\alpha}. \tag{2.280}$$

Similarly, the ensemble averaged time averaged MSD for large T and $\Delta \ll T$ is,

$$\langle \overline{\delta^2(\Delta)} \rangle = \frac{2D}{\gamma(2-\alpha)(3-\alpha)} \Delta^{3-\alpha}, \tag{2.281}$$

which is different from the ensemble averaged MSD in Eq. (2.280) by a constant multiplier. Although the difference between ensemble averaged MSD and ensemble averaged time averaged MSD is a constant, there are some intrinsic discrepancies between $0 < \alpha < 1$ and $1 < \alpha < 2$. For $1 < \alpha < 2$, the mean running time is finite and individual trajectories become self-averaging at sufficiently long (infinite) measurement time [Godec and Metzler (2013); Froemberg and Barkai (2013c); Wang et al. (2019a)]. While for $\alpha < 1$, the characteristic timescale is infinite, thus the individual time averaged MSD is irreproducible even for long measurement times.

2.3.2.3 The nonergodicity of scaled Brownian motion and heterogeneous diffusion processes

The deviations from Brownian motion are very ubiquitous in the natural world. Besides fractional Brownian motion and Lévy walk, another simple generalization of Brownian motion is that the diffusivity is not a constant, but dependent on the evolution time t or the position x. Denoting the diffusivity as $D(x, t)$, the corresponding overdamped Langevin equation is [Cherstvy and Metzler (2015a)]

$$\dot{x}(t) = \sqrt{D(x,t)}\xi(t), \tag{2.282}$$

where $\xi(t)$ is also the Gaussian white noise with zero mean and δ-correlation $\langle\xi(t_1)\xi(t_2)\rangle = \delta(t_1 - t_2)$. If the diffusivity is only time-dependent, then $D(x,t) \equiv D(t)$ and the process $x(t)$ described by Langevin equation (2.282) is named scaled Brownian motion [Jeon *et al.* (2014); Thiel and Sokolov (2014); Safdari *et al.* (2015)]. On the other hand, if the diffusivity is only position-dependent, then $D(x,t) \equiv D(x)$ and the process $x(t)$ is named heterogeneous diffusion process [Cherstvy *et al.* (2013); Cherstvy and Metzler (2013, 2014, 2015b); Leibovich and Barkai (2019); Wang *et al.* (2019c)]. They are quite different due to different diffusivity: the heterogeneous diffusion process is described by the Langevin equation (2.282) with a multiplicative noise while the time-dependence of the diffusivity in scaled Brownian motion implies its fundamentally non-stationary character.

Let us consider the scaled Brownian motion firstly by assuming

$$D(t) = 2\alpha K_\alpha t^{\alpha-1}, \tag{2.283}$$

which decreases or increases in time for $0 < \alpha < 1$ and $1 < \alpha < 2$, respectively. Based on the Langevin equation (2.282), the ensemble averaged MSD can be obtained as:

$$\langle x^2(t)\rangle = 2K_\alpha t^\alpha, \tag{2.284}$$

which displays the anomalous diffusion for $\alpha \neq 1$. Due to the δ-correlation of Gaussian white noise, it can be verified that the position correlation function is equivalent to the ensemble averaged MSD, i.e.,

$$\langle x(t_1)x(t_2)\rangle = \langle x^2(t_1)\rangle, \tag{2.285}$$

for $t_1 < t_2$. Combining it with the definition of time averaged MSD yields the mean of time averaged MSD

$$\langle\overline{\delta^2(\Delta)}\rangle \simeq 2K_\alpha \frac{\Delta}{T^{1-\alpha}}, \tag{2.286}$$

in the limit $\Delta \ll t$. The mean of time averaged MSD is normal with respect to the lag time Δ, which deviates from the anomalous ensemble averaged MSD and implies the ergodicity breaking of scaled Brownian motion [Jeon *et al.* (2014); Thiel and Sokolov (2014); Safdari *et al.* (2015)].

Although the scaled Brownian motion is nonergodic, its time averaged MSD has the property of self-averaging since the corresponding dimensionless variable $\xi = \overline{\delta^2(\Delta)}/\langle\overline{\delta^2(\Delta)}\rangle$ has a narrow distribution around $\xi = 1$ for large measurement time T. The width of the distribution ξ is characterized in terms of the EB parameter [Thiel and Sokolov (2014)]:

$$\text{EB} \simeq \begin{cases} 4I_\alpha (\Delta/T)^{2\alpha}, & 0 < \alpha < 1/2, \\ \frac{1}{3}(\Delta/T)\ln(T/\Delta), & \alpha = 1/2, \\ \frac{4\alpha^2}{3(2\alpha-1)}(\Delta/T), & \alpha > 1/2, \end{cases} \tag{2.287}$$

where the constant $I_\alpha = \int_0^1 dy \int_0^\infty dx [(x+1)^\alpha - (x+y)^\alpha]^2$. The EB parameter for the scaled Brownian motion decays to zero as $T \to \infty$.

Now we turn attention to the heterogeneous diffusion process with the diffusivity having the power-law form [Cherstvy *et al.* (2013); Cherstvy and Metzler (2013, 2014, 2015b); Leibovich and Barkai (2019); Wang *et al.* (2019c)]:

$$D(x) = 2D_0|x|^\alpha. \tag{2.288}$$

In the Stratonovich interpretation [Stratonovich (1966); Klimontovich (1990)], considering the substitution $y = \int^x dx'/\sqrt{D(x')}$, the process $y(t)$ is exactly the Brownian motion and the corresponding PDF of $y(t)$ is Gaussian. For the initial condition $p(x,0) = \delta(x)$, the PDF of heterogeneous diffusion process $x(t)$ is [Cherstvy *et al.* (2013)]

$$p(x,t) = \frac{|x|^{1/p-1}}{\sqrt{4\pi D_0 t}} \exp\left(-\frac{|x|^{2/p}}{(2/p)^2 D_0 t}\right) \tag{2.289}$$

with $p = 2/(2-\alpha)$. Therefore, the ensemble averaged MSD is

$$\langle x^2(t)\rangle = \frac{\Gamma(p+1/2)}{\pi^{1/2}}\left(\frac{2}{p}\right)^{2p}(D_0 t)^p, \tag{2.290}$$

which presents subdiffusion for $\alpha < 0$ and superdiffusion for $\alpha > 0$.

To further characterize the heterogeneous diffusion process $x(t)$, we provide the position autocorrelation for $t_1 < t_2$ [Cherstvy *et al.* (2013)]:

$$\langle x(t_1)x(t_2)\rangle = \frac{2^{p+1}D_0^p}{\sqrt{(\pi)}}\frac{\Gamma(p+1)\Gamma(p/2+1)}{p^{2p}\Gamma((p+1)/2)}t_1^{(p+1)/2}(t_2-t_1)^{(p-1)/2}$$

$$\times {}_2F_1\left(\frac{1-p}{2},\frac{p}{2}+1;\frac{3}{2};\frac{-t_1}{t_2-t_1}\right), \tag{2.291}$$

where $_2F_1(a, b, c, z)$ is the Gaussian hypergeometric function defined as

$$_2F_1(a, b, c, z) = \sum_{n=0}^{\infty} \frac{(a)_n (b)_n}{(c)_n} \frac{z^n}{n!}, \tag{2.292}$$

and $(q)_n$ is the (rising) Pochhammer symbol. Then the mean of time averaged MSD can be obtained as

$$\langle \overline{\delta^2(\Delta)} \rangle = \frac{\Gamma(p + 1/2)}{\pi^{1/2}} \left(\frac{2}{p} \right)^{2p} D_0^p \frac{\Delta}{T^{1-p}}, \tag{2.293}$$

which is normal for any α. The discrepancy between ensemble averaged MSD and the mean of time averaged MSD, i.e., $\langle \overline{\delta^2(\Delta)} \rangle \neq \langle x^2(\Delta) \rangle$ implies the ergodicity breaking of the heterogeneous diffusion process. Despite the same ergodicity breaking as scaled Brownian motion, the time averaged MSD of heterogeneous diffusion process is not self-averaging. The EB parameter of the latter cannot be obtained easily due to the multiplicative noise. However, the apparent diffusion coefficients and scaling exponents from the individual $\overline{\delta^2(\Delta)}$ can be investigated via fitting method in [Cherstvy et al. (2013)]. For small Δ, the $\overline{\delta^2(\Delta)}$ can be fitted by a linear law with diffusivity D_1. While for large Δ, the $\overline{\delta^2(\Delta)}$ is fitted by the nonlinear law $\overline{\delta^2(\Delta)} = D_\beta \Delta^\beta$. The distribution of the three parameters D_1, D_β, and β are fitted with the three-parameter gamma distribution $g(z) \simeq z^{\nu-1} e^{-z/b} e^{-a/z}$. The fittings show that the amplitude scatter presents a pronounced spread around the mean value $\xi = 1$ for both $\alpha < 0$ and $\alpha > 0$ [Cherstvy et al. (2013)].

The case of the diffusivity $D(x, t)$ depending on both position and time is investigated in [Cherstvy and Metzler (2015a)], where the underlying process combines the features of both scaled Brownian motion and heterogeneous diffusion processes. Therefore, the scaling exponent of the ensemble averaged MSD is shown to be the product of the critical exponents of the two parent processes, and describes both subdiffusive and superdiffusive systems [Cherstvy and Metzler (2015a)]. Besides the deterministic generalization of the diffusivity above, the diffusivity is sometimes parameterized as a variable random to model the effect of complex heterogeneous medium. One typical example is the Brownian non-Gaussian phenomenon, where the EAMSD is normal but the PDF is non-Gaussian [Wang et al. (2009); Toyota et al. (2011); e Silva et al. (2014); Bhattacharya et al. (2013); Samanta and Chakrabarti (2016)]. The effective idea of interpreting this novel phenomenon is to introduce a random diffusivity which is analogous to the concept of superstatistics [Beck (2001); Beck and Cohen (2003); Beck

(2006)]. It says that each one of particles moving in a complex heterogeneous environment has its own diffusivity [Wang *et al.* (2012); Hapca *et al.* (2009); Chubynsky and Slater (2014); Chechkin *et al.* (2017)]. In general, the idea of superstatistics has been applied to many dynamics: turbulent dispersion [Beck (2001)], renewal critical events in intermittent systems [Paradisi *et al.* (2015)], and different effective statistical mechanics with fluctuating intensive quantities [Beck (2001); Beck and Cohen (2003)]. To further describe the particle's stochastic motion in complex environments, the idea of random diffusivity has been applied to generalized Langevin equation [Ślęzak *et al.* (2018)], generalized grey Brownian motion [Sposini *et al.* (2018)], and fractional Brownian motion [Jain and Sebastian (2018); Maćkała and Magdziarz (2019)]. Besides, the exponential tail is found to be universal for short-time dynamics of the CTRW by using large deviation theory [Barkai and Burov (2020); Wang *et al.* (2020c)]. The ergodic property of a random diffusivity model has been partly discussed for some models, such as the model with local diffusivity fluctuating in time [Cherstvy and Metzler (2016)], the one with a power-law correlated fractional Gaussian noise [Wang *et al.* (2020a,b)], and the one with superstatistical, uncorrelated or correlated diffusivity [Wang and Chen (2021)].

At the end of this section, we list some common anomalous diffusion processes and their ergodicity or nonergodic behaviors. The parameter α in the table represents the exponent of power law distributed waiting time and H is the Hurst parameter. Abbreviation "WEB" means weak ergodicity breaking; "FBM" is for the fractional Brownian motion; "FLE" is for the fractional Langevin equation; and "TFBM" means the tempered fractional Brownian motion.

Stochastic Process	$\langle x^2(t) \rangle$	$\overline{\langle \delta^2(\Delta) \rangle}$	WEB	Reference
1. Lévy walk, $0 < \alpha < 1$	$\simeq B(\alpha)t^2$	$\simeq \frac{B(\alpha)}{1-\alpha}\Delta^2$	Yes	[Froemberg and Barkai (2013a,b,c)]
2. Lévy walk, $1 < \alpha < 2$	$\simeq C(\alpha)t^{3-\alpha}$	$\simeq \frac{C(\alpha)}{\alpha-1}\Delta^{3-\alpha}$	Yes	[Froemberg and Barkai (2013a,c)] [Godec and Metzler (2013)] [Akimoto (2012)]
3. Lévy flight	$= \infty$	$\simeq \Delta t^{2/\alpha - 1}$	Yes	[Vahabi et al. (2013)] [Froemberg and Barkai (2013b)]
4. FBM, $0 < H < 1$	$\simeq t^{2H}$	$\simeq \Delta^{2H}$	No	[Deng and Barkai (2009)] [Jeon and Metzler (2010, 2012)] [Kepten et al. (2011)]
5. FLE motion, $1/2 < H < 1$	$\simeq t^{2-2H}$	$\simeq \Delta^{2-2H}$	No	[Deng and Barkai (2009)] [Jeon and Metzler (2010, 2012)] [Kepten et al. (2011)]
6. TFBM, $H > 0$	$= C_t^2 t^{2H}$	$= C_\Delta^2 \Delta^{2H}$	No	[Meerschaert and Sabzikar (2013)] [Chen et al. (2017)] [Molina-Garcia et al. (2018)]
7. Subdiffusive CTRW	$\simeq t^\alpha$	$\simeq \Delta t^{\alpha-1}$	Yes	[Burov et al. (2011)] [Lubelski et al. (2008)] [He et al. (2008)]
8. Confined subdiffusive CTRW	$\simeq t^0$	$\simeq (\Delta/t)^{1-\alpha}$	Yes	[Burov et al. (2010)] [Jeon et al. (2011)] [Neusius et al. (2009)]
9. Scaled Brownian motion	$\simeq t^\alpha$	$\simeq \Delta t^{\alpha-1}$	Yes	[Jeon et al. (2014)] [Thiel and Sokolov (2014)] [Safdari et al. (2015)]
10. Heterogeneous diffusion process	$\simeq t^{2/(2-\alpha)}$	$\simeq \Delta t^{2/(2-\alpha)-1}$	Yes	[Cherstvy et al. (2013)] [Cherstvy and Metzler (2013, 2014, 2015b)]

Chapter 3

Functional Distributions

We have studied the PDF of particles' displacement and derived their governing equations (Fokker-Planck equations) in CTRW framework or Langevin picture in the last chapter. Based on the information contained in the displacement, we find that a large amount of processes display the anomalous and nonergodic behaviors. Sometimes, people are more interested in the functional of the sample path of a process in some specific applications. The governing equations of the PDF of the functionals are named as Feynman-Kac equation, which was firstly derived by Kac in 1949 for normal diffusion [Kac (1949)], influenced by Feynman's thesis about Schrödinger's equation. Since then, the functionals of the path of Brownian particle have been investigated in numerous studies. In recent years, as the accelerated development of anomalous and nonergodic processes in many fields, the Feynman-Kac equations governing the PDF of the functional of the anomalous path in complex systems have been derived, together with its applications, such as first-passage time and occupation time in a given domain.

3.1 Functional

For a stochastic process $x(t)$, its functional is defined as

$$A = \int_0^t U[x(t')]dt', \tag{3.1}$$

where $U(x)$ is a specified function. The functional A takes different values for different samples of $x(t)$. Therefore, the functional A is a random variable for any given time t. Influenced by Feynman's thesis about Schrödinger's equation, Kac derives the classical Feynman-Kac equation governing the PDF of functional A in 1949 for normal diffusion [Kac (1949)].

If we define the PDF of finding the functional taking value A at time t for the particles with initial position x_0 as $G_{x_0}(A, t)$, then its Laplace transform $G_{x_0}(\rho, t)$ $(A \to \rho)$ satisfies the Feynman-Kac equation

$$\frac{\partial G_{x_0}(\rho, t)}{\partial t} = K\frac{\partial^2 G_{x_0}(\rho, t)}{\partial x_0^2} - \rho U(x_0)G_{x_0}(\rho, t). \tag{3.2}$$

Here, we omit the Laplace notion $\hat{\ }$ with respect to the argument of ρ for convenience, since the Feynman-Kac equations are usually presented in the Laplace ρ space. Since the spatial derivatives in Eq. (3.2) are with respect to the initial position x_0, it is named as backward Feynman-Kac equation. The initial condition of Eq. (3.2) is $G_{x_0}(\rho, 0) = 1$, which can be extracted from the definition of functional in Eq. (3.1) that $G_{x_0}(A, 0) = \delta(A)$.

The $G_{x_0}(\rho, t)$ in Eq. (3.2) can be uniquely solved by giving an initial condition and a specific function $U(x)$. The different choices of $U(x)$ result in different meanings of functional A within numerous applications, such as probability theory, finance, data analysis, disordered systems, and computer science [Majumdar (2005)]. We show some typical examples below.

(1) The occupation time within a time window of size t [Lévy (1939)] is an important quantity of interest in probability theory, which is the time spent by a Brownian motion in some specific domain. In this case, $U(x)$ is just a δ-function or an indicator function $\chi(x)$.

(2) For fluctuating $(1+1)$-dimensional interfaces of the Edwards-Wilkinson [Edwards and Wilkinson (1982)] or the Kardar-Parisi-Zhang [Kardar *et al.* (1986)] varieties, the interface profile in stationary state is characterized by the one-dimensional Brownian motion in space [Barabasi and Stanley (1995)]. The fluctuations are captured by the PDF of the spatially averaged variance of height fluctuations, i.e., the PDF of functional $A = \frac{1}{L}\int_0^L B^2(x)dx$ where $B(x)$ is the Brownian motion in space to describe the deviation of the height from its spatial average and $U(x) = x^2$ now.

(3) In finance, the stock price $S(t)$ is usually modelled by the geometric Brownian motion [Yor (2000)], i.e., the exponential of a Brownian motion $S(t) = e^{-\beta x(t)}$ with a positive constant β. The Asian option price [Geman and Yor (1993)] depends on the time average of the stock's history price, i.e., $A = \int_0^t e^{-\beta x(t')}dt'$. Thus in this problem $U(x) = e^{-\beta x}$. In addition, the form of the integrated stock price has an interesting analogy in a disordered system where an overdamped particle moves in a random potential. The popular Sinai model [Sinai (1983)] describes the random potential as the trace of a random walker in space. If we

interpret the time t as the spatial distance, then $x(t)$ is the potential energy of the particle and $e^{-\beta x(t)}$ is the Boltzmann factor. Thus $A = \int_0^t e^{-\beta x(t')} dt'$ is just the partition function of the particle in the random potential [Comtet *et al.* (1998)].

3.2 Fractional Feynman-Kac Equation

It has been shown that functional has numerous applications in the last section. Nowadays, however, many underlying processes in the physical applications exhibit anomalous diffusion, which means the classical Feynman-Kac equation (3.2) is not valid any more for the anomalous cases. It is urgent to establish a set of theories of new Feynman-Kac equations for these anomalous diffusion processes.

There are many microscopic models to describe the motion of a single particle, with two typical ones being CTRW model and Langevin equation. There have been some attempts of deriving the Feynman-Kac equations from CTRW model [Turgeman *et al.* (2009); Carmi *et al.* (2010); Carmi and Barkai (2011); Wu *et al.* (2016); Wang and Deng (2018); Xu and Deng (2018a); Hou and Deng (2018)] and from Langevin equation [Cairoli and Baule (2015a, 2017); Wang *et al.* (2018)]. Fortunately, the obtained results are interesting and the ordinary time derivative in Eq. (3.2) is replaced by a fractional substantial derivative [Friedrich *et al.* (2006a)]. With some special value of parameters, Eq. (3.2) can be also recovered.

3.2.1 *Derivation in Continuous Time Random Walk Framework*

In [Turgeman *et al.* (2009)], the authors use the CTRW model as the underlying process exhibiting anomalous diffusion. In CTRW framework, the particles move on an infinite one-dimensional lattice with spacing a. The particles are only allowed to jump to the nearest neighbors with equal probability of left or right at each step. Waiting times between successive jumps are i.i.d. random variables, which are distributed in the power law form

$$\phi(\tau) \simeq \frac{B_\alpha \tau^{-(1+\alpha)}}{|\Gamma(-\alpha)|} \tag{3.3}$$

with B_α being a constant and $0 < \alpha < 1$. Under this condition, the mean waiting time is infinite which leads to subdiffusion behavior $\langle x^2(t) \rangle \propto t^\alpha$.

3.2.1.1 *Forward fractional Feynman-Kac equation*

Let $G(x, A, t)$ be the joint PDF of finding the particle at position x with the functional value A at time t. Using the method of transport equation in Sec. 2.1.1, we introduce an auxiliary function $Q_n(x, A, t)$ to represent the probability of the particle making its nth jump into (x, A) at time t.

Recalling the definition of functional $A = \int_0^t U[x(t')]dt'$, we can build the recursion relation for Q_n:

$$
\begin{aligned}
Q_{n+1}(x, A, t) =& \frac{1}{2} \int_0^t \phi(\tau) Q_n[x - a, A - \tau U(x - a), t - \tau]d\tau \\
&+ \frac{1}{2} \int_0^t \phi(\tau) Q_n[x + a, A - \tau U(x + a), t - \tau]d\tau.
\end{aligned}
\tag{3.4}
$$

Equation (3.4) can be interpreted as follows. For particle making its $(n + 1)$th jump into (x, A) at time t, it must be waiting at the nearest neighbors $x \pm a$ with equal probability $1/2$. The waiting time τ can be any value between 0 and t, and the functional value added within this waiting time is $\tau U(x \pm a)$. For $n = 0$, i.e., no jumps happened, there is $Q_0 = \delta(x)\delta(A)\delta(t)$.

Then, we build the relationship between $G(x, A, t)$ and its nearest flux Q_n as

$$
G(x, A, t) = \int_0^t \Phi(\tau) \sum_{n=0}^{\infty} Q_n[x, A - \tau U(x), t - \tau]d\tau,
\tag{3.5}
$$

where

$$
\Phi(t) = 1 - \int_0^t \phi(\tau)\, d\tau
\tag{3.6}
$$

is the survival probability. In Eq. (3.5), we assume the particle performs its last jump at time $t-\tau$ at position x. The particle stays at position x without jumping before time t. Throughout this waiting time, the functional value increases at the magnitude $A - \tau U(x)$ to the final value A at time t. The particle's waiting time exceeds τ, so we use the survival probability $\Phi(\tau)$ in the integrand.

As usual, we will apply the technique of Laplace and Fourier transform to solve the Eqs. (3.4) and (3.5). We assume the functional A is positive and use the Laplace transform $A \to \rho$. Otherwise, the Fourier transform would be used and the procedure is similar.

Performing Laplace transforms $A \to \rho, t \to \lambda$ and Fourier transform $x \to k$ on both sides of Eq. (3.4) gives

$$
\tilde{\hat{Q}}_{n+1}(k, \rho, \lambda) = \cos(ka)\hat{\phi}\left[\lambda + \rho U\left(-i\frac{\partial}{\partial k}\right)\right]\tilde{\hat{Q}}_n(k, \rho, \lambda).
\tag{3.7}
$$

Combining it with the initial condition $\tilde{\hat{Q}}_0(k, \rho, \lambda) = 1$, we have

$$\tilde{\hat{Q}}_n(k, \rho, \lambda) = \left[\cos(ka)\hat{\phi}\left[\lambda + \rho U\left(-i\frac{\partial}{\partial k}\right)\right]\right]^n. \tag{3.8}$$

On the other hand, taking the Laplace transform $A \to \rho$ on Eq. (3.5) yields

$$\int_0^\infty Q_n[x, A - U(x)\tau, t - \tau]e^{-\rho A}dA = e^{-\rho U(x)\tau}Q_n(x, \rho, t - \tau). \tag{3.9}$$

Then we take the Laplace transform $t \to \lambda$, utilize the convolution theorem of Laplace transform, and obtain

$$\hat{G}(x, \rho, \lambda) = \sum_{n=0}^\infty \hat{\Phi}(\lambda + \rho U(x))\hat{Q}_n(x, \rho, \lambda), \tag{3.10}$$

where $\hat{\Phi}(\lambda) = (1 - \hat{\phi}(\lambda))/\lambda$ is the Laplace transform of the survival probability $\Phi(t)$. Further performing the Fourier transform $x \to k$ on Eq. (3.10) and substituting the expression of Q_n in Eq. (3.8) into it, we obtain

$$\tilde{\hat{G}}(k, \rho, \lambda) = \sum_{n=0}^\infty \hat{\Phi}\left[\lambda + \rho U\left(-i\frac{\partial}{\partial k}\right)\right]\tilde{\hat{Q}}_n(k, \rho, \lambda)$$

$$= \frac{1 - \hat{\phi}[\lambda + \rho U(-i\frac{\partial}{\partial k})]}{\lambda + \rho U(-i\frac{\partial}{\partial k})} \cdot \frac{1}{1 - \cos(ka)\hat{\phi}[\lambda + \rho U(-i\frac{\partial}{\partial k})]}. \tag{3.11}$$

Note that the relationship between joint PDF and marginal PDF is

$$\tilde{\hat{G}}(k, \lambda) = \int_0^\infty \tilde{\hat{G}}(k, A, \lambda)dA = \tilde{\hat{G}}(k, \rho = 0, \lambda). \tag{3.12}$$

By taking $\rho = 0$ in Eq. (3.11), we obtain the marginal PDF of position x satisfies

$$\tilde{\hat{G}}(k, \lambda) = \frac{1 - \hat{\phi}(\lambda)}{\lambda}\frac{1}{1 - \cos(ka)\hat{\phi}(\lambda)}, \tag{3.13}$$

which is exactly the well-known Montroll-Weiss equation [Montroll and Weiss (1965); Metzler and Klafter (2000b)].

For the concerned power law waiting time distribution, the corresponding $\tilde{\hat{G}}(k, \rho, \lambda)$ can be obtained by using the asymptotic expansion of $\hat{\phi}(\lambda)$ for small λ, i.e.,

$$\hat{\phi}(\lambda) \simeq 1 - B_\alpha\lambda^\alpha. \tag{3.14}$$

Considering the long time limit of $\lambda, k \to 0$ in Eq. (3.11) and denoting the anomalous diffusion coefficient as

$$K_\alpha = \frac{a^2}{2B_\alpha}, \tag{3.15}$$

we obtain

$$\tilde{\hat{G}}(k, \rho, \lambda) \sim \left[\lambda + \rho U\left(-i\frac{\partial}{\partial k}\right)\right]^{\alpha-1} \frac{1}{K_\alpha k^2 + \left[\lambda + \rho U\left(-i\frac{\partial}{\partial k}\right)\right]^\alpha}. \tag{3.16}$$

Rearranging the above expression and inverting it back to the time-space domain result in the forward fractional Feynman-Kac equation

$$\frac{\partial G(x, \rho, t)}{\partial t} = K_\alpha \frac{\partial^2}{\partial x^2} \mathcal{D}_t^{1-\alpha} G(x, \rho, t) - \rho U(x) G(x, \rho, t), \tag{3.17}$$

where the fractional substantial derivative operator $\mathcal{D}_t^{1-\alpha}$ is defined in Eq. (2.163). In Laplace space, this operator in Eq. (3.17) satisfies $\mathcal{D}_t^{1-\alpha} G(x, \rho, t)$ $\to [\lambda + \rho U(x)]^{1-\alpha} \hat{G}(x, \rho, \lambda)$.

There are some remarks on the forward Feynman-Kac equation (3.17).

(1) For normal diffusion with $\alpha = 1$, the fractional equation (3.17) reduces to the classical Feynman-Kac equation

$$\frac{\partial G(x, \rho, t)}{\partial t} = K_1 \frac{\partial^2}{\partial x^2} G(x, \rho, t) - \rho U(x) G(x, \rho, t). \tag{3.18}$$

(2) The Feynman-Kac equation (3.17) governs the joint PDF of position x and functional A. By taking $\rho = 0$, it reduces to the fractional Fokker-Planck equation governing the PDF of position x

$$\frac{\partial G(x, t)}{\partial t} = K_\alpha \frac{\partial^2}{\partial x^2} D_t^{1-\alpha} G(x, t), \tag{3.19}$$

where $D_t^{1-\alpha}$ is the Riemann-Liouville fractional derivative operator ($D_t^{1-\alpha} G(x, t) \to \lambda^{1-\alpha} \hat{G}(x, \lambda)$ in Laplace space $t \to \lambda$).

(3) If the jump length is also power law distributed as $w(x) \propto |x|^{-1-\beta}$ with $0 < \beta < 2$, rather than a fixed length a, the corresponding characteristic function is $\tilde{w}(k) \simeq 1 - C_\beta |k|^\beta$ and the fractional Feynman-Kac equation is [Carmi and Barkai (2011)]

$$\frac{\partial}{\partial t} G(x, \rho, t) = K_{\alpha,\beta} \nabla_x^\beta \mathcal{D}_t^{1-\alpha} G(x, \rho, t) - \rho U(x) G(x, \rho, t), \tag{3.20}$$

where $K_{\alpha,\beta} = C_\beta / B_\alpha$ (units $m^\beta/\sec^\alpha$), and ∇_x^β is the Riesz spatial fractional derivative operator ($\nabla_x^\beta \to -|k|^\beta$ in Fourier $x \to k$ space).

3.2.1.2 Backward fractional Feynman-Kac equation

To obtain $G(A, t)$, the distribution of functional A, we need to solve Eq. (3.17) and then integrate the solution over all x to eliminate the dependence on the position x of the particle. The integration over x is not always easy and this step can be avoided by solving the backward Feynman-Kac equation, which governs the PDF $G_{x_0}(A, t)$ of A at time t given that the particle starts at x_0. Note that the initial position x_0 is a deterministic variable, not a random one.

In contrast to the case of deriving forward Feynman-Kac equation, the PDF $G_{x_0}(A, t)$ depends on the initial position x_0 instead of the final position x, and we should focus on the first step in the CTRW framework rather than the last step. More precisely, the particle jumps to either $x_0 + a$ or $x_0 - a$ with equal probabilities $1/2$ after the waiting time τ in the first step. Then, the remaining steps can be characterized through $G_{x_0 \pm a}(A - \tau U(x_0), t - \tau)$, which is

$$G_{x_0}(A, t) = \frac{1}{2} \int_0^t \phi(\tau) G_{x_0 + a}[A - \tau U(x_0), t - \tau] d\tau$$

$$+ \frac{1}{2} \int_0^t \phi(\tau) G_{x_0 - a}[A - \tau U(x_0), t - \tau] d\tau \qquad (3.21)$$

$$+ \Phi(t) \delta[A - U(x_0)t].$$

The last term in Eq. (3.21) denotes the particle staying at the initial position x_0 during the measurement time $[0, t]$.

Similar to the method of deriving forward Feynman-Kac equation, we take the Laplace transform $(A \to \rho, t \to \lambda)$ and Fourier transform $x_0 \to k_0$, and obtain

$$\hat{\tilde{G}}_{k_0}(\rho, \lambda) = \hat{\phi} \left[\lambda + \rho U \left(-i \frac{\partial}{\partial k_0} \right) \right] \cos(k_0 a) \hat{\tilde{G}}_{k_0}(\rho, \lambda)$$

$$+ \frac{1 - \hat{\phi} \left[\lambda + \rho U \left(-i \frac{\partial}{\partial k_0} \right) \right]}{\lambda + \rho U \left(-i \frac{\partial}{\partial k_0} \right)} \delta(k_0). \qquad (3.22)$$

We use the asymptotic expansion of $\hat{\phi}(\lambda)$ in Eq. (3.14), and obtain

$$\lambda \hat{\tilde{G}}_{k_0}(\rho, \lambda) - \delta(k_0) = -K_\alpha \left[\lambda + \rho U \left(-i \frac{\partial}{\partial k_0} \right) \right]^{1-\alpha} k_0^2 \hat{\tilde{G}}_{k_0}(\rho, \lambda)$$

$$- \rho U \left(-i \frac{\partial}{\partial k_0} \right) \hat{\tilde{G}}_{k_0}(\rho, \lambda). \qquad (3.23)$$

Inverting the above equation yields the backward fractional Feynman-Kac equation

$$\frac{\partial}{\partial t}G_{x_0}(\rho,t) = K_\alpha \mathcal{D}_t^{1-\alpha}\frac{\partial^2}{\partial x_0^2}G_{x_0}(\rho,t) - \rho U(x_0)G_{x_0}(\rho,t). \qquad (3.24)$$

There are some remarks on the backward Feynman-Kac equation (3.24).

(1) Taking $\alpha = 1$, the backward Feynman-Kac equation (3.24) reduces to the classical one Eq. (3.2).
(2) If the jump length is also power law distributed as $w(x) \propto |x|^{-1-\beta}$ with $0 < \beta < 2$, rather than a fixed length a, the corresponding backward fractional Feynman-Kac equation is [Carmi and Barkai (2011)]

$$\frac{\partial}{\partial t}G_{x_0}(\rho,t) = K_{\alpha,\beta}\mathcal{D}_t^{1-\alpha}\nabla_{x_0}^\beta G_{x_0}(\rho,t) - \rho U(x_0)G_{x_0}(\rho,t). \qquad (3.25)$$

(3) Comparing the forward fractional Feynman-Kac equation (3.17) and the backward one Eq. (3.24), we find the main change is the variable $x \to x_0$ and the operators $\frac{\partial^2}{\partial x^2}\mathcal{D}_t^{1-\alpha} \to \mathcal{D}_t^{1-\alpha}\frac{\partial^2}{\partial x_0^2}$. Note that the order of the operators cannot be changed casually since the fractional substantial derivative operator $\mathcal{D}_t^{1-\alpha}$ depends on the spatial variable.

3.2.1.3   *Other kinds of Feynman-Kac equations*

The underlying processes discussed above are assumed with power law distributed waiting time or jump length. Actually, the results can be extended to other kinds of processes with different distributions. The most typical one is with tempered power law distribution, which exponentially truncates the pure power law distribution. This truncation has the significant physical meanings in many situations, for example, the life or the motion space of a tracer particle is finite.

The method of deriving the fractional Feynman-Kac equation in the CTRW framework is still valid for these different processes. For example, we assume the waiting time is tempered power law distributed, which is denoted as $\phi(t,\mu)$ explicitly depending on the truncation parameter μ. The Laplace transform of $\phi(t,\mu)$ is, for small λ,

$$\hat{\phi}(\lambda,\mu) \simeq 1 - B_\alpha(\mu+\lambda)^\alpha + B_\alpha\mu^\alpha, \qquad (3.26)$$

which goes back to the one of power law distribution in Eq. (3.14) when $\mu = 0$.

Similar to the microscopic description in the Sec. 3.2.1, the particle still starts its movement at $x = x_0$. And it jumps to the nearest neighbors with equal probability $1/2$. The difference is that the waiting time distribution is taken as tempered power law distribution $\phi(t,\mu)$. With the new asymptotic expansion Eq. (3.26) and some lengthy calculations [Wu *et al.* (2016)], we obtain the forward tempered fractional Feynman-Kac equation

$$\frac{\partial}{\partial t}G(x,\rho,t) = \left[\mu^\alpha \mathcal{D}_t^{1-\alpha,\mu} - \mu\right]\left[G(x,\rho,t) - e^{-\rho U(x)t}\delta(x)\right]$$
$$- \rho U(x)G(x,\rho,t) + K_\alpha \frac{\partial^2}{\partial x^2}\mathcal{D}_t^{1-\alpha,\mu}G(x,\rho,t), \tag{3.27}$$

and the backward version

$$\frac{\partial}{\partial t}G_{x_0}(\rho,t) = \left[\mu^\alpha \mathcal{D}_t^{1-\alpha,\mu} - \mu\right]\left[G_{x_0}(\rho,t) - e^{-\rho U(x_0)t}\right]$$
$$- \rho U(x_0)G_{x_0}(\rho,t) + K_\alpha \mathcal{D}_t^{1-\alpha,\mu}\frac{\partial^2}{\partial x_0^2}G_{x_0}(\rho,t), \tag{3.28}$$

where the symbol $\mathcal{D}_t^{1-\alpha,\mu}$ is the tempered fractional substantial derivative defined as [Wu *et al.* (2016)]

$$\mathcal{D}_t^{1-\alpha,\mu}G(x,\rho,t) =$$
$$\frac{1}{\Gamma(\alpha)}\left[\frac{\partial}{\partial t} + \mu + \rho U(x)\right]\int_0^t \frac{e^{-(t-\tau)(\mu+\rho U(x))}}{(t-\tau)^{1-\alpha}}G(x,\rho,\tau)d\tau \tag{3.29}$$

and in Laplace λ space

$$\mathcal{D}_t^{1-\alpha,\mu}G(x,\rho,t) \to [\lambda + \mu + \rho U(x)]^{1-\alpha}\hat{G}(x,\rho,\lambda). \tag{3.30}$$

More generally, there may be some kind of external force acting on the particles. In CTRW framework, the force changes the probabilities of jumping left or right. They are not equal any more and now denoted as $L(x)$ and $R(x)$, respectively. The sum of $L(x)$ and $R(x)$ is still equal to unit, but their difference will depend on the external force $F(x)$ explicitly. We assume the system is coupled to a heat bath at temperature $\mathcal{T}$ and it stays in a detailed balance. In this case, there exists

$$L(x)\exp\left[-\frac{V(x)}{k_B\mathcal{T}}\right] = R(x-a)\exp\left[-\frac{V(x-a)}{k_B\mathcal{T}}\right], \tag{3.31}$$

where $V(x)$ represents the potential, that is, $F(x) = -V'(x)$. For small a, using the expansion of exponential function leads to

$$L(x) \simeq 1 - \frac{aF(x)}{2k_B\mathcal{T}} \quad \text{and} \quad R(x) \simeq 1 + \frac{aF(x)}{2k_B\mathcal{T}}. \tag{3.32}$$

In this case, due to the asymmetry of the probability of jumping left and right, the first order of k (Fourier transform $x \to k$) remains in the derivations, which is equivalent to the first order of derivative with respect of x

in time domain. The forward tempered fractional Feynman-Kac equation becomes

$$\frac{\partial}{\partial t}G(x,\rho,t) = \left[\mu^\alpha \mathcal{D}_t^{1-\alpha,\mu} - \mu\right]\left[G(x,\rho,t) - e^{-\rho U(x)t}\delta(x)\right]$$

$$- \rho U(x)G(x,\rho,t) + K_\alpha\left[\frac{\partial^2}{\partial x^2} - \frac{\partial}{\partial x}\frac{F(x)}{k_B\mathcal{T}}\right]\mathcal{D}_t^{1-\alpha,\mu}G(x,\rho,t),$$

(3.33)

and the backward tempered fractional Feynman-Kac equation is

$$\frac{\partial}{\partial t}G_{x_0}(\rho,t) = \left[\mu^\alpha \mathcal{D}_t^{1-\alpha,\mu} - \mu\right]\left[G_{x_0}(\rho,t) - e^{-\rho U(x_0)t}\right]$$

$$- \rho U(x_0)G_{x_0}(\rho,t) + K_\alpha\mathcal{D}_t^{1-\alpha,\mu}\left[\frac{\partial^2}{\partial x_0^2} + \frac{F(x_0)}{k_B\mathcal{T}}\frac{\partial}{\partial x_0}\right]G_{x_0}(\rho,t).$$

(3.34)

3.2.2 Derivation in Langevin Picture

Compared with the CTRW model, there are some cases where Langevin picture is a more natural choice, e.g., the Langevin equation with multiplicative noise and an external force field

$$\dot{x}(t) = f(x(t),t) + g(x(t),t)\xi(t),$$
(3.35)

where $x(t)$ is the particle coordinate, $f(x,t)$ is the force field, $\xi(t)$ is the noise resulting from a fluctuating environment, and $g(x,t)$ is the multiplicative noise coefficient. The multiplicative noise is aimed to describe the heterogeneous dynamical behavior, which is particularly remarkable in biological systems. The cytoplasm of biological cells is always crowded with various obstacles, including proteins, nucleic acids, ribosomes, the cytoskeleton, as well as internal membranes compartmentalizing the cell [Zimmerman and Minton (1993); Zhou *et al.* (2008)]. The nonuniform distribution of crowders in the cytoplasm provides the heterogeneous media for tracer particles of different sizes in it. This kind of heterogeneous diffusion process can be also realized in the experiments on eukaryotic cells [Kühn *et al.* (2011)] and a local variation of the temperature in thermophoresis experiments [Maeda *et al.* (2012); Mast *et al.* (2013)].

Here, we choose $\xi(t)$ to be the Lévy noise, which is the formal time derivative of the corresponding Lévy process $\eta(t)$. That is to say, the increment $\delta\eta(t) = \eta(t+\tau) - \eta(t)$ of $\eta(t)$ could be defined as the time integral of $\xi(t)$, i.e., $\delta\eta(t) = \int_t^{t+\tau}\xi(t')dt'$. Similarly, the increment $\delta x(t) = x(t+\tau) - x(t)$ of the particle trajectory undergoing the Langevin system Eq. (3.35) during a time small interval τ satisfies

$$\delta x(t) \simeq f(x(t),t)\tau + g(x(t),t)\delta\eta(t),$$
(3.36)

which defines the meaning of equation (3.35) in the Itô interpretation [Itô (1950); Risken (1989)]. The particle location $x(t)$ only relies on the previous increments of $\eta(t)$ and thus it is independent on the increment $\delta\eta(t)$ because of the independence of the increments of Lévy process in non-overlapping intervals. The stationary increment of the Lévy process imples that $\delta\eta(t)$ has the same distribution as $\eta(\tau)$ with characteristic function denoted by [Applebaum (2009)]:

$$\langle e^{-ik\eta(\tau)} \rangle = e^{\tau\phi_0(k)}, \tag{3.37}$$

where the Lévy exponent $\phi_0(k)$ has the specific forms of $\phi_0(k) = -k^2$ for Gaussian white noise and $\phi_0(k) = -|k|^\beta$ $(0 < \beta < 2)$ for non-Gaussian β-stable Lévy noise.

3.2.2.1 *Forward fractional Feynman-Kac equation*

Consistent to the notations in the previous sections, we denote the functional $A = \int_0^t U[x(t')]dt'$ and $G(x, A, t)$ as the joint PDF of position x and functional A at time t. The Fourier transform $x \to k, A \to \rho$ of $G(x, A, t)$ is

$$\tilde{G}(k, \rho, t) = \langle e^{-ikx(t)}e^{-i\rho A(t)} \rangle. \tag{3.38}$$

Similar to the increment $\delta x(t)$ in Eq. (3.36), one has the increment

$$\delta A(t) = A(t + \tau) - A(t) \simeq U(x(t))\tau \tag{3.39}$$

during the small time interval τ. Then we consider the increment of $G(x, A, t)$ in Fourier space, which can be written as

$$\begin{aligned} \delta\tilde{G}(k, \rho, t) :&= \tilde{G}(k, \rho, t + \tau) - \tilde{G}(k, \rho, t) \\ &= \langle e^{-ikx(t+\tau)-i\rho A(t+\tau)} \rangle - \langle e^{-ikx(t)-i\rho A(t)} \rangle. \end{aligned} \tag{3.40}$$

Substituting the increments $\delta x(t)$ and $\delta A(t)$ into Eq. (3.40) and taking $\tau \to 0$ give

$$\begin{aligned} \delta\tilde{G}(k, \rho, t) = &\langle e^{-ikx(t)-i\rho A(t)}(e^{-ikg(x(t),t)\delta\eta(t)} - 1) \rangle \\ &- ik\tau \langle e^{-ikx(t)-i\rho A(t)} f(x(t), t) \rangle \\ &- i\rho\tau \langle e^{-ikx(t)-i\rho A(t)} U(x(t)) \rangle. \end{aligned} \tag{3.41}$$

Note that the angle brackets in the first term in Eq. (3.41) denote the average with the joint PDF $G(x, A, t)$ and the PDF of the noise increment $\delta\eta(t)$ since $\delta\eta(t)$ is independent of particle trajectory $x(t)$. Utilizing the characteristic function of the noise increment $\delta\eta(t)$ in Eq. (3.37), we have

$$\lim_{\tau \to 0} \frac{1}{\tau} \langle (e^{-ikg(x(t),t)\delta\eta(t)} - 1) \rangle = \phi_0(kg(x(t), t)). \tag{3.42}$$

It can be seen that the second and third terms in Eq. (3.41) are exactly the Fourier transforms of some kind of compound functions with respect to $G(x, A, t)$, i.e.,

$$ik\langle e^{-ikx(t)-i\rho A(t)} f(x(t), t)\rangle = \mathcal{F}_x \mathcal{F}_A \left\{ \frac{\partial}{\partial x} f(x, t) G(x, A, t) \right\}, \qquad (3.43)$$

and

$$i\rho \langle e^{-ikx(t)-i\rho A(t)} U(x(t))\rangle = i\rho \mathcal{F}_x \mathcal{F}_A \left\{ U(x) G(x, A, t) \right\}. \qquad (3.44)$$

Up to now, the terms on the right-hand side of Eq. (3.41) have been evaluated in Eqs. (3.42), (3.43) and (3.44), respectively. Then dividing Eq. (3.41) by τ and taking the limit $\tau \to 0$, we can obtain the forward Feynman-Kac equation in Fourier space:

$$\begin{aligned}
\frac{\partial \tilde{G}(k, \rho, t)}{\partial t} &= \mathcal{F}_x \{\phi_0(kg(x, t)) G(x, \rho, t)\} \\
&\quad - \mathcal{F}_x \left\{ \frac{\partial}{\partial x} f(x, t) G(x, \rho, t) + i\rho U(x) G(x, \rho, t) \right\}.
\end{aligned} \qquad (3.45)$$

For some specific problem with Gaussian white noise or Lévy noise, we can determine the Fourier symbol $\phi_0(kg(x, t))$ and obtain the corresponding forward Feynman-Kac equation in x space.

There are some remarks on the forward fractional Feynman-Kac equation (3.45).

(1) *Generalized Fokker-Planck equation.* Let $\rho = 0$ in Eq. (3.45). In this case, $G(x, \rho = 0, t) = \int_0^\infty G(x, A, t) dA$ reduces to $G(x, t)$, the marginal PDF of finding the particle at position x at time t. Correspondingly, the forward Feynman-Kac equation (3.45) reduces to the generalized Fokker-Planck equation [Denisov *et al.* (2009)], where three kinds of noises (Gaussian white noise, Poisson white noise and Lévy stable noise) are considered for the specific forms of this equation.

(2) *Gaussian white noise.* If $\xi(t)$ is the Gaussian white noise in Eq. (3.45), for arbitrary $f(x, t)$ and $g(x, t)$, we get the forward Feynman-Kac equation:

$$\begin{aligned}
\frac{\partial G(x, \rho, t)}{\partial t} &= \left[-\frac{\partial}{\partial x} f(x, t) + \frac{\partial^2}{\partial x^2} g^2(x, t) \right] G(x, \rho, t) \\
&\quad - i\rho U(x) G(x, \rho, t).
\end{aligned} \qquad (3.46)$$

(3) *Non-Gaussian β-stable noise.* If $\xi(t)$ is the non-Gaussian β-stable noise in Eq. (3.45), for arbitrary $f(x,t)$ and $g(x,t)$, the forward Feynman-Kac equation becomes

$$\frac{\partial G(x,\rho,t)}{\partial t} = \left[-\frac{\partial}{\partial x} f(x,t) + \nabla_x^\beta |g(x,t)|^\beta \right] G(x,\rho,t)$$
$$- i\rho U(x) G(x,\rho,t),$$

(3.47)

where ∇_x^β is the Riesz space fractional derivative operator with Lévy exponent $-|k|^\beta$. This equation extends Eq. (3.46) to the case corresponding to Lévy stable noise, denoting the power law distributed jump length in CTRWs.

(4) *A positive functional.* If the functional A is positive at any time t, the Fourier transform $A \to \rho$ will be replaced by the Laplace transform $G(x,\rho,t) = \int_0^\infty e^{-\rho A} G(x,A,t) dA$. Eventually, the forward Feynman-Kac equation corresponding to Eq. (3.45) is obtained by replacing $i\rho$ with ρ.

3.2.2.2 *Backward fractional Feynman-Kac equation*

The backward fractional Feynman-Kac equation can be also derived from Langevin picture. Similar to the derivations from CTRW framework, the underlying process we consider should be time homogenous. So we assume that the stochastic process is

$$\dot{x}(t) = f(x(t)) + g(x(t))\xi(t), \tag{3.48}$$

where $\xi(t)$ is also a Lévy noise. Compared with the model Eq. (3.35), f and g do not explicitly depend on t. Otherwise, the time-dependent force field (or the multiplicative term) induces different displacement for a particle located at the same position but different time. In this case, it is difficult to let the functional A only depend on initial position x_0 without using the information of the whole path $x(t)$.

Let $G_{x_0}(A,t)$ be the PDF of functional A at time t, given that the process has started at x_0. We should figure out how functional A depends on initial position x_0. Different from the increment δA considered in the forward Feynman-Kac equation, we should build the relationship between A and x_0 as, during the small time interval τ,

$$A(t+\tau)|_{x_0} = \int_0^\tau U(x(t'))dt' + \int_\tau^{t+\tau} U(x(t'))dt'$$
$$= U(x_0)\tau + A(t)|_{x(\tau)},$$

(3.49)

where $A(t+\tau)|_{x_0}$ denotes the functional A at time $t+\tau$ with the initial position x_0. Letting $t=0$ in Eq. (3.36), $x(\tau)$ can be approximated as

$$x(\tau) = x_0 + f(x_0)\tau + g(x_0)\eta(\tau). \tag{3.50}$$

The expression of $G_{x_0}(A,t)$ in Fourier space is

$$G_{x_0}(\rho,t) = \langle e^{-i\rho A(t)|_{x_0}} \rangle. \tag{3.51}$$

Then

$$G_{x_0}(\rho,t+\tau) = \langle\langle e^{-i\rho A(t)|_{x(\tau)}} \rangle\rangle e^{-i\rho U(x_0)\tau} \tag{3.52}$$

by using Eq. (3.49). Since $A(t)|_{x(\tau)}$ denotes the functional A at time t with the initial position $x(\tau)$, it is independent of the event before $x(\tau)$, e.g., $\eta(\tau)$. So the internal angle bracket in Eq. (3.52) denotes the average of $A(t)|_{x(\tau)}$ while the external one the average of $\eta(\tau)$. Then the increment $\delta G_{x_0}(\rho,t)$ can be expressed as

$$\begin{aligned} \delta G_{x_0}(\rho,t) : &= G_{x_0}(\rho,t+\tau) - G_{x_0}(\rho,t) \\ &= \langle\langle e^{-i\rho A(t)|_{x(\tau)}} \rangle\rangle e^{-i\rho U(x_0)\tau} - \langle e^{-i\rho A(t)|_{x_0}} \rangle. \end{aligned} \tag{3.53}$$

Taking $\tau \to 0$, omitting the higher order terms of τ, we get

$$\delta G_{x_0}(\rho,t) = \langle\langle e^{-i\rho A(t)|_{x(\tau)}} \rangle\rangle - \langle e^{-i\rho A(t)|_{x_0}} \rangle - i\rho U(x_0)\tau \langle e^{-i\rho A(t)|_{x_0}} \rangle, \tag{3.54}$$

where the last term equals to $-i\rho U(x_0)\tau G_{x_0}(\rho,t)$. Next, we will deal with the first two terms in the right-hand side of Eq. (3.54) carefully and omit the higher order of τ. Taking Fourier transform $x_0 \to k_0$ in Eq. (3.54), then $\langle e^{-i\rho A(t)|_{x_0}} \rangle$ becomes $G_{k_0}(\rho,t)$. But for $\langle\langle e^{-i\rho A(t)|_{x(\tau)}} \rangle\rangle$, it is not easy to get the form in Fourier space.

Hence, we firstly take $g(x) \equiv 1$, i.e., the noise in this system is additive noise. Denote $T_\eta = \langle e^{-i\rho A(t)|_{x(\tau)}} \rangle$. Since $g(x) \equiv 1$, Eq. (3.50) becomes

$$x(\tau) = x_0 + f(x_0)\tau + \eta(\tau), \tag{3.55}$$

where $f(x_0)$ only depends on the initial position x_0. Because of this dependence, when we consider the Fourier transform, $x(\tau)$ is not a simple shift of x_0. Therefore, we firstly write the Fourier transform $(x_0 \to k_0)$ of $\langle T_\eta \rangle$ as

$$\mathcal{F}_{x_0}\{\langle T_\eta \rangle\} = \left\langle \int_{-\infty}^{\infty} e^{-ik_0 x(\tau)} T_\eta e^{ik_0(f(x_0)\tau + \eta(\tau))} dx_0 \right\rangle. \tag{3.56}$$

Then turning dx_0 into $dx(\tau)$, one arrives at

$$\mathcal{F}_{x_0}\{\langle T_\eta \rangle\} = \Big\langle \int_{-\infty}^{\infty} e^{-ik_0 x(\tau)} T_\eta e^{ik_0(f(x_0)\tau+\eta(\tau))} dx(\tau) \Big\rangle$$

$$- \Big\langle \int_{-\infty}^{\infty} e^{-ik_0 x(\tau)} T_\eta e^{ik_0(f(x_0)\tau+\eta(\tau))} \frac{df(x_0)}{dx_0} \tau dx_0 \Big\rangle. \tag{3.57}$$

Since all x_0 and $f(x_0)$ are multiplied by τ in Eq. (3.57), replacing all x_0 by $x(\tau)$ in Eq. (3.57) yields higher-order terms of τ, which can be omitted. Then we utilize the asymptotics $e^{ik_0 f(x_0)\tau} \simeq 1 + ik_0 f(x_0)\tau$. The first term on the right-hand side of Eq. (3.57) reduces to

$$\Big\langle \int_{-\infty}^{\infty} e^{-ik_0 x(\tau)} T_\eta e^{ik_0 \eta(\tau)} dx(\tau) \Big\rangle + ik_0 \tau \Big\langle \int_{-\infty}^{\infty} e^{-ik_0 x(\tau)} T_\eta f(x(\tau)) dx(\tau) \Big\rangle, \tag{3.58}$$

where the latter term above equals to

$$\tau \mathcal{F}_{x_0}\Big\{ \frac{\partial}{\partial x_0} f(x_0) G_{x_0}(\rho, t) \Big\}. \tag{3.59}$$

On the other hand, omitting the exponential term and replacing x_0 by $x(\tau)$ on the second term on the right-hand side of Eq. (3.57) yield

$$-\tau \Big\langle \int_{-\infty}^{\infty} e^{-ik_0 x(\tau)} T_\eta \frac{df(x(\tau))}{dx(\tau)} dx(\tau) \Big\rangle = -\tau \mathcal{F}_{x_0}\Big\{ \frac{df(x_0)}{dx_0} G_{x_0}(\rho, t) \Big\}. \tag{3.60}$$

Therefore, the Fourier transform of $\langle\langle e^{-i\rho A(t)|x(\tau)}\rangle\rangle - \langle e^{-i\rho A(t)|x_0}\rangle$ in Eq. (3.54), replacing $x(\tau)$ by y, reduces to

$$\Big\langle \int_{-\infty}^{\infty} e^{-ik_0 y} T_\eta (e^{ik_0 \eta(\tau)} - 1) dy \Big\rangle + \tau \mathcal{F}\Big\{ f(x_0) \frac{\partial G_{x_0}(\rho, t)}{\partial x_0} \Big\}, \tag{3.61}$$

i.e.,

$$\tau \phi_0(-k_0) G_{k_0}(\rho, t) + \tau \mathcal{F}_{x_0}\Big\{ f(x_0) \frac{\partial G_{x_0}(\rho, t)}{\partial x_0} \Big\} \tag{3.62}$$

on account of Eq. (3.42). Dividing Eq. (3.54) by τ and taking $\tau \to 0$, we obtain the backward Feynman-Kac equation in Fourier space:

$$\frac{\partial \tilde{G}_{k_0}(\rho, t)}{\partial t} = \phi_0(-k_0) \tilde{G}_{k_0}(\rho, t)$$

$$+ \mathcal{F}_{x_0}\Big\{ f(x_0) \frac{\partial G_{x_0}(\rho, t)}{\partial x_0} - i\rho U(x_0) G_{x_0}(\rho, t) \Big\}. \tag{3.63}$$

There are some remarks on the backward Feynman-Kac equation (3.63).

(1) If the noise $\xi(t)$ is Gaussian white noise, then $\phi_0(-k_0) = -k_0^2$ and we get the backward Feynman-Kac equation:

$$\frac{\partial G_{x_0}(\rho, t)}{\partial t} = \frac{\partial^2}{\partial x_0^2} G_{x_0}(\rho, t)$$
$$+ f(x_0) \frac{\partial}{\partial x_0} G_{x_0}(\rho, t) - i\rho U(x_0) G_{x_0}(\rho, t), \tag{3.64}$$

which is the same as the backward Feynman-Kac equation proposed in [Carmi and Barkai (2011)] with $\alpha = 1$ in CTRW framework. Here, α is the exponent characterizing the waiting time distribution in CTRWs or the PDF of subordinator in Langevin system.

(2) If the noise $\xi(t)$ is non-Gaussian β-stable noise, i.e., $\phi_0(-k_0) = -|k_0|^\beta$, then the backward Feynman-Kac equation becomes

$$\frac{\partial G_{x_0}(\rho, t)}{\partial t} = \nabla_{x_0}^\beta G_{x_0}(\rho, t)$$
$$+ f(x_0) \frac{\partial}{\partial x_0} G_{x_0}(\rho, t) - i\rho U(x_0) G_{x_0}(\rho, t), \tag{3.65}$$

which is an extension for the backward Feynman-Kac equation derived in CTRW framework [Carmi *et al.* (2010)], in which jump length obeys heavy-tailed distribution but without a force field $f(x)$.

(3) In the case that $g(x)$ is not a constant, we assume $\xi(t)$ to be Gaussian white noise and the backward Feynman-Kac equation is

$$\frac{\partial G_{x_0}(\rho, t)}{\partial t} = g^2(x_0) \frac{\partial^2}{\partial x_0^2} G_{x_0}(\rho, t)$$
$$+ f(x_0) \frac{\partial}{\partial x_0} G_{x_0}(\rho, t) - i\rho U(x_0) G_{x_0}(\rho, t), \tag{3.66}$$

which goes back to Eq. (3.64) when $g(x_0) \equiv 1$. See the detailed derivations in [Wang *et al.* (2018)].

3.2.2.3 *Coupled Langevin equation*

The Langevin equation (3.35) evolves in the deterministic time variable t, which avoids the possible trap events in practical problems. For the latter case, we can replace the time variable t by a process $t(s)$ with positive values. The common choice is a subordinator, which is a positive nondecreasing one-dimensional Lévy process. Then the subordinated stochastic process could be described by the coupled Langevin equation:

$$\dot{x}(s) = f(x(s), t(s)) + g(x(s), t(s))\xi(s),$$
$$\dot{t}(s) = \theta(s). \tag{3.67}$$

Here we adopt the fully skewed α-stable Lévy noise $\theta(s)$ with $0 < \alpha < 1$, which is independent of the arbitrary Lévy noise $\xi(s)$. Then the combined process is defined as $y(t) = x(s(t))$ with the inverse α-stable subordinator $s(t)$, which is the first-passage time of the α-stable subordinator $\{t(s), s \geq 0\}$ and defined [Piryatinska *et al.* (2005); Magdziarz *et al.* (2007)] as $s(t) = \inf_{s>0}\{s : t(s) > t\}$. Note that the time-dependent force f and multiplicative noise term g should depend on the physical time $t(s)$, rather than the operational time s, due to a physical interpretation [Magdziarz *et al.* (2008); Heinsalu *et al.* (2007)].

Since $y(t) = x(s(t))$, we can build the Langevin equation of $y(t)$ from Eq. (3.67) as:

$$\dot{y}(t) = f(y(t), t)\dot{s}(t) + g(y(t), t)\xi(s(t))\dot{s}(t). \tag{3.68}$$

Similar to Eq. (3.36), with the Itô interpretation, the increment of $y(t)$ reads

$$\delta y(t) = f(y(t), t)\delta s(t) + g(y(t), t)\delta\eta(s(t)), \tag{3.69}$$

where $\delta s(t) = s(t + \tau) - s(t)$ and $\delta\eta(s(t)) = \eta(s(t + \tau)) - \eta(s(t))$. Next, similar to Eq. (3.41), we obtain the increment of $G(y, W, t)$ in Fourier space $(y \to k, W \to \rho)$:

$$\delta\tilde{G}(k, \rho, t) = \langle e^{-iky(t)-i\rho W(t)}(e^{-ikg(y(t),t)\delta\eta(s(t))} - 1)\rangle$$

$$- ik\langle e^{-iky(t)-i\rho W(t)}f(y(t), t)\delta s(t)\rangle \tag{3.70}$$

$$- i\rho\tau\langle e^{-iky(t)-i\rho W(t)}U(y(t))\rangle,$$

where the first term on the right-hand side can be reduced to

$$\langle e^{-iky(t)-i\rho W(t)}\phi_0(kg(y(t), t))\,\delta s(t)\rangle \tag{3.71}$$

as usual due to the characteristic function of $\delta\eta(t)$ in Eq. (3.37). So dividing Eq. (3.70) by τ and taking the limit $\tau \to 0$, we obtain

$$\frac{\partial}{\partial t}\tilde{G}(k, \rho, t) = \langle e^{-iky(t)-i\rho W(t)}\phi_0(kg(y(t), t))\dot{s}(t)\rangle$$

$$- ik\langle e^{-iky(t)-i\rho W(t)}f(y(t), t)\dot{s}(t)\rangle$$

$$- i\rho\langle e^{-iky(t)-i\rho W(t)}U(y(t))\rangle \tag{3.72}$$

$$=: Q_1 + Q_2 + Q_3.$$

It is obvious that the inverse Fourier transform $(k \to y)$ of Q_3 is $-i\rho U(y)G(y, \rho, t)$. But for Q_1 and Q_2, they look a little bit difficult due to the new term $\dot{S}(t)$ compared with Eq. (3.41). Note that the angle brackets in Q_1 denote the average of two kinds of independent stochastic processes with the joint PDF $G(y(t), W(t), t)$ and Lévy α-stable noise $\theta(t)$ on which $s(t)$ depends. In order to deal with the term Q_1, we first add a technical δ-function $\delta(y - y(t))$ in it and get

$$Q_1 = \int_{-\infty}^{\infty} e^{-iky} \phi_0(kg(y,t)) \langle e^{-i\rho W(t)} \delta(y - y(t)) \dot{s}(t) \rangle dy. \tag{3.73}$$

Then applying the technique in [Cairoli and Baule (2017)] of rewriting the functional $W(t)$ as a subordinated process:

$$W(t) = V(s(t)), \qquad V(s) = \int_0^s U(x(s')) \theta(s') ds'. \tag{3.74}$$

Substituting $y(t) = x(s(t))$ and $W(t) = V(s(t))$ into Q_1 gives the middle term of Q_1 as

$$\langle e^{-i\rho V(s(t))} \delta(y - x(s(t))) \dot{s}(t) \rangle = \int_0^{\infty} \langle e^{-i\rho V(s)} \delta(y - x(s)) \delta(t - T(s)) \rangle ds. \tag{3.75}$$

Taking Laplace transform $(t \to \lambda)$ of Eq. (3.75), we obtain

$$\hat{Q}_1(\lambda) = \int_{-\infty}^{\infty} e^{-iky} \phi_0(kg(y,t)) \int_0^{\infty} \langle e^{-i\rho V(s) - \lambda t(s)} \delta(y - x(s)) \rangle ds\, dy. \tag{3.76}$$

On the other hand, $G(y, \rho, t)$ can be rewritten as:

$$\begin{aligned} G(y, \rho, t) &= \langle e^{-i\rho V(s(t))} \delta(y - x(s(t))) \rangle \\ &= \int_0^{\infty} \langle e^{-i\rho V(s)} \delta(s - s(t)) \delta(y - x(s)) \rangle ds. \end{aligned} \tag{3.77}$$

So its Laplace transform $(t \to \lambda)$ is

$$\hat{G}(y, \rho, \lambda) = \int_0^{\infty} \langle e^{-i\rho V(s) - \lambda t(s)} \theta(s) \delta(y - x(s)) \rangle ds. \tag{3.78}$$

The characteristic function of the Lévy process $t(s)$ in Eq. (3.67) is

$$\langle e^{-\lambda t(s)} \rangle = e^{-s\lambda^{\alpha}}. \tag{3.79}$$

Then we use the independence between $\theta(s)$ and $x(s)$ and perform the average with respect to $\theta(s)$ first. Thus the part of the integrand of Eq.

(3.78) becomes

$$\langle e^{-i\rho V(s)-\lambda t(s)}\theta(s)\rangle = \Big\langle \theta(s)e^{-\int_0^s[\lambda+i\rho U(x(r))]\theta(r)dr}\Big\rangle$$

$$= -\frac{1}{\lambda+i\rho U(x(s))}\frac{\partial}{\partial s}\Big\langle e^{-\int_0^s[\lambda+i\rho U(x(r))]\theta(r)dr}\Big\rangle$$

$$= -\frac{1}{\lambda+i\rho U(x(s))}\frac{\partial}{\partial s}e^{-\int_0^s[\lambda+i\rho U(x(r))]^\alpha dr} \qquad (3.80)$$

$$= [\lambda+i\rho U(x(s))]^{\alpha-1}\Big\langle e^{-\int_0^s[\lambda+i\rho U(x(r))]\theta(r)dr}\Big\rangle$$

$$= [\lambda+i\rho U(x(s))]^{\alpha-1}\langle e^{-i\rho V(s)-\lambda t(s)}\rangle,$$

which yields

$$\hat{G}(y,\rho,\lambda) = [\lambda+i\rho U(y)]^{\alpha-1}\int_0^\infty \langle e^{-i\rho V(s)-\lambda t(s)}\delta(y-x(s))\rangle ds. \qquad (3.81)$$

Comparing Eq. (3.76) with Eq. (3.81), we find that

$$Q_1(\lambda) = \int_{-\infty}^\infty e^{-iky}\phi_0(kg(y,t))[\lambda+i\rho U(y)]^{1-\alpha}G(y,\rho,\lambda)dy. \qquad (3.82)$$

Taking the inverse Laplace transform ($\lambda \to t$), we obtain

$$Q_1 = \int_{-\infty}^\infty e^{-iky}\phi_0(kg(y,t))\mathcal{D}_t^{1-\alpha}G(y,\rho,t)dy. \qquad (3.83)$$

As for Q_2, it can be obtained similarly, i.e.,

$$Q_2 = -ik\int_{-\infty}^\infty e^{-iky}f(y,t)\mathcal{D}_t^{1-\alpha}G(y,\rho,t)dy. \qquad (3.84)$$

Finally, substituting Eqs. (3.83) and (3.84) into Eq. (3.72), we obtain the forward Feynman-Kac equation in Fourier space:

$$\frac{\partial\tilde{G}(k,\rho,t)}{\partial t} = \mathcal{F}_y\{\phi_0(kg(y,t))\mathcal{D}_t^{1-\alpha}G(y,\rho,t)\}$$
$$- \mathcal{F}_y\left\{\frac{\partial}{\partial y}f(y,t)\mathcal{D}_t^{1-\alpha}G(y,\rho,t)+i\rho U(y)G(y,\rho,t)\right\}. \qquad (3.85)$$

An alternative method to derive the forward Feynman-Kac equation (3.85) corresponding the coupled Langevin equation (3.67) is to use the Itô formula [Øksendal (2005)] in Ref. [Cairoli and Baule (2017)], which studied the basic mathematic knowledge of Lévy processes, semimartingales and their stochastic calculus. In addition, it also extends the waiting time to be arbitrarily distributed, such as tempered power law distribution. The different waiting time distribution contributes to a different fractional substantial derivative operator.

3.2.3 *Multiple Internal States with Anisotropic Diffusion*

The underlying processes considered in the previous Secs. 3.2.1 and 3.2.2 are in one dimension. For high dimensional processes, by contrast, the discrepancy of the particles' motion in different directions is common in natural world, which is named as anisotropic diffusion. On the other hand, all the models mentioned above are for the diffusion with single internal state, implying that the processes have the same distributions of waiting time and jump length at each step. As a kind of inhomogeneous motion, multiple internal states have been demonstrated to be much applicable. In this subsection, we will extend the processes discussed previously to high dimension by studying the anisotropic fractional Laplacian operator characterizing the anisotropic motion with long jumps, and deriving the corresponding fractional Feynman-Kac equations with multiple internal states.

The Lévy-Khintchine formula [Applebaum (2009)] is convenient to deal with the stochastic process in high dimension, where the characteristic function of a Lévy process has a specific form in Eq. (1.93). Now we consider the compound Poisson process in n dimensions, in which Poisson process is taken as a renewal process. Let Poisson process $N(t)$ satisfy $P\{N(t) = n\} = \frac{t^n}{n!}e^{-t}$, where the mean number of jumps per unit time is unit. Then the compound Poisson process is defined as $\mathbf{X}(t) = \sum_{j=0}^{N(t)} \mathbf{X}_j$, where $\mathbf{X}_j$ are i.i.d. random variables. The characteristic function of $\mathbf{X}(t)$ has a specific form as [Deng *et al.* (2018b)]

$$\langle e^{i\mathbf{k}\cdot\mathbf{X}} \rangle = \int_{\mathbb{R}^n} e^{i\mathbf{k}\cdot\mathbf{X}} p(\mathbf{X}, t) d\mathbf{X} = e^{t(\Phi_0(\mathbf{k})-1)}, \qquad (3.86)$$

where $\Phi_0(\mathbf{k}) = \langle e^{i\mathbf{k}\cdot\mathbf{X}_j} \rangle$, $j = 0, 1, \cdots, N(t)$. Denoting the probability measure of the jump length $\mathbf{X}_j$ by $\nu(d\mathbf{Y})$, we have

$$\Phi_0(\mathbf{k}) - 1 = \int_{\mathbb{R}^n\backslash\{0\}} \left[e^{i\mathbf{k}\cdot\mathbf{Y}} - 1 - i\mathbf{k}\cdot\mathbf{Y}_{\chi_{\{|\mathbf{Y}|<1\}}} \right] \nu(d\mathbf{Y}), \qquad (3.87)$$

which is the same as the Lévy-Khintchine formula in Eq. (1.94) by taking $\mathbf{a} = 0$ and $\mathbf{b} = 0$. Considering the definition of Fourier transform and Eq. (3.86), we have

$$\tilde{p}(\mathbf{k}, t) = e^{t(\Phi_0(\mathbf{k})-1)}, \qquad (3.88)$$

which implies that the equation in $\mathbf{k}$ space is

$$\frac{\partial \tilde{p}(\mathbf{k}, t)}{\partial t} = (\Phi_0(\mathbf{k}) - 1)\tilde{p}(\mathbf{k}, t). \qquad (3.89)$$

Here, the term $i\mathbf{k}\cdot\mathbf{Y}$ aims to overcome the possible divergence of the integral of Eq. (3.87) because of the possible strong singularity of $\nu(d\mathbf{Y})$ at zero for the case of anomalous diffusion. For an isotropic β-stable anomalous diffusion process in n dimensions, its distribution of jump length is $c_\beta r^{-n-\beta}$, which means that

$$\nu(d\mathbf{Y}) = c_\beta |\mathbf{Y}|^{-n-\beta} d\mathbf{Y}. \tag{3.90}$$

When $0 < \beta < 1$, the term $i\mathbf{k} \cdot \mathbf{Y}$ can be omitted due to weak singularity (the integral in Eq. (3.87) is convergent at origin). If $1 \le \beta < 2$, though the singularity is strong, this term can also be omitted due to the possible symmetry of the Lévy measure $\nu(d\mathbf{Y})$, i.e., $\nu(d\mathbf{Y}) = \nu(-d\mathbf{Y})$ (the integral in Eq. (3.87) at origin can be understood in the sense of Cauchy principal value). Therefore, if $1 \le \beta < 2$ meets with the asymmetry of $\nu(d\mathbf{Y})$, this term is required. Based on the analyses above, we will keep the term $i\mathbf{k} \cdot \mathbf{Y}$ formally for $0 < \beta < 2$ in the following, though it vanishes in some appropriate situations.

Two special cases are the isotropic one in Eq. (3.90) and the horizontal-vertical one

$$\begin{aligned}
\nu(d\mathbf{Y}) = {} & c_{\beta_1} |y_1|^{-1-\beta_1} \delta(y_2)\delta(y_3) \cdots \delta(y_n) d\mathbf{Y} \\
& + c_{\beta_2} |y_2|^{-1-\beta_2} \delta(y_1)\delta(y_3) \cdots \delta(y_n) d\mathbf{Y} + \cdots \\
& + c_{\beta_n} |y_n|^{-1-\beta_n} \delta(y_1)\delta(y_2) \cdots \delta(y_{n-1}) d\mathbf{Y},
\end{aligned} \tag{3.91}$$

where $\beta_i \in (0,2)$ and y_i is the component of $\mathbf{Y}$, i.e., $\mathbf{Y} = [y_1, y_2, \cdots, y_n]^T$. Their corresponding macroscopic equations are

$$\begin{aligned}
\frac{\partial p(\mathbf{X}, t)}{\partial t} &= \Delta^{\beta/2} p(\mathbf{X}, t) \\
&= -c_{n,\beta} \, \mathrm{P.V.} \int_{\mathbb{R}^n} \frac{p(\mathbf{X}, t) - p(\mathbf{Y}, t)}{|\mathbf{X} - \mathbf{Y}|^{n+\beta}} d\mathbf{Y}
\end{aligned} \tag{3.92}$$

and

$$\frac{\partial p(\mathbf{X}, t)}{\partial t} = (\Delta_{x_1}^{\beta_1/2} + \Delta_{x_2}^{\beta_2/2} + \cdots + \Delta_{x_n}^{\beta_n/2}) p(\mathbf{X}, t), \tag{3.93}$$

respectively, where $\Delta_{x_i}^{\beta_i/2}$ is the fractional Laplacian in $\mathbb{R}^1$ with respect to x_i. Besides the two cases, there are also a large number of irregular motions the microscopic particles perform. In general, we call it anisotropy. With the aid of Lévy-Khintchine formula in Eq. (1.94), the concrete form of $\nu(d\mathbf{Y})$ can be given.

Following Eqs. (3.89) and (3.87), with inverse Fourier transform, we have

$$\frac{\partial p(\mathbf{X}, t)}{\partial t} = \int_{\mathbb{R}^n \setminus \{0\}} [p(\mathbf{X}-\mathbf{Y}) - p(\mathbf{X}) + (\mathbf{Y} \cdot \nabla_{\mathbf{X}} p(\mathbf{X}))_{\chi_{[|\mathbf{Y}|<1]}}] \nu(d\mathbf{Y}), \tag{3.94}$$

where $\nabla_{\mathbf{X}} = [\partial_{x_1}, \partial_{x_2}, \cdots, \partial_{x_n}]^T$. Taking

$$\nu(d\mathbf{Y}) = \frac{1}{|\Gamma(-\beta)|} \frac{m(\mathbf{Y})}{|\mathbf{Y}|^{n+\beta}} d\mathbf{Y} \tag{3.95}$$

gives

$$\frac{\partial p(\mathbf{X},t)}{\partial t} = \frac{1}{|\Gamma(-\beta)|} \int_{\mathbb{R}^n \setminus \{0\}} \left[p(\mathbf{X}-\mathbf{Y}) - p(\mathbf{X}) + (\mathbf{Y} \cdot \nabla_{\mathbf{X}} p(\mathbf{X}))\chi_{[|\mathbf{Y}|<1]} \right]$$
$$\times \frac{m(\mathbf{Y})}{|\mathbf{Y}|^{n+\beta}} d\mathbf{Y}. \tag{3.96}$$

Here, we introduce the directional measure function $m(\mathbf{Y})$ to characterize the anisotropic feature of the particles' motion. Therefore, the meaning of $m(\mathbf{Y})$ can be presented clearly by transforming the diffusion equation into polar coordinate system. In the two and three dimensional cases, Eq. (3.96) can be written as, respectively,

$$\frac{\partial p(\mathbf{X},t)}{\partial t} = \frac{1}{|\Gamma(-\beta)|} \int_0^\infty \int_0^{2\pi} \left[p(x_1 - r\cos(\theta), x_2 - r\sin(\theta)) - p(x_1, x_2) \right.$$
$$\left. + \left(r\cos(\theta)\frac{\partial p}{\partial x_1} + r\sin(\theta)\frac{\partial p}{\partial x_2} \right)_{\chi_{[r<1]}} \right] \frac{m(\theta)}{r^{1+\beta}} d\theta dr \tag{3.97}$$

and

$$\frac{\partial p(\mathbf{X},t)}{\partial t}$$
$$= \frac{1}{|\Gamma(-\beta)|} \int_0^\infty \int_0^\pi \int_0^{2\pi} \left[p(x_1 - r\sin(\theta)\cos(\phi), x_2 - r\sin(\theta)\sin(\phi), \right.$$
$$x_3 - r\cos(\theta)) - p(x_1, x_2, x_3)$$
$$\left. + \left(r\sin(\theta)\cos(\phi)\frac{\partial p}{\partial x_1} + r\sin(\theta)\sin(\phi)\frac{\partial p}{\partial x_2} + r\cos(\theta)\frac{\partial p}{\partial x_3} \right)_{\chi_{[r<1]}} \right]$$
$$\times \frac{m(\theta,\phi)\sin\theta}{r^{1+\beta}} d\phi d\theta dr, \tag{3.98}$$

where the directional measure $m(\theta)$ or $m(\theta,\phi)$ specifies the distribution of particles spreading in the radial direction of $\mathbf{Y}$; among them, $m(\theta)$ is defined on $[0, 2\pi]$, satisfying $\int_0^{2\pi} m(\theta)d\theta = 1$, while $m(\theta,\phi)$ is defined on a $[0, \pi] \times [0, 2\pi]$ rectangular domain, satisfying $\int_0^\pi \int_0^{2\pi} m(\theta,\phi)d\phi d\theta = 1$. The situation becomes much simple if the particles move in one dimension. It is like the biased CTRW model with asymmetric probability of jumping left or right.

For the tempered Lévy flight, we can describe the movement of microscopic particles and derive the macroscopic equations by defining

$$\nu(d\mathbf{Y}) = \frac{1}{|\Gamma(-\beta)|} \frac{m(\mathbf{Y})}{e^{\mu|\mathbf{Y}|}|\mathbf{Y}|^{n+\beta}} d\mathbf{Y}; \tag{3.99}$$

and Eq. (3.94) becomes

$$\frac{\partial p(\mathbf{X}, t)}{\partial t} = \frac{1}{|\Gamma(-\beta)|} \int_{\mathbb{R}^n \setminus \{0\}} \left[p(\mathbf{X} - \mathbf{Y}) - p(\mathbf{X}) + (\mathbf{Y} \cdot \nabla_{\mathbf{X}} p(\mathbf{X}))_{\chi_{[|\mathbf{Y}|<1]}} \right]$$
$$\times \frac{m(\mathbf{Y})}{e^{\mu|\mathbf{Y}|}|\mathbf{Y}|^{n+\beta}} d\mathbf{Y}. \tag{3.100}$$

We write Eqs. (3.96) and (3.100), respectively, as

$$\frac{\partial p(\mathbf{X}, t)}{\partial t} = \Delta_m^{\beta/2} p(\mathbf{X}, t) \quad \text{and} \quad \frac{\partial p(\mathbf{X}, t)}{\partial t} = \Delta_m^{\beta/2, \mu} p(\mathbf{X}, t), \tag{3.101}$$

where the notation $\Delta_m^{\beta/2}$ ($\Delta_m^{\beta/2,\mu}$) denotes the anisotropic (tempered) fractional Laplacian in $\mathbb{R}^n$; and their definitions are the right-hand sides of Eqs. (3.96) and (3.100).

Different from Eqs. (3.96) and (3.100), an alternative definition of the anisotropic (tempered) fractional Laplacians can be given in Fourier space:

$$\mathcal{F}[\Delta_m^{\beta/2} p(\mathbf{X}, t)] = (-1)^{\lceil \beta \rceil} \left[\int_{|\boldsymbol{\phi}|=1} (-i\mathbf{k} \cdot \boldsymbol{\phi})^\beta m(\boldsymbol{\phi}) d\boldsymbol{\phi} \right] \tilde{p}(\mathbf{k}, t) \tag{3.102}$$

and

$$\mathcal{F}[\Delta_m^{\beta/2, \mu} p(\mathbf{X}, t)] = (-1)^{\lceil \beta \rceil} \left[\int_{|\boldsymbol{\phi}|=1} \left((\mu - i\mathbf{k} \cdot \boldsymbol{\phi})^\beta - \mu^\beta \right) m(\boldsymbol{\phi}) d\boldsymbol{\phi} \right] \tilde{p}(\mathbf{k}, t), \tag{3.103}$$

where $\lceil \beta \rceil$ denotes the smallest integer that is bigger than or equal to β. Equation (3.102) has been given in [Meerschaert et al. (1999)]. It seems that these definitions are natural for the study of the governing equations, since the symbol $(-i\mathbf{k} \cdot \boldsymbol{\phi})^\beta$ for $\beta \in (0, 1) \cup (1, 2)$ denotes β-order fractional directional derivative. Now we consider the question of when the two ways of defining the operators are equivalent. To establish the relationship between them, we focus on two cases:

- **Case I**: $0 < \beta < 1$ or m is symmetric. Recall that here the third term in Eqs. (3.96) and (3.100) can be deleted,

$$\Delta_m^{\beta/2} p(\mathbf{X}, t) = \frac{1}{|\Gamma(-\beta)|} \int_{\mathbb{R}^n \setminus \{0\}} [p(\mathbf{X} - \mathbf{Y}) - p(\mathbf{X})] \frac{m(\mathbf{Y})}{|\mathbf{Y}|^{n+\beta}} d\mathbf{Y},$$
(3.104)

$$\Delta_m^{\beta/2, \mu} p(\mathbf{X}, t) = \frac{1}{|\Gamma(-\beta)|} \int_{\mathbb{R}^n \setminus \{0\}} [p(\mathbf{X} - \mathbf{Y}) - p(\mathbf{X})] \frac{m(\mathbf{Y})}{e^{\mu |\mathbf{Y}|} |\mathbf{Y}|^{n+\beta}} d\mathbf{Y}.$$
(3.105)

- **Case II**: $1 < \beta < 2$ and m is asymmetric. Recall that the integrals in Eqs. (3.96) and (3.100) without the third terms can be understood in the Hadamard sense [Samko *et al.* (1993)], i.e.,

$$\Delta_m^{\beta/2} p(\mathbf{X}, t)$$

$$= \text{p.f.} \frac{1}{|\Gamma(-\beta)|} \int_{\mathbb{R}^n \setminus \{0\}} [p(\mathbf{X} - \mathbf{Y}) - p(\mathbf{X})] \frac{m(\mathbf{Y})}{|\mathbf{Y}|^{n+\beta}} d\mathbf{Y}$$

$$= \frac{1}{|\Gamma(-\beta)|} \int_{\mathbb{R}^n \setminus \{0\}} [p(\mathbf{X} - \mathbf{Y}) - p(\mathbf{X}) + (\mathbf{Y} \cdot \nabla_{\mathbf{X}} p(\mathbf{X}))] \frac{m(\mathbf{Y})}{|\mathbf{Y}|^{n+\beta}} d\mathbf{Y},$$
(3.106)

and

$$\Delta_m^{\beta/2, \mu} p(\mathbf{X}, t)$$

$$= \text{p.f.} \frac{1}{|\Gamma(-\beta)|} \int_{\mathbb{R}^n \setminus \{0\}} [p(\mathbf{X} - \mathbf{Y}) - p(\mathbf{X})] \frac{m(\mathbf{Y})}{e^{\mu |\mathbf{Y}|} |\mathbf{Y}|^{n+\beta}} d\mathbf{Y}$$

$$= \frac{1}{|\Gamma(-\beta)|} \int_{\mathbb{R}^n \setminus \{0\}} [p(\mathbf{X} - \mathbf{Y}) - p(\mathbf{X}) + (\mathbf{Y} \cdot \nabla_{\mathbf{X}} p(\mathbf{X}))]$$

$$\times \frac{m(\mathbf{Y})}{e^{\mu |\mathbf{Y}|} |\mathbf{Y}|^{n+\beta}} d\mathbf{Y} - \frac{1}{|\Gamma(-\beta)|} \Gamma(1 - \beta) \mu^{\beta-1} (\mathbf{b} \cdot \nabla_{\mathbf{X}} p(\mathbf{X})),$$
(3.107)

where $\mathbf{b} = \int_{|\boldsymbol{\phi}|=1} \boldsymbol{\phi} \, m(\boldsymbol{\phi}) d\boldsymbol{\phi}$.

In Case II, since the high singularity makes the integral divergent, we use the notation p.f. to denote its finite part in the Hadamard sense.

Let $m(\mathbf{Y})$ be any directional measure on unit sphere and $\mu \geq 0$. The definitions of the anisotropic (tempered) fractional Laplacians $\Delta_m^{\beta/2, \mu}$ in both Case I and Case II are, respectively, equivalent to $\Delta_m^{\beta/2, \mu}$ in Eqs. (3.102) and (3.103) in $\mathbb{R}^n$. The detailed proofs are presented in Ref. [Deng *et al.* (2020)].

We have just defined the anisotropic (tempered) fractional Laplacian by extending the Lévy measure $\nu(d\mathbf{Y})$ with different probability distribution in different directions. More generally, another two variables jump length exponent β and truncation exponent μ can be also generalized to be anisotropic, i.e., $\beta(\boldsymbol{\phi})$ and $\mu(\boldsymbol{\phi})$, or abused by $\beta(\mathbf{Y})$ and $\mu(\mathbf{Y})$ similar to $m(\mathbf{Y})$. The corresponding forward and backward Feynman-Kac equations governing the anisotropic (tempered) Lévy flight can be obtained by replacing the classical spatial derivative operators in Eqs. (3.17) and (3.24) by the anisotropic (tempered) fractional Laplacian $\Delta_m^{\beta/2,\mu}$.

Now, we consider the case with multiple internal states, and present the corresponding fractional Fokker-Planck and Feynman-Kac equations with the spatial operators being the anisotropic (tempered) fractional Laplacian $\Delta_m^{\beta/2,\mu}$. In fact, multiple internal states have many applications, e.g., the particles moving in multiphase viscous liquid composed of materials with different chemical properties. In fact, the case of two internal states is considered in [Godec and Metzler (2017); Pollak and Talkner (1993)] with applications, including trapping in amorphous semiconductors, electronic burst noise, movement in systems with fractal boundaries, the digital generation of $1/f$ noise, and ionic currents in cell membranes; Niemann *et al.* [Niemann *et al.* (2006)] investigate in detail a stochastic signal with multiple states, in which each state has an associated joint distribution for the signal's intensity and its holding time. Xu and Deng [Xu and Deng (2018a)] extended the Fokker-Planck and Feynman-Kac equations to the cases with multiple temporal internal states.

Now the first thing is to make it clear what multiple internal states mean. The motion of particles is characterized by waiting time ξ and jump length η in CTRW framework. Assume the process only has three different possibilities of distributions of ξ or η at each step. We call it three internal states $S1$, $S2$ and $S3$, as in Fig. 3.1. The information contained in each internal state $Si\,(i = 1, 2, 3)$ is the distributions of ξ and η at current step. More general models may contain more information and more internal states. In one step, each possibility of the three will yield the next step still with three different possibilities. So step after step, a Markov chain is formed. As long as the initial distribution $|\text{init}\rangle$ and transition matrix M are given, the distribution of internal states of nth step can be easily obtained, denoted by $(M^T)^{n-1}|\text{init}\rangle$. Here, the element m_{ij} of the matrix M denotes the transition probability from state i to state j, and the notations bras $\langle\cdot|$ and kets $|\cdot\rangle$ denote the row and column vectors, respectively.

The number of the internal states are taken as N for fractional Fokker-Planck and Feynman-Kac equations, the derivation processes of which are similar to the ones given in [Xu and Deng (2018a)]. We denote the column vector by capital letter and its components by lowercase letters,

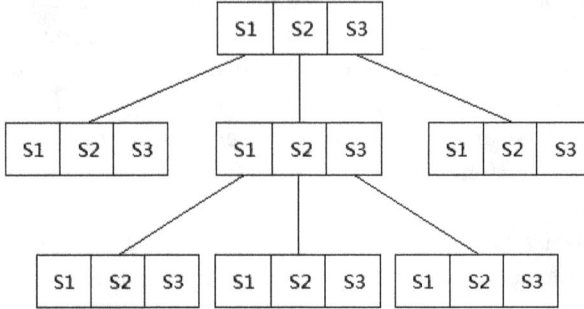

Fig. 3.1: Three internal states in each step. Each internal state of $S1$, $S2$, and $S3$ contains different distributions of waiting time ξ or jump length η.

e.g., $|G(\mathbf{X}, t)\rangle$ with its components $g^i(\mathbf{X}, t), i = 1, 2, \cdots, N$ being the PDF of finding the particle, at time t, position $\mathbf{X}$ in n dimensional space and internal state i. Then define the waiting time distribution matrix $\Phi(t) = \text{diag}(\psi^1(t), \psi^2(t), \cdots, \psi^N(t))$ and the jump length one $\Lambda(\mathbf{X}) = \text{diag}(w^1(\mathbf{X}), w^2(\mathbf{X}), \cdots, w^N(\mathbf{X}))$, where $\psi^i(t)$ and $w^i(\mathbf{X})$ are, respectively, the PDFs of waiting time and jump length at the ith internal state.

Let $|Q_n(\mathbf{X}, t)\rangle$ be composed by $q_n^i(\mathbf{X}, t), i = 1, 2, \cdots, N$, representing the PDF of the particle that just arrives at position $\mathbf{X}$, time t, and ith internal state, after n steps. Thus the matrix of survival probability is

$$W(t) = \text{diag}\left(\int_t^\infty \psi^1(\tau)d\tau, \cdots, \int_t^\infty \psi^N(\tau)d\tau\right) = I - \int_0^t \Phi(\tau)d\tau,$$

$$\tag{3.108}$$

where I denotes the identity matrix. This indicates that the Laplace transform of $W(t)$ is

$$\hat{W}(\lambda) = \frac{I - \hat{\Phi}(\lambda)}{\lambda}. \tag{3.109}$$

For G and Q, there exists

$$|G(\mathbf{X}, t)\rangle = \int_0^t W(\tau) \sum_{n=0}^\infty |Q_n(\mathbf{X}, t - \tau)\rangle d\tau. \tag{3.110}$$

On the other hand, for each component q_n^i of Q_n, we have

$$q_n^i(\mathbf{X}, t) = \sum_{j=1}^N \int_0^t \int_{\mathbb{R}^n} m_{ji} \Lambda(\mathbf{X} - \mathbf{Y}) \Phi(t - \tau) q_{n-1}^j(\mathbf{Y}, \tau) d\mathbf{Y} d\tau. \tag{3.111}$$

Thus Q satisfies

$$|Q_n(\mathbf{X}, t)\rangle = \int_0^t \int_{\mathbb{R}^n} M^T \Lambda(\mathbf{X} - \mathbf{Y}) \Phi(t - \tau) |Q_{n-1}(\mathbf{Y}, \tau)\rangle d\mathbf{Y} d\tau. \quad (3.112)$$

Performing Fourier-Laplace transform on Eqs. (3.110) and (3.112) leads to

$$|\tilde{\hat{G}}(\mathbf{k}, \lambda)\rangle = \frac{I - \hat{\Phi}(\lambda)}{\lambda} [I - M^T \tilde{\Lambda}(\mathbf{k}) \hat{\Phi}(\lambda)]^{-1} |\text{init}\rangle. \quad (3.113)$$

The Fokker-Planck equation can be obtained by applying inverse Fourier-Laplace transform on Eq. (3.113). Here, we take the waiting time distributions as asymptotic power laws, i.e., in Laplace space $\hat{\Phi}(\lambda) \sim I - \text{diag}(\lambda^{\alpha_1}, \cdots, \lambda^{\alpha_N})$, $0 < \alpha_1, \cdots, \alpha_N < 1$. As for jump lengths, they obey the Lévy distributions, i.e., in Fourier space, each component of $\tilde{\Lambda}(\mathbf{k})$ is the form of Eq. (3.103) with particular β_i and μ_i. Then, the Fokker-Planck equation with N internal states is

$$M^T \frac{\partial}{\partial t} |G(\mathbf{X}, t)\rangle = (M^T - I) \text{diag}(D_t^{1-\alpha_1}, \cdots, D_t^{1-\alpha_N}) |G(\mathbf{X}, t)\rangle$$
$$+ M^T \text{diag}(D_t^{1-\alpha_1} \Delta_m^{\beta_1/2, \mu_1}, \cdots, D_t^{1-\alpha_N} \Delta_m^{\beta_N/2, \mu_N}) |G(\mathbf{X}, t)\rangle, \quad (3.114)$$

where $D_t^{1-\alpha_i}$ is the Riemann-Liouville fractional derivative operator and $\Delta_m^{\beta_i/2, \mu_i}$ denotes the anisotropic (tempered) fractional Laplacian.

For Feynman-Kac equations, we define the functional $A = \int_0^t U(\mathbf{X}(\tau)) d\tau$, where U is a prespecified function. Let $G(\mathbf{X}, A, t)$ be the PDF of the functional A and position $\mathbf{X}$ and $G(\mathbf{X}, \rho, t)$ be the Fourier transform from A to ρ. Then the forward Feynman-Kac equation is

$$M^T \frac{\partial}{\partial t} |G(\mathbf{X}, \rho, t)\rangle = (M^T - I) \text{diag}(\mathcal{D}_t^{1-\alpha_1}, \cdots, \mathcal{D}_t^{1-\alpha_N}) |G(\mathbf{X}, \rho, t)\rangle$$
$$+ M^T \text{diag}(\Delta_m^{\beta_1/2, \mu_1} \mathcal{D}_t^{1-\alpha_1}, \cdots, \Delta_m^{\beta_N/2, \mu_N} \mathcal{D}_t^{1-\alpha_N}) |G(\mathbf{X}, \rho, t)\rangle \quad (3.115)$$
$$+ i\rho U(\mathbf{X}) M^T |G(\mathbf{X}, \rho, t)\rangle,$$

and the backward version is

$$M^T \frac{\partial}{\partial t} |G_{\mathbf{X}_0}(\rho, t)\rangle = (M^T - I) \text{diag}(\mathcal{D}_t^{1-\alpha_1}, \cdots, \mathcal{D}_t^{1-\alpha_N}) |G_{\mathbf{X}_0}(\rho, t)\rangle$$
$$+ M^T \text{diag}(\mathcal{D}_t^{1-\alpha_1} \Delta_{m,\mathbf{X}_0}^{\beta_1/2, \mu_1}, \cdots, \mathcal{D}_t^{1-\alpha_N} \Delta_{m,\mathbf{X}_0}^{\beta_N/2, \mu_N}) |G_{\mathbf{X}_0}(\rho, t)\rangle \quad (3.116)$$
$$+ i\rho U(\mathbf{X}_0) M^T |G_{\mathbf{X}_0}(\rho, t)\rangle.$$

3.3   Functional Distribution Governed by Feynman-Kac Equation

Since we have derived the forward and backward Feynman-Kac equation in the previous sections, we can obtain the distribution of functional A by specifying the function $U(x)$ and solve the Feynman-Kac equation. Now, we will present the typical three examples for the functionals—occupation time in half-space, first-passage time, the area under the random walk curve—in the following.

3.3.1   *Occupation Time in Half-Space*

The occupation time of a particle in the positive half-space is defined as

$$T_+ = \int_0^t \Theta[x(\tau)]d\tau, \tag{3.117}$$

where $\Theta(x) = 1$ for $x > 0$ and $\Theta(x) = 0$ for $x < 0$. Equation (3.117) implies the function $U(x) = \Theta(x)$. We first consider the CTRW model with power law distributed waiting time in Eq. (3.14). We further substitute $U(x) = \Theta(x)$ into the backward fractional Feynman-Kac equation (3.24) and obtain

$$\frac{\partial}{\partial t}G_{x_0}(\rho, t) = \begin{cases} K_\alpha D_t^{1-\alpha}\frac{\partial^2}{\partial x_0^2}G_{x_0}(\rho, t) & x_0 < 0 \\ K_\alpha \mathcal{D}_t^{1-\alpha}\frac{\partial^2}{\partial x_0^2}G_{x_0}(\rho, t) - \rho G_{x_0}(\rho, t) & x_0 > 0. \end{cases} \tag{3.118}$$

Taking Laplace transform $(t \to \lambda)$ yields

$$\lambda\hat{G}_{x_0}(\rho, \lambda) - 1 = \begin{cases} K_\alpha\lambda^{1-\alpha}\frac{\partial^2}{\partial x_0^2}\hat{G}_{x_0}(\rho, \lambda) & x_0 < 0 \\ K_\alpha(\lambda + \rho)^{1-\alpha}\frac{\partial^2}{\partial x_0^2}\hat{G}_{x_0}(\rho, \lambda) - \rho\hat{G}_{x_0}(\rho, \lambda) & x_0 > 0. \end{cases} \tag{3.119}$$

The remaining thing is to solve the second order ordinary differential equations (3.119) with respect to x_0.

Since the particle might move in an unbounded space or an bounded domain with reflecting boundary conditions, we discuss these two cases separately. For the former one without any boundary condition, solving Eq. (3.119) leads to

$$\hat{G}_{x_0}(\rho, \lambda) = \begin{cases} C_0 \exp[x_0\lambda^{\alpha/2}/\sqrt{K_\alpha}] + 1/\lambda & x_0 < 0 \\ C_1 \exp[-x_0(\lambda + \rho)^{\alpha/2}/\sqrt{K_\alpha}] + 1/(\lambda + \rho) & x_0 > 0, \end{cases} \tag{3.120}$$

where we have used the natural boundary condition that $\hat{G}_{x_0}(\rho, \lambda)$ is finite for $|x_0| \to \infty$. Some physical intuitions can be verified from Eq. (3.120).

For $x_0 \to -\infty$, the particle will never arrive at $x > 0$ and thus $G_{x_0}(T_+, t) = \delta(T_+)$ and $\hat{G}_{x_0}(\rho, \lambda) = 1/\lambda$. On the other hand, for $x_0 \to \infty$, the particle will stay at $x > 0$ all the time and thus $G_{x_0}(T_+, t) = \delta(T_+ - t)$ and $\hat{G}_{x_0}(\rho, \lambda) = 1/(\lambda + \rho)$. The coefficients C_0 and C_1 can be solved by using the continuity of $\hat{G}_{x_0}(\rho, \lambda)$ and its first derivative at $x_0 = 0$. More precisely, we have

$$\begin{cases} C_0 + \lambda^{-1} = C_1 + (\lambda + \rho)^{-1}, \\ C_0 \lambda^{\alpha/2} = -C_1 (\lambda + \rho)^{\alpha/2}, \end{cases} \tag{3.121}$$

and thus

$$\begin{cases} C_0 = -\dfrac{\rho(\lambda+\rho)^{\alpha/2-1}}{\lambda[\lambda^{\alpha/2}+(\lambda+\rho)^{\alpha/2}]}, \\ C_1 = \dfrac{\rho \lambda^{\alpha/2-1}}{(\lambda+\rho)[\lambda^{\alpha/2}+(\lambda+\rho)^{\alpha/2}]}. \end{cases} \tag{3.122}$$

Substituting C_0 and C_1 into Eq. (3.120) and assuming the particle starts at $x_0 = 0$, we obtain

$$\hat{G}_0(\rho, \lambda) = \frac{\lambda^{\alpha/2-1} + (\lambda + \rho)^{\alpha/2-1}}{\lambda^{\alpha/2} + (\lambda + \rho)^{\alpha/2}}. \tag{3.123}$$

Taking inverse transform, the PDF of the occupation fraction

$$p_+ := \frac{T_+}{t} \tag{3.124}$$

is obtained as the symmetric Lamperti PDF:

$$f(p_+) = \frac{\sin(\pi\alpha/2)}{\pi} \frac{(p_+)^{\alpha/2-1}(1 - p_+)^{\alpha/2-1}}{(p_+)^\alpha + (1 - p_+)^\alpha + 2(p_+)^{\alpha/2}(1 - p_+)^{\alpha/2} \cos(\pi\alpha/2)}. \tag{3.125}$$

For $\alpha = 1$, Eq. (3.125) recovers to the well-known arcsine law of Lévy [Majumdar (2005); Majumdar and Comtet (2002); Barkai (2006)]

$$f(p_+) = \frac{1}{\pi} \frac{1}{\sqrt{p_+(1 - p_+)}}. \tag{3.126}$$

The inverse transform of Eq. (3.123) is technical, where the formulas in [Godrèche and Luck (2001)] should be used. More precisely, by writing the $\hat{G}_0(\rho, \lambda)$ in the scaling form as

$$\hat{G}_0(\rho, \lambda) = \frac{1}{\lambda} g\left(\frac{\rho}{\lambda}\right) \tag{3.127}$$

in the regime $\lambda, \rho \to 0$ with ρ/λ arbitrary, we obtain the scaling function $g(\cdot)$. Then the PDF of the occupation fraction p_+ is

$$f(p_+) = -\frac{1}{\pi p_+} \lim_{\epsilon \to 0} \Im g\left(-\frac{1}{p_+ + i\epsilon}\right). \tag{3.128}$$

Comparing the Eqs. (3.125) and (3.126), we find that $f(p_+)$ has two peaks at $p_+ = 0$ and $p_+ = 1$ for $0 < \alpha \le 1$, which implies that the particle tends to spend most of the time at either $x > 0$ or $x < 0$. This result is adverse to the naive expectation that the particle spends about half the time at $x > 0$ or $x < 0$.

Now we consider the second example that the particle's motion is restricted to a box $[-L, L]$. In this case, Eq. (3.123) is also valid, but with a reflecting boundary condition

$$\frac{\partial}{\partial x_0} \hat{G}_{x_0}(\rho, \lambda) \bigg|_{x_0 = \pm L} = 0. \tag{3.129}$$

Under this condition, the solution G is

$$\hat{G}_{x_0}(\rho, \lambda) = \begin{cases} C_0 \cosh\left[(L + x_0)\frac{\lambda^{\alpha/2}}{\sqrt{K_\alpha}}\right] + \frac{1}{\lambda} & x_0 < 0 \\ C_1 \cosh\left[(L - x_0)\frac{\lambda^{\alpha/2}}{\sqrt{K_\alpha}}\right] + \frac{1}{\lambda + \rho} & x_0 > 0. \end{cases} \tag{3.130}$$

Matching G_{x_0} and its derivative at $x_0 = 0$ solves the coefficients C_0 and C_1. With some calculations, we have

$$\hat{G}_0(\rho, \lambda) = \frac{\lambda^{\alpha/2-1} \tanh[(\lambda\tau)^{\alpha/2}] + (\lambda + \rho)^{\alpha/2-1} \tanh[(\lambda + \rho\tau)^{\alpha/2}]}{\lambda^{\alpha/2} \tanh[(\lambda\tau)^{\alpha/2}] + (\lambda + \rho)^{\alpha/2} \tanh[(\lambda + \rho\tau)^{\alpha/2}]}, \tag{3.131}$$

where we defined $\tau^\alpha := L^2/K_\alpha$. Since it cannot be directly inverted, we consider the long time asymptotics $\lambda, \rho \to 0$, and obtain

$$\hat{G}_0(\rho, \lambda) \simeq \frac{\lambda^{\alpha-1} + (\lambda + \rho)^{\alpha-1}}{\lambda^\alpha + (\lambda + \rho)^\alpha}. \tag{3.132}$$

Similar to the procedure of the case without boundary condition, the PDF of the occupation fraction p_+ here is

$$f(p_+) = \frac{\sin(\pi\alpha)}{\pi} \frac{(p_+)^{\alpha-1}(1 - p_+)^{\alpha-1}}{(p_+)^{2\alpha} + (1 - p_+)^{2\alpha} + 2\cos(\pi\alpha)(p_+)^\alpha(1 - p_+)^\alpha}, \tag{3.133}$$

still a Lamperti's PDF. Compared with the result in Eq. (3.125) with natural boundary condition, it just replaces α with 2α here for reflecting boundary condition.

This Lamperti's PDF implies that the two peaks are still $p_+ = 1$ and $p_+ = 0$ for the reflecting boundary case. However, the situation becomes different as $\alpha \to 1$. When $\alpha = 1$, Eq. (3.133) reduces to $f(p_+) = \delta(p_+ - 1/2)$, a δ-function. This is an expected result based on the ergodicity of the normal diffusion.

More generally, if the particles move in a heterogenous environment, which is described by the Langevin equation with multiplicative noise Eq. (3.35), then we will apply the similar procedure to the backward fractional Feynman-Kac equation (3.66). Now, performing the Laplace transform on Eq. (3.66) yields

$$\lambda \hat{G}_{x_0}(\rho, \lambda) - 1 = g^2(x_0) \frac{\partial^2}{\partial x_0^2} \hat{G}_{x_0}(\rho, \lambda) + f(x_0) \frac{\partial}{\partial x_0} \hat{G}_{x_0}(\rho, \lambda)$$
$$- \rho U(x_0) \hat{G}_{x_0}(\rho, \lambda). \tag{3.134}$$

To consider the effect of multiplicative noise, we specify $f(x_0) = 0$ and

$$g(x_0) = aL - x_0 \tag{3.135}$$

with $a > 1$ to keep $g(x_0)$ positive. The constant aL in $g(x_0)$ measures the intensity of the additive component of the random force. Hence, Eq. (3.134) becomes

$$(aL - x_0)^2 \frac{\partial^2 \hat{G}_{x_0}(\rho, \lambda)}{\partial x_0^2} - (\lambda + \rho U(x_0)) \hat{G}_{x_0}(\rho, \lambda) = -1. \tag{3.136}$$

By using a variable substitution $y = aL - x_0 > 0$, we can obtain the celebrated Euler equation, i.e.,

$$y^2 \frac{\partial^2 \hat{\bar{G}}_y(\rho, \lambda)}{\partial y^2} - (\lambda + \rho \bar{U}(y)) \hat{\bar{G}}_y(\rho, \lambda) = -1. \tag{3.137}$$

This equation can be solved by another variable substitution $y = e^z$. Finally, we get the solutions of Eq. (3.134) in two half-spaces, i.e.,

$$\hat{G}_{x_0}(\rho, \lambda) = \begin{cases} C_1(aL - x_0)^{z_1} + C_2(aL - x_0)^{z_2} + \frac{1}{\lambda + \rho} & x_0 > 0 \\ C_3(aL - x_0)^{z_3} + C_4(aL - x_0)^{z_4} + \frac{1}{\lambda} & x_0 < 0, \end{cases} \tag{3.138}$$

where

$$z_{1,2} = \frac{1 \mp \sqrt{1 + 4(\lambda + \rho)}}{2}, \quad z_{3,4} = \frac{1 \mp \sqrt{1 + 4\lambda}}{2}. \tag{3.139}$$

The two reflecting boundary conditions in Eq. (3.129), together with two other conditions ($\hat{G}_{x_0}(\rho, \lambda)$ and its derivative are continuous at $x_0 = 0$) can determine the four coefficients C_i ($i = 1, \cdots, 4$) in Eq. (3.138) uniquely. Then we get the final solution $\hat{G}_{x_0}(\rho, \lambda)$ at $x_0 = 0$:

$$\hat{G}_0(\rho, \lambda) = \frac{\rho}{\lambda(\rho + \lambda)} \cdot \frac{F_1 F_2}{F_3 F_4 - F_1 F_2} + \frac{1}{\lambda}, \tag{3.140}$$

where

$$F_1 = a^{z_4} - \frac{z_4}{z_3}(a+1)^{z_4-z_3}a^{z_3},$$

$$F_2 = z_2[a^{z_2} - (a-1)^{z_2-z_1}a^{z_1}],$$

$$F_3 = z_4[a^{z_4} - (a+1)^{z_4-z_3}a^{z_3}], \qquad (3.141)$$

$$F_4 = a^{z_2} - \frac{z_2}{z_1}(a-1)^{z_2-z_1}a^{z_1}.$$

Equation (3.140) is the PDF of T_+ in Laplace space, but it cannot be inverted directly. Nevertheless, we can find the scaling relation $\langle T_+ \rangle \simeq t$ from Eq. (3.140). Using the technique of Laplace transform and the formula

$$\langle T_+(\lambda) \rangle = -\left.\frac{\partial \hat{G}_0(\rho, \lambda)}{\partial \rho}\right|_{\rho=0}, \qquad (3.142)$$

one arrives at

$$\langle T_+(\lambda) \rangle = -\frac{1}{\lambda^2} \cdot \left.\frac{F_1 F_2}{F_3 F_4 - F_1 F_2}\right|_{\rho=0}. \qquad (3.143)$$

For long time, i.e., $\lambda \ll 1$, $(z_1 = z_3 \sim -s, z_2 = z_4 \sim 1)$,

$$\langle T_+(t) \rangle \simeq \frac{a+1}{2a}t. \qquad (3.144)$$

For short time, i.e., $\lambda \gg 1$, $(z_1 = z_3 \sim -\sqrt{s}, z_2 = z_4 \sim \sqrt{s})$,

$$\langle T_+(t) \rangle \simeq \frac{1}{2}t. \qquad (3.145)$$

It can be seen that for both long time and short time, $\langle T_+(t) \rangle$ scales asymptotically as t, which is also verified in Fig. 3.2. Four curves evolute from $t/2$ to $\frac{a+1}{2a}t$ for the case $g(x) = aL - x$ or $\frac{a-1}{2a}t$ for the case $g(x) = aL + x$. Therefore, it is natural to consider the PDF of the occupation fraction $p_+ \equiv T_+/t$.

For long time, i.e., $\lambda \ll 1$, together with $\rho \ll 1$ due to the scale of $T_+(t)$, we have $z_1 \sim -(\lambda+\rho)$, $z_2 \sim 1$, $z_3 \sim -\lambda$, $z_4 \sim 1$ from Eq. (3.139) and $F_1 \sim (a+1)/\lambda$, $F_2 \sim 1$, $F_3 \sim -1$, $F_4 \sim (a-1)/(\lambda+\rho)$. Then we obtain the asymptotic expression of Eq. (3.140):

$$\hat{G}_0(\rho, \lambda) \simeq \frac{2a}{2a\lambda + (a+1)\rho}. \qquad (3.146)$$

By inverting the scaling form of a double Laplace transform in [Godrèche and Luck (2001)], after some calculations, using the nascent δ-function:

$$\lim_{\epsilon \to 0} \frac{\epsilon}{\pi(x^2 + \epsilon^2)} = \delta(x), \qquad (3.147)$$

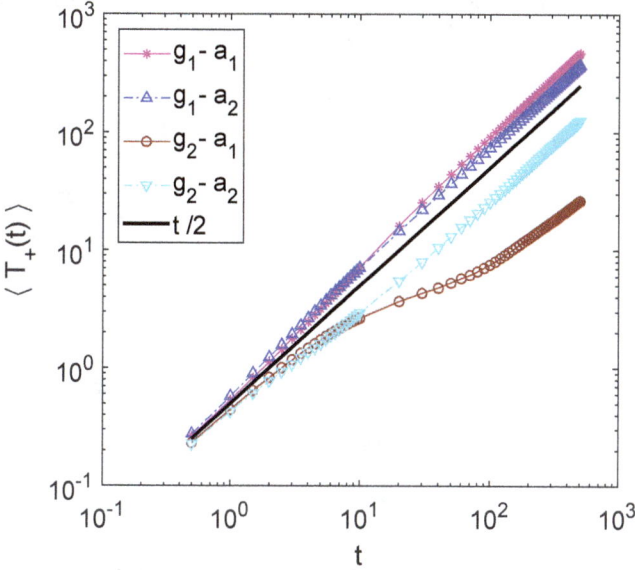

Fig. 3.2: The mean value of the occupation time T_+ in positive half-space for a particle moving in the box $[-1,1]$. Here g_1 represents the case $g(x) = aL - x$ and g_2 represents the case $g(x) = aL + x$. The other parameters are $a_1 = 1.1$ and $a_2 = 2$. Four kinds of different cases ($g_1 - a_1$ with star-markers; $g_1 - a_2$ with triangle markers; $g_2 - a_1$ with inverted-triangle markers; $g_2 - a_2$ with circle markers) are simulated with 1000 trajectories and the total time $T = 500$.

we obtain the PDF of p_+:

$$f(p_+) \simeq \frac{r}{p_+} \cdot \delta(p_+ - r) \overset{d}{=} \delta(p_+ - r), \qquad (3.148)$$

where $r = \frac{a+1}{2a}$ and $\overset{d}{=}$ denotes identical distribution. Note that the PDF of p_+ in Eq. (3.148) is normalized. Especially, p_+ reduces to a deterministic event for large t, occurring at r with probability 1. But the value r depends on a. When a is sufficiently large, this value will approach $\frac{1}{2}$ (see the curve for $a = 20$ which has a peak at $\frac{1}{2}$ in Fig. 3.3). This phenomenon has an intuitive explanation that in this case the multiplicative noise term approximates an additive noise term aL and thus it is consistent with the case of $\alpha = 1$ in [Carmi and Barkai (2011)]. On the other hand, when a is small and close to 1, the value r is near 1, which means that the particle stays in positive half-plane all the time. This phenomenon results from the multiplicative noise term. We simulate $f(p_+)$ with $a = 2$ and it has a peak at $\frac{a+1}{2a}$ for $g(x) = aL - x$ (see g_1- LT in Fig. 3.3) and a peak at $\frac{a-1}{2a}$ for $g(x) = aL + x$ (see g_2- LT in Fig. 3.3).

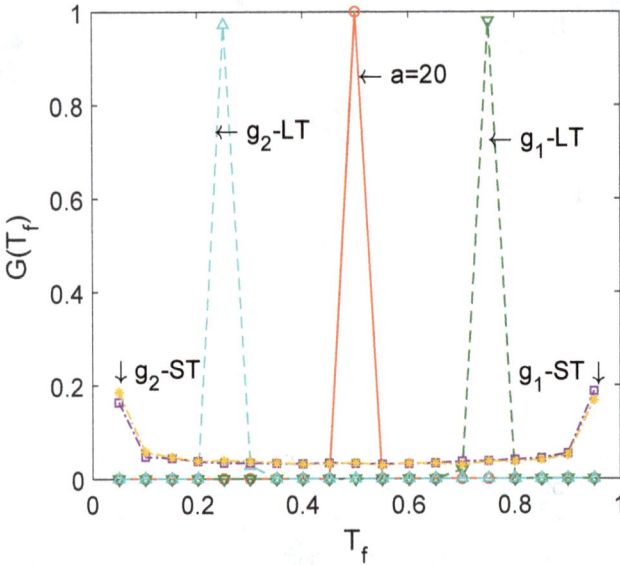

Fig. 3.3: PDF of the occupation fraction T_f in positive half-space for a particle moving in the box $[-1, 1]$. Here the PDF of T_f for long times ($T = 500$) and short times ($T = 0.01$) are shown together, recorded as "LT" and "ST", respectively. 1000 trajectories are used. In this figure, g_1 and g_2 represent the cases $g(x) = aL - x$ and $g(x) = aL + x$, respectively. The solid line denotes $G(T_f)$ in long times with $a = 20$ as well as $g = aL \pm x$ (the lines coincide for the cases $g = g_1$ and $g = g_2$). Except for the solid line, the other lines represent the case $a = 2$, but g or times are different.

For short times, i.e., $\lambda \gg 1$, we have $z_1 \sim -\sqrt{\lambda + \rho}$, $z_2 \sim \sqrt{\lambda + \rho}$, $z_3 \sim -\sqrt{\lambda}$, $z_4 \sim \sqrt{\lambda}$ from Eq. (3.139) and $F_1 \sim (a + 1)^{2\sqrt{\lambda}} a^{-\sqrt{\lambda}}$, $F_2 \sim \sqrt{\lambda + \rho} a^{\sqrt{\lambda + \rho}}$, $F_3 \sim -\sqrt{\lambda}(a + 1)^{2\sqrt{\lambda}} a^{-\sqrt{\lambda}}$, $F_4 \sim a^{\sqrt{\lambda + \rho}}$, which result in the asymptotic expression of Eq. (3.140):

$$\hat{G}_0(\rho, \lambda) \simeq -\frac{\rho}{\lambda(\rho + \lambda)} \cdot \frac{\sqrt{\lambda + \rho}}{\sqrt{\lambda} + \sqrt{\lambda + \rho}} + \frac{1}{\lambda}. \qquad (3.149)$$

Then we obtain the PDF of p_+:

$$f(p_+) \simeq \frac{1}{\pi} \cdot \frac{1}{\sqrt{x}\sqrt{1-x}}, \tag{3.150}$$

which coincides exactly with the first Lévy arcsine law as the functional of Brownian motion [Majumdar (2005)] (Generalized arcsine laws for fractional Brownian motion have been discussed in [Sadhu *et al.* (2018)]). This result is as expected since for short times the particle does not interact with the boundaries and behaves like a free particle. Furthermore, if the time t is sufficiently small, such that $x \ll aL$, then the multiplicative noise term approximates an additive noise term aL, so the PDFs of occupation fractions p_+ in cases $g(x) = aL \pm x$ all become the Lamperti PDF and present a symmetric curve with two peaks at $p_+ = 0$ and $p_+ = 1$ (see g_2- ST and g_1- ST in Fig. 3.3). Though $x \ll aL$, there is still a slight difference between two kinds of the multiplicative noises $g(x) = aL \pm x$. Therefore, the two curves in Fig. 3.3 look a little skew to one side (0 or 1).

For both long times and short times, in another perspective, the particle driven by the multiplicative noise term $g(x) = aL - x$ is more likely to move to the positive half-space since the distribution of p_+ has a larger proportion on the right side of 0.5 in Fig. 3.3. On the contrary, for $g(x) = aL + x$, the distribution of p_+ concentrates on the left side of 0.5. This phenomenon can be explained by the corresponding Fokker-Planck equation (3.46). Taking $\alpha = 1$, $\rho = 0$, $f(x,t) = 0$ in Eq. (3.46) and using the notation $g'(x) = dg(x)/dx$, one obtains

$$\begin{aligned}
\frac{\partial G(x,t)}{\partial t} &= \frac{\partial^2}{\partial x^2} g^2(x) G(x,t) \\
&= g^2(x)\frac{\partial^2 G(x,t)}{\partial x^2} + 4g(x)g'(x)\frac{\partial G(x,t)}{\partial x} \\
&\quad + 2\Big(g'(x)^2 + g(x)g''(x)\Big)G(x,t),
\end{aligned} \tag{3.151}$$

where the coefficient $E := 4g(x)g'(x)$ in front of the first-order derivative of $G(x,t)$ is called noise induced drift [Coffey *et al.* (2004)]. The cases of $g(x) = aL - x$ and $g(x) = aL + x$ lead to $E = 4(x - aL) < 0$ and $E = 4(x + aL) > 0$, respectively, which means that the multiplicative noise term $g(x) = aL - x$ induces a positive drift while $g(x) = aL + x$ a negative drift. Furthermore, as time goes on, the curve of the distribution of occupation fraction T_f in Fig. 3.3 changes from concave upwards to convex upwards, which is similar to the case of Brownian motion. The reason is that by a variable substitution, the Langevin equation for $x(t)$

with multiplicative noise can be turned into another Langevin equation for a new process $y(t)$ with an additive noise (i.e., $y(t)$ is the Brownian motion with a drift); see [Dhar and Majumdar (1999); Cherstvy *et al.* (2013)] for more details.

3.3.2   First-Passage Time

A good beginning of the application of occupation time in the previous part is to consider a problem of the first-passage time t_f. We still consider two cases of natural boundary condition and reflecting boundary condition.

For the former case, we discuss the first-passage time t_f denoting the time when a particle starting at $x_0 = -b$ first hits $x = 0$. The distribution of first-passage time t_f for anomalous paths can be obtained from the backward fractional Feynman-Kac equation by using an identity due to Kac [Kac (1951)]:

$$P(t_f > t) = P\left(\max_{0 \leq \tau \leq t} x(\tau) < 0\right) = \lim_{\rho \to \infty} G_{x_0}(\rho, t), \qquad (3.152)$$

where $G_{x_0}(\rho, t)$ is the Laplace transform of the PDF of functional T_+ representing the occupation time on the positive half-space in the Sec. 3.3.1, i.e.,

$$G_{x_0}(\rho, t) = \int_0^\infty e^{-\rho T_+} G_{x_0}(T_+, t) dT_+. \qquad (3.153)$$

If the particle has crossed $x = 0$ at time t, we have $T_+ > 0$ and $e^{-\rho T_+} = 0$ for $\rho \to \infty$. Then two sides of Eq. (3.152) equal to 0. Otherwise $T_+ = 0$ and $e^{-\rho T_+} = 1$ make two sides equal 1.

Based on the analyses above, we just need to take $x_0 = -b$ in Eq. (3.123), and obtain

$$\hat{G}_{-b}(\rho, \lambda) = \frac{1}{\lambda}\left[1 - e^{-\frac{b}{\sqrt{K_\alpha}}\lambda^{\alpha/2}} \frac{\rho(\lambda + \rho)^{\alpha/2-1}}{\lambda^{\alpha/2} + (\lambda + \rho)^{\alpha/2}}\right]. \qquad (3.154)$$

Then taking the limit of infinite ρ leads to the right-hand side of Eq. (3.152)

$$\lim_{\rho \to \infty} \hat{G}_{-b}(\rho, \lambda) = \frac{1}{\lambda}\left(1 - e^{-\frac{b}{\sqrt{K_\alpha}}\lambda^{\alpha/2}}\right). \qquad (3.155)$$

Defining $\tau_f = (b^2/K_\alpha)^{1/\alpha}$ and performing inverse Laplace transform, we obtain

$$P\{t_f > t\} = \lim_{\rho \to \infty} G_{-b}(\rho, t) = 1 - \int_0^t \frac{1}{\tau_f} L_{\alpha/2}\left(\frac{\tau}{\tau_f}\right) d\tau, \qquad (3.156)$$

where $L_{\alpha/2}(\cdot)$ is the one-sided Lévy distribution of order $\alpha/2$, whose Laplace transform is $\hat{L}_{\alpha/2}(\lambda) = e^{-\lambda^{\alpha/2}}$. Equation (3.156) implies the PDF of the first-passage time t_f is

$$f(t_f) = \frac{1}{\tau_f} L_{\alpha/2}\left(\frac{t_f}{\tau_f}\right). \tag{3.157}$$

The long time behavior of $f(t_f)$ can be obtained by taking $\lambda \to 0$ in Eq. (3.155) and inverse Laplace transform:

$$f(t_f) \simeq \frac{b}{|\Gamma(-\frac{\alpha}{2})|\sqrt{K_\alpha}} t_f^{-1-\alpha/2}. \tag{3.158}$$

For $\alpha = 1$, the famous $t^{-3/2}$ decay law of first-passage time of Brownian motion is obtained [Redner (2001)].

For another case where the particle is restricted in a box $[-L, L]$, we assume the particle starts at the point $x_0 = -bL$ ($0 < b < 1$). With similar calculations, there is [Carmi and Barkai (2011)]

$$\lim_{\rho \to \infty} \hat{G}_{-bL}(\rho, \lambda) = \frac{1}{\lambda}\left(1 - \frac{\cosh\left[(1-b)L\frac{\lambda^{\alpha/2}}{\sqrt{K_\alpha}}\right]}{\cosh\left(L\frac{\lambda^{\alpha/2}}{\sqrt{K_\alpha}}\right)}\right). \tag{3.159}$$

Then the PDF of first-passage time is, in Laplace λ space

$$\hat{f}(\lambda) = \frac{\cosh\left[(1-b)L\frac{\lambda^{\alpha/2}}{\sqrt{K_\alpha}}\right]}{\cosh\left(L\frac{\lambda^{\alpha/2}}{\sqrt{K_\alpha}}\right)}. \tag{3.160}$$

For long times, the small λ asymptotics is

$$\hat{f}(\lambda) \simeq 1 - \frac{b(2-b)}{2K_\alpha}L^2\lambda^\alpha, \tag{3.161}$$

the inverse of which is

$$f(t_f) \simeq \frac{b(2-b)L^2}{2K_\alpha|\Gamma(-\alpha)|} t_f^{-1-\alpha}. \tag{3.162}$$

Similar to the result of occupation fraction, the exponent of the PDF $f(t_f)$ of first-passage time is also twice that with natural boundary conditions in Eq. (3.158).

When $\alpha = 1$ in Eq. (3.161), we have

$$\hat{f}(\lambda) \simeq 1 - \frac{b(2-b)}{2K_1}L^2\lambda \simeq e^{-\frac{b(2-b)}{2K_1}L^2\lambda} \tag{3.163}$$

and

$$f(t_f) \simeq \delta\left(t_f - \frac{b(2-b)}{2K_1}L^2\right). \tag{3.164}$$

The PDF of the first-passage time becomes a δ-function for Brownian motion, which means that the first-passage time is a deterministic event, occurring at $\frac{b(2-b)}{2K_1}L^2$ with probability 1.

Furthermore, if the particles move in heterogenous media described by the Langevin equation (3.35) with multiplicative noise, the similar result can be obtained by taking $x_0 = -bL$ in Eq. (3.138):

$$\hat{G}_{-bL}(\rho, \lambda) = \frac{\rho}{\lambda(\rho + \lambda)} \cdot \frac{F_{1b}F_2}{F_3F_4 - F_1F_2} + \frac{1}{\lambda}, \tag{3.165}$$

where

$$F_{1b} = (a+b)^{z_4} - \frac{z_4}{z_3}(a+1)^{z_4-z_3}(a+b)^{z_3}. \tag{3.166}$$

For the long time limit ($\lambda \to 0$), we have

$$z_1 \sim -\sqrt{\rho}, z_2 \sim \sqrt{\rho}, z_3 \sim -\lambda, z_4 \sim 1, \tag{3.167}$$

and thus

$$\lim_{\rho \to \infty} \hat{G}_{-bL}(\rho, \lambda) \simeq \ln\left(1 + \frac{b}{a}\right) - \frac{b}{1+a} =: C_{ab}. \tag{3.168}$$

Considering the first-passage time PDF satisfying $f(t) = \frac{\partial}{\partial t}[1 - P(t_f > t)]$, we have the PDF of t_f in Laplace λ space

$$\hat{f}(\lambda) \simeq 1 - C_{ab}\lambda \simeq e^{-C_{ab}\lambda}, \tag{3.169}$$

and thus

$$f(t_f) \simeq \delta(t_f - C_{ab}). \tag{3.170}$$

The PDF of first-passage time is also a δ-function, similar to Eq. (3.164); see the distribution of first passage time t_f in Fig. 3.4. Furthermore, for $0 < b < 1 < a$, C_{ab} is monotonously increasing with respect to b but decreasing with respect to a, being the same as physical intuition.

3.3.3   *Area under Random Walk Curve*

Now we turn to another functional $A_x = \int_0^t x(t')dt'$ ($U(x) = x$), denoting the total area under the curve of trajectory $x(t)$ [Baule and Friedrich (2006); Grebenkov (2007)]. This functional A_x is also related to the phase accumulated by spins in an NMR experiment [Grebenkov (2007)]. If we further assume that the particle starts and ends at the origin $x(0) = x(t) = 0$ but stays positive in between, this motion is called excursion [Majumdar (2005)]. The question about the area under Brownian excursion has been studied quite extensively by mathematicians [Darling (1983); Louchard

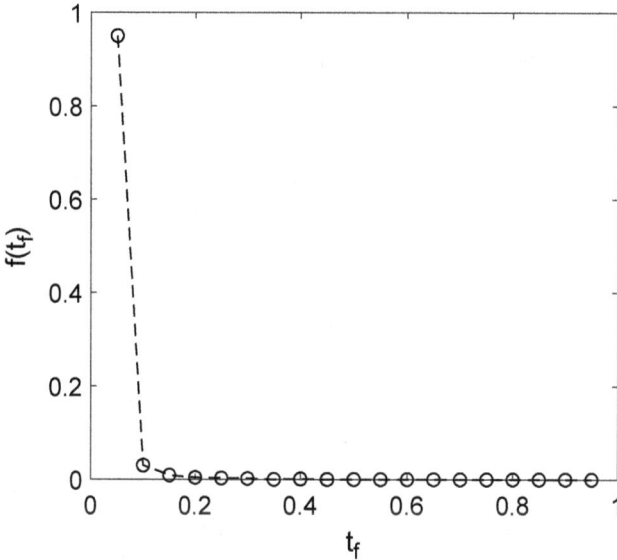

Fig. 3.4: The PDF of the first-passage time of a particle in the box $[-1, 1]$ starting at $-b$ and reaching $x = 0$ for the first time. We simulate it with 1000 trajectories and the total time $T = 10$. The parameters are $b = 1/2$, $a = 2$, and then $C_{ab} = 0.0565$, which is consistent with the curve that has a peak near 0.0565 in the figure.

(1984); Takacs (1995); Flajolet *et al.* (1998); Flajolet and Louchard (2001)]. Recently, the applications of Brownian excursion are further studied [Gall (1991); Majumdar and Comtet (2005)] and even extended to the Bessel excursion for anomalous diffusion [Barkai *et al.* (2014)].

Similar to the previous sections, we firstly consider the random walk with power law distributed waiting times. Since the area A_x is not necessarily positive, we use the Fourier transform instead of Laplace transform ($A_x \to \rho$). Besides, the backward fractional Feynman-Kac equation (3.24) with $U(x) = x$ is not easy to be solved. So we resort to the forward one in Eq. (3.17) and obtain

$$\lambda \hat{G}(x, \rho, \lambda) - \delta(x) = K_\alpha \frac{\partial^2}{\partial x^2} (\lambda + i\rho x)^{1-\alpha} \hat{G}(x, \rho, \lambda) - i\rho x \hat{G}(x, \rho, \lambda). \quad (3.171)$$

Here, $\hat{G}(x, \rho, \lambda)$ is the Fourier-Laplace transform of $G(x, A_x, t)$ and the initial position is assumed as $x_0 = 0$. Integrating Eq. (3.171) over all x

yields

$$\lambda \hat{G}(\rho, \lambda) - 1 = -i\rho \int_{-\infty}^{\infty} x\hat{G}(x, \rho, \lambda)dx. \tag{3.172}$$

Then using the formulas

$$\langle A_x^2 \rangle(\lambda) = \int_{-\infty}^{\infty} -\frac{\partial^2}{\partial \rho^2} \hat{G}(x, \rho, \lambda)\bigg|_{\rho=0} dx \tag{3.173}$$

and

$$\langle A_x \rangle(\lambda) = \int_{-\infty}^{\infty} -i\frac{\partial}{\partial \rho} \hat{G}(x, \rho, \lambda)\bigg|_{\rho=0} dx, \tag{3.174}$$

taking the derivatives with respect to ρ and substituting $\rho = 0$ in Eq. (3.172), we obtain

$$\lambda \langle A_x^2 \rangle(\lambda) = 2\langle x A_x \rangle(\lambda). \tag{3.175}$$

On the other hand, considering the definition of functional A_x, there is

$$\frac{d}{dt}\langle x A_x \rangle = \frac{d}{dt}\left\langle x \int_0^t x(\tau)d\tau \right\rangle = \langle x^2 \rangle \tag{3.176}$$

and

$$\lambda \langle x A_x \rangle(\lambda) = \langle x^2 \rangle(\lambda). \tag{3.177}$$

Based on these relations and $\langle x^2 \rangle(\lambda) = 2K_\alpha \lambda^{-1-\alpha}$, we finally obtain

$$\langle A_x^2 \rangle(t) = \frac{4K_\alpha}{\Gamma(3+\alpha)}t^{2+\alpha}. \tag{3.178}$$

If the underlying process is Brownian motion with $\alpha = 1$, we have

$$\langle A_x^2 \rangle(t) = \frac{2}{3}K_1 t^3, \tag{3.179}$$

which is consistent to the scaling relation $x(t) \propto t^{1/2}$ and

$$A_x = \int_0^t x(\tau)d\tau \propto t^{3/2}. \tag{3.180}$$

Here we study the functional A_x in the Langevin system containing a force field and non-Gaussian β-stable noise. Similarly, the analytical solutions of $G_{x_0}(\rho, t)$ in Eq. (3.65) cannot be easily obtained due to the Riesz space fractional derivative operator ∇_x^β, we resort to the forward Feynman-Kac equation (3.47) by integrating the solution $G(x, \rho, t)$ over x with initial position x_0 to get the marginal PDF of $G_{x_0}(\rho, t)$.

Let us consider the particle confined in a harmonic potential $V(x,t) = bx^2/2$ with $b > 0$, (i.e., $f(x,t) = -\partial V(x,t)/\partial x = -bx$). We further assume $g(x,t) \equiv 1$, $U(x) = x$, $\alpha = 1$, and obtain the forward Feynman-Kac equation (3.45) as

$$\frac{\partial \tilde{G}(k,\rho,t)}{\partial t} + (bk - \rho)\frac{\partial}{\partial k}\tilde{G}(k,\rho,t) = \phi_0(k)\tilde{G}(k,\rho,t). \tag{3.181}$$

Its general solution has the form as [Polyanin *et al.* (2002)]:

$$\tilde{G}(k,\rho,t) = \exp\left[\int_0^k \frac{\phi_0(z)}{bz - \rho}dz + c_1\right] \cdot \Psi\left[\frac{1}{b}\ln|bk - \rho| - t + c_2\right], \tag{3.182}$$

where c_1, c_2 are constants and $\Psi(x)$ is an arbitrary function. Since we have assumed that the particle starts its movement from the origin, the initial condition $\tilde{G}(k,\rho,0) = 1$ holds. Therefore, we have

$$\Psi\left[\frac{1}{b}\ln|bk - \rho| + c_2\right] = \exp\left[-\int_0^k \frac{\phi_0(z)}{bz - \rho}dz - c_1\right]. \tag{3.183}$$

Then replacing k by $l(k) := \frac{bk-\rho}{be^{bt}} + \frac{\rho}{b}$, Eq. (3.183) becomes

$$\Psi\left[\frac{1}{b}\ln|bk - \rho| - t + c_2\right] = \exp\left[-\int_0^{l(k)} \frac{\phi_0(z)}{bz - \rho}dz - c_1\right]. \tag{3.184}$$

Substituting this result into Eq. (3.182) gives

$$\tilde{G}(k,\rho,t) = \exp\left[\int_{l(k)}^k \frac{\phi_0(z)}{bz - \rho}dz\right]. \tag{3.185}$$

By taking $k = 0$, we can get the PDF of functional A_x in Fourier space ($A_x \to \rho$):

$$G(\rho,t) := \tilde{G}(k,\rho,t)|_{k=0} = \exp\left[\int_{\frac{\rho}{b}(1-e^{-bt})}^0 \frac{\phi_0(z)}{bz - \rho}dz\right]. \tag{3.186}$$

The one having not been determined in Eq. (3.186) is $\phi_0(\cdot)$. We will discuss the specific dynamical behavior of functional A_x for Lévy particles by substituting $\phi_0(k) = -|k|^\beta$ into Eq. (3.186). Considering a variable substitution $z = \frac{\rho}{b}(1 - e^{-bt})y$, Eq. (3.186) can be represented in the form

$$\ln G(\rho,t) = -C_b(t)\left(\frac{1 - e^{-bt}}{b}\right)^{\beta+1}|\rho|^\beta, \tag{3.187}$$

where $C_b(t)$ is independent of ρ [Gradshteyn *et al.* (1980)]:

$$C_b(t) = \int_0^1 \frac{y^\beta}{1 - (1 - e^{-bt})y} dy$$
$$= B(\beta + 1, 1) \cdot {}_2F_1(1, \beta + 1; \beta + 2; 1 - e^{-bt}), \tag{3.188}$$

where ${}_2F_1(\cdot)$ is the hypergeometric function. It can be seen from Eq. (3.187) that the functional A_x also obeys Lévy β-stable distribution. Next what we need to pay attention to is the coefficient in front of $|\rho|^\beta$ in Eq. (3.187).

For long times $t \to \infty$, we cannot omit the term e^{-bt} in the integrand of $C_b(t)$. Instead, we should evaluate it more carefully. Since the integrand monotonically decreases with the increase of β. For the two critical cases $\beta = 0$ and $\beta = 2$, the integral of $C_b(t)$ can be calculated analytically. Afterwards we take the limit $t \to \infty$ and find that both of them have the asymptotic result bt. Therefore, for any $0 \leq \beta \leq 2$, it holds that

$$C_b(t) = \int_0^1 \frac{y^\beta}{1 - (1 - e^{-bt})y} dy \simeq bt. \tag{3.189}$$

Substituting it into Eq. (3.187) makes

$$G(\rho, t) \simeq \exp(-b^{-\beta} t |\rho|^\beta) \qquad \text{as} \quad t \to \infty. \tag{3.190}$$

For short times $t \to 0$, ${}_2F_1(1, \beta + 1; \beta + 2; 1 - e^{-bt}) \sim 1$, and thus

$$G(\rho, t) \simeq \exp\left(-\frac{t^{\beta+1}}{\beta + 1} |\rho|^\beta\right) \qquad \text{as} \quad t \to 0. \tag{3.191}$$

For the special case $\beta = 2$, i.e., Gaussian white noise, by the formula

$$\langle A_x^2 \rangle = \frac{\partial^2}{\partial \rho^2} G(\rho, t) \bigg|_{\rho=0}, \tag{3.192}$$

we get

$$\langle A_x^2 \rangle \simeq 2b^{-2} t, \qquad \text{as} \quad t \to \infty, \tag{3.193}$$

and

$$\langle A_x^2 \rangle \simeq \frac{2}{3} t^3, \qquad \text{as} \quad t \to 0, \tag{3.194}$$

which can be verified by numerical simulations. In Fig. 3.5, the functional A_x exhibits a crossover from t^3 to t. For short times, since the particle begins its movement from the origin, i.e., $x \ll 1$, the effect of force ($f = -bx$) can be omitted. As time goes on, this effect is getting bigger and bigger, and eventually produces the multi-scale phenomenon for long times. On the contrary, for the force-free case (i.e., $b = 0$), it is equivalent to that

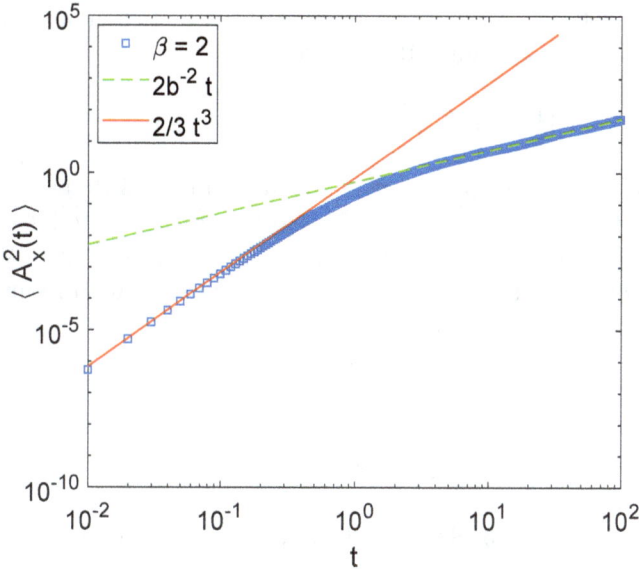

Fig. 3.5: Second moment $\langle A_x^2 \rangle$ of the area under the trajectory curve with $\beta = 2$ and $b = 2$, generated with 1000 trajectories and the total time $T = 100$. The circle-markers denote the simulation results. The dotted line denotes the theoretical result $\langle A_x^2 \rangle \simeq \frac{2}{3} t^3$ for short time while the solid line represents $\langle A_x^2 \rangle \simeq \frac{2}{b^2} t$ for long time. This figure shows a crossover of scaling regimes from t^3 to t.

$b \to 0$ for any t in Eq. (3.187). Then only the single-scale phenomenon $\langle A_x^2 \rangle \simeq \frac{2}{3} t^3$ can be observed, which is consistent with [Carmi *et al.* (2010)] by taking $\alpha = 1$ there.

As for the general case $0 < \beta < 2$, the MSD of A_x diverges [Metzler and Klafter (2000b)]: $\langle A_x^2 \rangle \to \infty$. But for a particle with non-diverging mass, a finite velocity of propagation exists, making long instantaneous jumps impossible. Their fractional moments can be written as

$$\langle |A_x|^\delta \rangle \propto \tilde{t}^{\delta/\beta}, \tag{3.195}$$

where $0 < \delta < \beta < 2$. Based on Eqs. (3.190) and (3.191), we find that $\tilde{t}$ should be $t^{\beta+1}$ for short times and t for long times in Eq. (3.195). So we rescale the fractional moments and get the pseudo second moment $[A_x^2] \propto \tilde{t}^{2/\beta}$. An alternative method is to consider the $(A_x - t)$ scaling relations, or to measure the width of the PDF $G(A_x, t)$ rather than its variance [Metzler and Klafter (2000b)]. More precisely, enclose the particle in an imaginary growing box [Jespersen *et al.* (1999)] and define

$$\langle A_x^2 \rangle_L := \int_{-L_0 t^{1/\beta}}^{L_0 t^{1/\beta}} A_x^2 G(A_x, t) dA_x \simeq \tilde{t}^{2/\beta}, \qquad (3.196)$$

where L_0 is chosen to adapt the scaling regimes in Eqs. (3.190) and (3.191), i.e., for long time $L_0 = \sqrt{2b^{-\beta}}$ while for short time $L_0 = \sqrt{2/(1+\beta)}$. This has been implemented numerically and can be seen in Fig. 3.6. We take β to be 1.4 or 0.7 and $b = 2$. The markers denote simulation results while the solid lines are the theoretical ones

$$\langle A_x^2 \rangle \simeq 2b^{-\beta} t^{2/\beta}, \qquad \text{as} \quad t \to \infty, \qquad (3.197)$$

and

$$\langle A_x^2 \rangle \simeq \frac{2}{\beta+1} t^{2(\beta+1)/\beta}, \qquad \text{as} \quad t \to 0, \qquad (3.198)$$

which go back to Eqs. (3.193) and (3.194), respectively, when $\beta = 2$.

3.3.4 *Other Functionals*

Apart from the three kinds of functionals discussed above, there are still various functionals which can be studied by solving Feynman-Kac equation. We list some of them in the following and briefly introduce the relationship between them and Feynman-Kac equations. See more details and discussions in the reference [Carmi *et al.* (2010)].

The first one is the total residence time of the particle in some interval. It is like the occupation time in half-space in Sec. 3.3.1. But now, the concerned domain is assumed as $[-b, b]$, being more general. The corresponding functional is defined as $A = \int_0^t U[x(\tau)]d\tau$, where

$$U(x) = \begin{cases} 1 & |x| < b, \\ 0 & |x| > b. \end{cases} \qquad (3.199)$$

Then we solve the backward fractional Feynman-Kac equation to obtain the PDF of functional A as the procedure for occupation time in Sec. 3.3.1.

The second one is the particle's maximal displacement $x_m := \max_{0 \le \tau \le t} x(\tau)$. It is a random variable and has been studied in many references [Comtet and Majumdar (2005); Majumdar *et al.* (2008); Schehr and Le-Doussal (2010); Tejedor *et al.* (2010)]. For this functional, we can still use the identity in Eq. (3.152)

$$P(x_m < b) = \lim_{\rho \to \infty} G_{x_0}(\rho, t), \qquad (3.200)$$

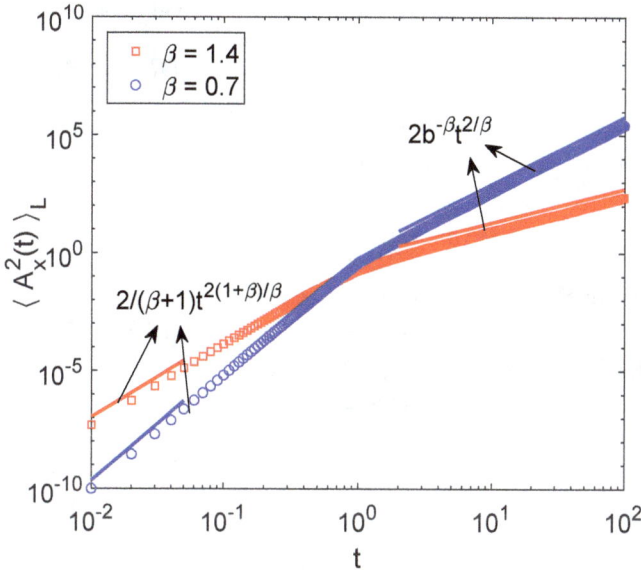

Fig. 3.6: Pseudo second moment $\langle A_x^2 \rangle_L$ by a cut-off approach, generated with 1000 trajectories and the total time $T = 100$ with $\beta = 1.4, 0.7$, and $b = 2$. The circle-markers and square-markers denote the simulation results of $\beta = 1.4$ and $\beta = 0.7$, respectively. The solid lines are the theoretical results with slope $2(\beta+1)/\beta$ for short time and $2/\beta$ for long time. It shows that for different Lévy β-stable noise ($0 < \beta < 2$), there is a crossover of scaling regimes from $t^{2(\beta+1)/\beta}$ to $t^{2/\beta}$.

where $G_{x_0}(\rho, t)$ is the Laplace transform of the PDF of functional $T_+ = \int_0^t U[x(\tau)]d\tau$, and

$$U(x) = \begin{cases} 0 & x < b, \\ 1 & x > b. \end{cases} \quad (3.201)$$

Now, $x_0 = 0$ and the right-hand side of Eq. (3.200) can be obtained from backward fractional Feynman-Kac equation, similar to the part of first-passage time. Then take the derivative with respect to b on both sides of Eq. (3.200) yields the PDF of maximal displacement.

Another one is the survival probability of the particle in a medium where the particle is absorbed at rate R in some interval. The survival probability of the particle, S, is related to the total time T at this interval through $S = e^{-RT}$. Thus, the first moment of survival probability is

$$\langle S \rangle = \int_0^\infty e^{-RT} G_{x_0}(T, t) dT =: \hat{G}_{x_0}(R, t), \qquad (3.202)$$

where the right-hand side can be obtained from backward fractional Feynman-Kac equation as the previous sections.

3.4 Klein-Kramers Equation

Another form of the Fokker-Planck equation in phase space is often called the Kramers equation or Klein-Kramers equation [Risken (1989)], originally derived [Chandrasekhar (1943)] by Klein in 1921. In general, the Fokker-Planck equation in the context of a dynamical system, where the motion in the absence of heat bath is governed by Hamilton's equations with a separable and additive Hamiltonian comprising the sum of the kinetic and potential energies, is known as the Klein-Kramers equation. Consider a Brownian particle moves in a potential $V(x)$, with the corresponding Langevin equation

$$\frac{dx(t)}{dt} = v(t), \quad m\frac{dv(t)}{dt} = -\frac{d}{dx}V[x(t)] - \gamma v(t) + \xi(t), \qquad (3.203)$$

where the Gaussian white noise satisfies

$$\langle \xi(t) \rangle = 0, \quad \langle \xi(t_1)\xi(t_2)) \rangle = 2kT\gamma\delta(t_1 - t_2). \qquad (3.204)$$

Let $G(x, v, t)$ be the joint PDF of the position x and velocity v. Then $G(x, v, t)$ satisfies the Klein-Kramers equation [Kramers (1940); Coffey *et al.* (2004)]:

$$\frac{\partial G}{\partial t} + v\frac{\partial G}{\partial x} - \frac{1}{m}\frac{\partial V}{\partial x}\frac{\partial G}{\partial v} = \frac{\gamma}{m}\left(\frac{\partial}{\partial v}vG + \frac{kT}{m}\frac{\partial^2}{\partial v^2}G\right). \qquad (3.205)$$

In contrast with Brownian motion, the Klein-Kramers equations corresponding to the anomalous diffusion processes are more complex and usually together with the fractional derivative operators. We will introduce the fractional Klein-Kramers equations without or with an external force in order in the following.

3.4.1 *Force-Free Case*

The anomalous diffusion process can be described by a set of Langevin equation:

$$\frac{d}{dt}x(t) = v(t), \quad \frac{d}{ds}v(s) = -\gamma v(s) + \xi(s), \quad \frac{d}{ds}t(s) = \eta(s), \qquad (3.206)$$

where γ is the friction coefficient, $\xi(s)$ is the Gaussian white noise satisfying $\langle \xi(s_1)\xi(s_2) \rangle = 2D_v\delta(s_1 - s_2)$, and $t(s)$ is the α-dependent subordinator $(0 < \alpha < 2)$ with Laplace exponent

$$\Phi(\lambda) \simeq \begin{cases} \mu_\alpha \lambda^\alpha & 0 < \alpha < 1, \\ \mu_1 \lambda - \mu_\alpha \lambda^\alpha & 1 < \alpha < 2, \end{cases} \tag{3.207}$$

introduced in Sec. 2.2.3. The case of $0 < \alpha < 1$ has been considered in [Cairoli and Baule (2017)]. Fortunately, the method in [Cairoli and Baule (2017); Wang *et al.* (2018)] can be applied to any subordinator with Laplace exponent $\Phi(\lambda)$, which only makes a difference in fractional substantial derivative operator proposed by [Friedrich *et al.* (2006a)].

The fractional Klein-Kramers equation corresponding to Eq. (3.206) can be directly obtained from the forward Feynman-Kac equation in [Cairoli and Baule (2017); Wang *et al.* (2018)] (see also [Cairoli and Baule (2015a)]), since $x(t) = \int_0^t v(t')dt'$ could be interpreted as a functional of $v(t)$. Therefore, for the force-free case, it is convenient to derive the Feynman-Kac equation of velocity process by using the method introduced in Sec. 3.2.

Denote the joint PDF of position $x(t)$ and velocity $v(t)$ as $G(x, v, t)$. Based on the analyses in Sec. 3.2, the forward fractional Feynman-Kac equation (3.17) with $0 < \alpha < 1$ is

$$\frac{\partial}{\partial t}\tilde{G}(\rho, v, t) = \mu_\alpha \mathcal{L}_{FP}\mathcal{D}_t^{1-\alpha}\tilde{G}(\rho, v, t) - i\rho v\tilde{G}(\rho, v, t), \tag{3.208}$$

where $\tilde{G}(\rho, v, t)$ is the Fourier transform of $G(x, v, t)$ $(x \to \rho)$, $\mathcal{D}_t^{1-\alpha}$ is the fractional substantial derivative operator, and $\mathcal{L}_{FP}$ is the Fokker-Planck collision operator

$$\mathcal{L}_{FP} = \gamma\frac{\partial}{\partial v}v + D_v\frac{\partial^2}{\partial v^2}. \tag{3.209}$$

Then performing inverse Fourier transform on Eq. (3.208), we obtain the fractional Klein-Kramers equation

$$\left[\frac{\partial}{\partial t} + v\frac{\partial}{\partial x}\right]G(x, v, t) = \mu_\alpha \mathcal{L}_{FP}\mathcal{D}_t^{1-\alpha}G(x, v, t). \tag{3.210}$$

Note that $\mathcal{D}_t^{1-\alpha}$ in Eq. (3.210) comes from the inverse Fourier-Laplace transform $(\rho \to x, \lambda \to t)$ of the symbol

$$\frac{\lambda + i\rho v}{\Phi_0(\lambda + i\rho v)} = (\lambda + i\rho v)^{1-\alpha}. \tag{3.211}$$

With the new $\Phi(\lambda)$ in Eq. (3.207) for the case of $1 < \alpha < 2$, we have

$$\frac{\lambda + i\rho v}{\Phi(\lambda + i\rho v)} = \frac{1}{\mu_1 - \mu_\alpha(\lambda + i\rho v)^{\alpha-1}} \simeq \frac{1}{\mu_1} + \frac{\mu_\alpha}{\mu_1^2}(\lambda + i\rho v)^{\alpha-1}, \tag{3.212}$$

as $\lambda \to 0$ and $\rho \to 0$. Taking the inverse Fourier-Laplace transform, we get the operator

$$\check{\mathcal{D}}_t^{\alpha-1} := \frac{1}{\mu_1} + \frac{\mu_\alpha}{\mu_1^2} \mathcal{D}_t^{\alpha-1} \qquad (3.213)$$

and obtain the fractional Klein-Kramers equation in the case of $1 < \alpha < 2$,

$$\left[\frac{\partial}{\partial t} + v \frac{\partial}{\partial x} \right] G(x, v, t) = \mathcal{L}_{\text{FP}} \check{\mathcal{D}}_t^{\alpha-1} G(x, v, t). \qquad (3.214)$$

Integrating over the position x, or performing the Fourier transform $(x \to \rho)$ together with taking $\rho = 0$, the fractional Fokker-Planck equation governing the PDF of velocity v

$$\frac{\partial}{\partial t} G(v, t) = \mathcal{L}_{\text{FP}} \check{D}_t^{1-\alpha} G(v, t) \qquad (3.215)$$

is obtained, where

$$\check{D}_t^{1-\alpha} = \begin{cases} \mu_\alpha D_t^{1-\alpha} & 0 < \alpha < 1, \\ \frac{1}{\mu_1} + \frac{\mu_\alpha}{\mu_1^2} D_t^{\alpha-1} & 1 < \alpha < 2, \end{cases} \qquad (3.216)$$

and $D_t^{\alpha-1}$ is the Riemann-Liouville fractional derivative operator [Podlubny (1999)] with Laplace symbol $\lambda^{\alpha-1}$. The corresponding equation governing the PDF of positive x cannot be easily obtained by the similar procedure, since v is embedded into the fractional substantial derivative operator $\check{\mathcal{D}}_t^{\alpha-1}$, where the time t and position x are coupled with each other.

Different from the Langevin equation (3.206), another kind of commonly considered coupled Langevin system is

$$\frac{d}{ds} x(s) = v(s), \quad \frac{d}{ds} v(s) = -\gamma v(s) + \xi(s), \quad \frac{d}{ds} t(s) = \eta(s), \quad (3.217)$$

where position x and velocity v are both subordinated. The corresponding fractional Klein-Kramers equation has been discussed in [Metzler and Klafter (2000a)] for the α-stable subordinator $t(s)$ with $0 < \alpha < 1$ and in [Wang *et al.* (2019b)] for subordinator $t(s)$ with $1 < \alpha < 2$. Here the fractional Klein-Kramers equation will be different from Eq. (3.214). Denote the joint PDF of position x and velocity v in operational time as $G_0(x, v, s)$ and the one in physical time $G(x, v, t)$. Then $G_0(x, v, s)$ solves the Klein-Kramers equation [Coffey *et al.* (2004)]

$$\left[\frac{\partial}{\partial s} + v \frac{\partial}{\partial x} \right] G_0(x, v, s) = \mathcal{L}_{\text{FP}} G_0(x, v, s). \qquad (3.218)$$

Using the relation

$$p(x, v, t) = \int_0^\infty ds \, p_0(x, v, s) h(s, t), \qquad (3.219)$$

where $h(s,t)$ is the PDF of inverse subordinator $s(t)$ introduced in Sec. 1.6.2, we have

$$\frac{\partial}{\partial t}G(x,v,t) = \left[-v\frac{\partial}{\partial x} + \mathcal{L}_{FP}\right]\check{D}_t^{\alpha-1}G(x,v,t). \qquad (3.220)$$

This is the fractional Klein-Kramers equation governing the joint PDF of position-velocity of the Langevin equation (3.217). Note that in this case, the Newton relation does not hold between $x(t)$ and $v(t)$ and Galilean invariance is violated [Metzler and Klafter (2000b); Eule and Friedrich (2009)].

Integrating over the position x on Eq. (3.220), we get the same equation governing the PDF of velocity $v(t)$ as Eq. (3.215). This is reasonable since the only difference between the Langevin equations (3.206) and (3.217) is the dependence of position $x(t)$ on velocity $v(t)$. But here, we can also derive the equation governing the PDF of position $x(t)$ by integrating Eq. (3.220) over $\int dv$ and $\int v\,dv$, and combining the two resulted equations. Assuming $1 < \alpha < 2$, with $\langle v^2(t)\rangle \simeq D_v/\gamma$ for long times in the case of $\gamma \neq 0$, this procedure yields the fractional diffusion equation of $G(x,t)$:

$$\frac{\partial^2}{\partial t^2}G(x,t) + \gamma\left(\frac{1}{\mu_1}\frac{\partial}{\partial t} + \frac{\mu_\alpha}{\mu_1^2}D_t^\alpha\right)G(x,t)$$
$$= \frac{D_v}{\gamma}\frac{\partial^2}{\partial x^2}\left(\frac{1}{\mu_1} + \frac{\mu_\alpha}{\mu_1^2}D_t^{\alpha-1}\right)^2 G(x,t), \qquad (3.221)$$

which becomes, in the long time or high-friction limit,

$$\frac{\partial}{\partial t}G(x,t) = \frac{D_v}{\gamma^2\mu_1}\frac{\partial^2}{\partial x^2}G(x,t). \qquad (3.222)$$

It can be seen that in the long time limit, the Langevin equation (3.217) undergoes normal diffusion, with the odd-order moments vanishing and even-order moments

$$\langle x^{2n}(t)\rangle \simeq \frac{(2n)!}{n!}\left(\frac{D_v}{\gamma^2\mu_1}\right)^n t^n. \qquad (3.223)$$

Another way of deriving the moments of $x(t)$ in Eq. (3.223) is based on the Gaussian distribution of the original process of $x(s)$ in operational time. For a Gaussian process, its PDF can be completely determined from the knowledge of its mean and variance. Based on the correlation function of v in operational time s, we calculate the second moment of $x(s)$ in operational time for model described by Eq. (3.217):

$$\langle x^2(s)\rangle = \int_0^s\int_0^s ds_1 ds_2 \,\langle v(s_1)v(s_2)\rangle \simeq \frac{2D_v}{\gamma^2}s. \qquad (3.224)$$

Since the mean value of $v(s)$ is zero, the motion is unbiased and the odd-order moments of $x(s)$ are zero; the even-order moments are

$$\langle x^{2n}(s)\rangle \simeq \frac{(2n)!}{n!}\left(\frac{D_v}{\gamma^2}\right)^n s^n. \tag{3.225}$$

Then using the relation

$$\langle x^{2n}(t)\rangle = \int_0^\infty ds\, \langle x^{2n}(s)\rangle h(s,t), \tag{3.226}$$

and the asymptotic expression

$$\Phi(\lambda) \simeq \mu_1\lambda, \quad \text{as } \lambda \to 0, \tag{3.227}$$

one can also get the result in Eq. (3.223). In the long time, the Langevin equation (3.217) coupled with α-dependent subordinator $(1 < \alpha < 2)$ still exhibits normal diffusion as in the operational time Eq. (3.225), although this subordinator might change the PDF of $x(t)$ and $v(t)$ in the Langevin system.

3.4.2   Force Case

The fractional Klein-Kramers equation with an external force $f(x)$ cannot be obtained easily from the fractional Feynman-Kac equation since the force depends on the position $x(t)$ explicitly so that the acceleration and the velocity also depends on the position. Fortunately, by using a CTRW-like approach, Friedrich and co-workers [Friedrich *et al.* (2006a,b)] were able to rigorously derive the fractional Klein-Kramers equation

$$\left[\frac{\partial}{\partial t} + v\frac{\partial}{\partial x} + f(x)\frac{\partial}{\partial v}\right] G(x,v,t) = \mathcal{L}_{\text{FP}}\mathcal{D}_t^{1-\alpha}G(x,v,t) \tag{3.228}$$

for particles subject to a deterministic acceleration and a series of random kicks. Equation (3.228) recovers the force-free case Eq. (3.218) when $f(x) = 0$.

In their dynamical model, both deterministic and stochastic elements are contained. The motion of a test particle is assumed to satisfy the time evolution equations:

$$\frac{d}{dt}x(t) = v(t), \qquad \frac{\partial}{\partial t}v(t) = f(x). \tag{3.229}$$

The latter equation describes the effects on acceleration contributed by the external force $f(x)$. However, the test particle is subject to a random collision with the surrounding light particles in a thermal bath which change its velocity abruptly from time to time. Then consider the particle located in the volume element dx' about x' and in the time interval $[t', t + dt']$ changes its velocity to a new value which lies in the velocity space element dv' about v'. This probability is denoted by $\eta(x', v', t')dx'dv'dt'$. Assume the time difference between the successive collisions is $\tau = t - t'$ with the PDF $\psi(\tau)$. It means that the particle will undergo a further transition to a state with the velocity v at the position x. Denote these procedures as one step with the conditional probability $\xi(x, v, \tau; x', v')dxdvd\tau$. Assume this quantity can be rewritten in the form

$$\xi(x, v, \tau; x', v') = \psi(\tau)\delta(x - X_\tau(x', v')) \int dv'' F(v; v'')\delta(v'' - V_\tau(x', v')).$$
(3.230)

Here, $F(v, v'')dv$ gives the probability that the particle's velocity transits from v'' to the element dv about v in velocity space. Moreover,

$$x = X_\tau(x', v') \qquad v = V_\tau(x', v') \tag{3.231}$$

are the solutions (at time t) of Eq. (3.229) with initial values x' and v' (at time t'). Therefore, the $\psi(\tau)$ in Eq. (3.230) characterizes the stochastic element in this model while other parts in Eq. (3.230) the deterministic element. Then assuming the distribution function $\xi(x, v, \tau; x', v')$ is statistically independent from the particle's path and using Eq. (3.230), we obtain the integral equation of $\eta(x, v, t)$, i.e.,

$$\eta(x, v, t) - G_0(x, v)\delta(t)$$

$$= \int_0^t dt' \int_{-\infty}^\infty dv' \int_{-\infty}^\infty dx' \xi(x, v, t - t'; x', u')\eta(x', v', t') \tag{3.232}$$

$$= \int_0^t dt' \psi(t - t') \int_{-\infty}^\infty dv' F(v; v')\mathcal{P}^{t,t'}\eta(x, v', t').$$

Here $G_0(x, v)$ is the initial condition, and the deterministic evolution is described by the Perron-Frobenius operator

$$\mathcal{P}^{t,t'}\eta(x, v, t') = \exp\left(-(t - t')\left[v\frac{\partial}{\partial x} + f(x)\frac{\partial}{\partial v}\right]\right)\eta(x, v, t')$$

$$= \int_{-\infty}^\infty dv' \int_{-\infty}^\infty dx' \delta(x - X_\tau(x', v'))\delta(v - V_\tau(x', v'))$$

$$\times \eta(x', v', t'). \tag{3.233}$$

In addition to the integral equation of $\eta(x, v, t)$, the relationship between it and the joint PDF $G(x, v, t)$ is also needed:

$$G(x, v, t) = \int_0^t dt' \Psi(t - t')\mathcal{P}^{t,t'}\eta(x, v, t'), \tag{3.234}$$

where $\Psi(\tau)$ denotes the survival probability that no random collision occurs within the time interval τ. Based on the Eqs. (3.232) and (3.234), the equation of $G(x, v, t)$ can be obtained with the technique of Laplace transform:

$$\left[\frac{\partial}{\partial t} + v \frac{\partial}{\partial x} + f(x) \frac{\partial}{\partial v} \right] G(x, v, t)$$

$$= \int_0^t dt' \Omega(t - t') \int_{-\infty}^{\infty} dv' F(v; v') \mathcal{P}^{t,t'} G(x, v', t') \qquad (3.235)$$

$$- \int_0^t dt' \Omega(t - t') \mathcal{P}^{t,t'} G(x, v, t'),$$

where the Laplace transform of the kernel $\Omega(t)$ $(t \to \lambda)$ satisfies

$$\hat{\Omega}(\lambda) = \frac{\lambda \hat{\phi}(\lambda)}{1 - \hat{\phi}(\lambda)}. \qquad (3.236)$$

From the aspect of kinetic theory, Eq. (3.235) can be interpreted as follows. The left-hand side shows that this system is subject to Eq. (3.229). The right-hand side represents a collision operator which consists of a source and a sink. The kernel $\Omega(t)$ is related with the survival probability $\Psi(t)$ as

$$\frac{d\Psi(t)}{dt} = -\int_0^t dt' \Omega(t - t') \Psi(t'). \qquad (3.237)$$

The collision operator on the right-hand side of Eq. (3.235) is highly nonlocal in space and time. This nonlocality can be viewed as a retardation effect and it is closely related to the fact that Eq. (3.235) is Galilean invariant, provided $F(v; v')$ depends only on the velocity difference $v - v'$.

For the concrete case of Gaussian distributed [Friedrich *et al.* (2006a,b)]

$$F(v; v') = \left(\frac{\Lambda}{4\pi\kappa} \right)^{3/2} \exp \left[-\frac{(v - v' + \gamma v'/\Lambda)^2}{4\kappa/\Lambda} \right], \qquad (3.238)$$

which can be derived in the spirit of the Rayleigh model for Brownian motion [van Kampen (1992)], we have

$$\int_{-\infty}^{\infty} dv' F(v; v') g(v') - g(v) = \Lambda^{-1} \mathcal{L}_{\text{FP}} g(v) \qquad (3.239)$$

and

$$\lim_{\Lambda \to \infty} \frac{\Omega(t)}{\Lambda} = \omega(t) \qquad (3.240)$$

for large values of parameter Λ. Then Eq. (3.235) takes the form

$$\left[\frac{\partial}{\partial t} + v\frac{\partial}{\partial x} + f(x)\frac{\partial}{\partial v}\right] G(x,v,t) = \mathcal{L}_{FP} \int_0^t dt' \omega(t-t') \mathcal{P}^{t,t'} G(x,v,t'). \tag{3.241}$$

For the power law form of kernel

$$\omega(t) = \frac{1}{\Gamma(\alpha)} \frac{\partial}{\partial t} \frac{1}{t^{1-\alpha}} \propto -\frac{1}{t^{2-\alpha}} \tag{3.242}$$

in long time limit with $0 < \alpha \leq 1$, Eq. (3.241) can be reduced to

$$\left[\frac{\partial}{\partial t} + v\frac{\partial}{\partial x} + f(x)\frac{\partial}{\partial v}\right] G(x,v,t) = \mathcal{L}_{FP} \mathcal{D}_{t,f}^{1-\alpha} G(x,v,t), \tag{3.243}$$

where the symbol $\mathcal{D}_{t,f}^{1-\alpha}$ is fractional substantial derivative defined in Laplace space as

$$\mathcal{L}[\mathcal{D}_{t,f}^{1-\alpha} G(x,v,t)](\lambda) = \left[\lambda + v\frac{\partial}{\partial x} + f(x)\frac{\partial}{\partial v}\right]^{1-\alpha} \hat{G}(x,v,\lambda). \tag{3.244}$$

The definition of the fractional substantial derivative containing an external force in time domain is

$$\mathcal{D}_{t,f}^{1-\alpha} G(x,v,t) = \frac{1}{\Gamma(\alpha)} \left[\frac{\partial}{\partial t} + v\frac{\partial}{\partial x} + f(x)\frac{\partial}{\partial v}\right]$$
$$\times \int_0^t \frac{dt'}{(t-t')^{1-\alpha}} e^{-(t-t')[v\frac{\partial}{\partial x} + f(x)\frac{\partial}{\partial v}]} G(x,v,t'). \tag{3.245}$$

When $\alpha = 1$, Eq. (3.243) recovers the classical Klein-Kramers equation (3.205).

Besides the fractional Klein-Kramers equation (3.243), there are another two kinds of generalizations for anomalous diffusion processes. For example, by means of the non-Markovian generalization of the Chapman-Kolmogorov equation, [Metzler and Klafter (2000a,c); Metzler (2000)] proposed the fractional Klein-Kramers equation

$$\frac{\partial}{\partial t} G(x,v,t) = \left[-v\frac{\partial}{\partial x} - f(x)\frac{\partial}{\partial v} + \mathcal{L}_{FP}\right] \gamma_\alpha D_t^{1-\alpha} G(x,v,t), \tag{3.246}$$

where γ_α is a damping coefficient with unit $[\gamma_\alpha] = s^{\alpha-1}$ and $D_t^{1-\alpha}$ is the fractional Riemann-Liouville derivative. Another kind of fractional Klein-Kramers equation is proposed by [Barkai and Silbey (2000)]:

$$\left[\frac{\partial}{\partial t} + v\frac{\partial}{\partial x} + f(x)\frac{\partial}{\partial v}\right] G(x,v,t) = \mathcal{L}_{FP} \gamma_\alpha D_t^{1-\alpha} G(x,v,t). \tag{3.247}$$

The difference between Eqs. (3.243) and (3.247) is embodied by the fractional operators. Compared with Eq. (3.247), the fractional substantial derivative operator in Eq. (3.243) characterizes the retardation effects in the models of Friedrich and co-workers [Friedrich *et al.* (2006a,b)].

The three kinds of fractional Klein-Kramers equations (3.243), (3.246) and (3.247) all recover the classical Klein-Kramers equation (3.205) when $\alpha = 1$. Their main difference is embodied by the various fractional operators in equations even without an external force. For comparison of the three equations, the corresponding Langevin equations for the three models without an external force are proposed in [Eule *et al.* (2007)].

For the force-free case, by integrating the Eqs. (3.243), (3.246) and (3.247) over x or performing Fourier transform $x \to \rho$ together with taking $\rho = 0$, we can obtain the same Fokker-Planck equation of velocity process as

$$\frac{\partial}{\partial t}G(v,t) = \mathcal{L}_{\mathrm{FP}}\gamma_\alpha D_t^{1-\alpha}G(v,t). \tag{3.248}$$

In spirit of Fogedby [Fogedby (1994)], the velocity process in Eq. (3.248) can be described by a Langevin equation coupled with a subordinator:

$$\frac{d}{ds}v(s) = -\gamma v(s) + \xi(s), \quad \frac{d}{ds}t(s) = \eta(s), \tag{3.249}$$

where $\xi(s)$ is Gaussian white noise satisfying $\langle \xi(s_1)\xi(s_2)\rangle = \delta(s_1 - s_2)$ and $t(s)$ is the α-stable subordinator. The next important thing is to define the position $x(t)$. Obviously, different relationships between $x(t)$ and $v(t)$ contribute to different fractional Klein-Kramers equations. In [Eule *et al.* (2007)], it has been pointed out that the Langevin system corresponding to fractional Klein-Kramers equation (3.243) is

$$\frac{d}{dt}x(t) = v(t), \quad \frac{d}{ds}v(s) = -\gamma v(s) + \xi(s), \quad \frac{d}{ds}t(s) = \eta(s). \tag{3.250}$$

For another two cases, the Langevin systems are

$$\frac{d}{ds}x(s) = v(s), \quad \frac{d}{ds}v(s) = -\gamma v(s) + \xi(s), \quad \frac{d}{ds}t(s) = \eta(s), \tag{3.251}$$

and

$$\frac{d}{ds}x(s) = v(s)\eta(s), \quad \frac{d}{ds}v(s) = -\gamma v(s) + \xi(s), \quad \frac{d}{ds}t(s) = \eta(s), \tag{3.252}$$

for the Eqs. (3.246) and (3.247), respectively. The derivations based on the first two Langevin systems in Eqs. (3.250) and (3.251) have been partially presented in Sec. 3.4.1 while the relationships between the last Langevin system in Eq. (3.252) and the fractional Klein-Kramers equation has been investigated in [Eule *et al.* (2007)].

3.5 First-Passage Time and Mean Exit Time

One of the most concerns of random walks and diffusion is the first-passage event. Typical examples include integrate-and-fire neurons, in which a neuron fires only when a fluctuating voltage level first reaches a specified level, fluorescence quenching, in which light emission of a fluorescent particle stops when it reacts with a quencher, and the execution of buy/sell a stock when the price first reaches a threshold. More detailed illustrations are presented in the book [Redner (2001)], which provides a comprehensive overview of first-passage time.

To appreciate the essential characterizations of first-passage event, we employ it to the random walks, exhibiting normal or anomalous diffusion. Assume the particle starts at x_0. Denote $f(x_0, t)$ as the PDF of first-passage time when the particle reaches a certain level for the first time, and first exit time when leaving a certain interval for the first time. We derive the macroscopic equations governing the PDF $f(x_0, t)$ or the moments of first-passage time, the most typical one of which is mean first exit time.

3.5.1 *First-Passage Properties for Uncoupled Langevin Equation*

In the past few decades, most of the research works on the first-passage time or mean first exit time, appearing in the mathematical, physical, chemical and engineering literatures [Benjacob *et al.* (1982); Bobrovsky and Schuss (1982); Carmeli and Nitzan (1983); Day (1990); Gardiner (1983)], are for the uncoupled Langevin equation:

$$\frac{d}{dt}x(t) = F(x) + \sqrt{2D(x)}\xi(t), \qquad (3.253)$$

where $\xi(t)$ is the Gaussian white noise satisfying $\langle \xi(t_1)\xi(t_2) \rangle = \delta(t_1 - t_2)$, $F(x)$ is the external force and $\sqrt{2D(x)}$ is the multiplicative noise term interpreted in the Itô sense. For a general stochastic differential equation, the Itô prescription is mathematically and technically the most satisfactory [Gardiner (1983)], while the Stratonovich prescription has the advantage of ordinary chain rule formulas under a transformation [Risken (1989); Øksendal (2005)] and of yielding physically correct results especially for the noise with infinitely short correlation times [West *et al.* (1979)]. It is well-known that the processes $x(t)$ described by Eq. (3.253) is Markovian. It means that $x(t + dt)$ only depends on the state at the previous time step $x(t)$. If we consider the conditional probability density $p(x, t|x_0, t_0)$ that the particle is at position x at time t provided that it started at x_0 at time t_0. The condition probability density $p(x, t|x_0, t_0)$ satisfies the Chapman-Kolmogorov equation

$$p(x, t|x_0, t_0) = \int_{-\infty}^{\infty} p(x, t|x', t')p(x', t'|x_0, t_0)dx' \qquad (3.254)$$

for any $t' \in [t_0, t]$. From the Chapman-Kolmogorov equation (3.254), we can also derive the Fokker-Planck equation

$$\frac{\partial}{\partial t}p(x, t|x_0, t_0) = \mathcal{L}_x p(x, t|x_0, t_0), \qquad (3.255)$$

where

$$\mathcal{L}_x = -\frac{\partial}{\partial x}F(x) + \frac{\partial^2}{\partial x^2}D(x) \qquad (3.256)$$

is the Fokker-Planck operator on variable x.

Then we derive the equation governing the PDF $f(x_0, t)$ of first-passage time from the Chapman-Kolmogorov equation (3.254). Differentiating on both sides of Eq. (3.254) with respect to t' yields

$$0 = \int_{-\infty}^{\infty} \frac{\partial p(x, t|x', t')}{\partial t'}p(x', t'|x_0, t_0) + p(x, t|x', t')\frac{\partial p(x', t'|x_0, t_0)}{\partial t'}dx'. \qquad (3.257)$$

Then we denote the two terms in Eq. (3.257) as I and II, respectively, and deal with them separately. For the first term I, we use the property of time-translation invariance and obtain

$$\frac{\partial}{\partial t'}p(x, t|x', t') = \frac{\partial}{\partial t'}p(x-x', t-t') = -\frac{\partial}{\partial t}p(x-x', t-t') = -\frac{\partial}{\partial t}p(x, t|x', t'), \qquad (3.258)$$

which implies

$$I = \int_{-\infty}^{\infty} -\frac{\partial p(x, t|x', t')}{\partial t}p(x', t'|x_0, t_0)dx'. \qquad (3.259)$$

For the second term II, we have

$$II = \int_{-\infty}^{\infty} p(x, t|x', t')\mathcal{L}_{x'}p(x', t'|x_0, t_0)dx'$$

$$= \int_{-\infty}^{\infty} p(x', t'|x_0, t_0)\mathcal{L}_{x'}^*p(x, t|x', t')dx', \qquad (3.260)$$

where

$$\mathcal{L}_x^* = F(x)\frac{\partial}{\partial x} + D(x)\frac{\partial^2}{\partial x^2} \qquad (3.261)$$

is the conjugate operator of the Fokker-Planck operator $\mathcal{L}_x$ in Eq. (3.256). Combining the two terms I and II, we obtain

$$\int_{-\infty}^{\infty} \left[\frac{\partial p(x,t|x',t')}{\partial t} - \mathcal{L}_{x'}^* p(x,t|x',t') \right] p(x',t'|x_0,t_0)dx' = 0. \quad (3.262)$$

Since the term $p(x',t'|x_0,t_0)$ in Eq. (3.262) is arbitrary with respect to t', we obtain the backward Fokker-Planck equation:

$$\frac{\partial p(x,t|x_0,0)}{\partial t} = \mathcal{L}_{x_0}^* p(x,t|x_0,0), \quad (3.263)$$

by writing x' and t' as x_0 and 0, respectively.

Next, let us consider a semi-infinite problem, i.e., the first-passage time that a particle starts at $x_0 > 0$ and reaches $x = 0$. For this problem, we set the absorbing boundary condition at $x = 0$. With this boundary condition, the survival probability $S(x_0,t)$, namely, the probability that a particle starts at x_0 and keeps in the positive half-space throughout the time interval $[0,t]$, satisfies [Gardiner (1983); Deng *et al.* (2017); Wang *et al.* (2019c)]

$$P\{\tau \geq t\} = S(x_0,t) = \int_0^{\infty} p(x,t|x_0,0)dx, \quad (3.264)$$

where τ is the exact time that the particle leaves the positive half-space. From the second equality in Eq. (3.264), we find that $S(x_0,t)$ also satisfies the backward Fokker-Planck equation (3.263) with the same boundary condition, i.e.,

$$\begin{cases} \frac{\partial}{\partial t}S(x_0,t) = \mathcal{L}_{x_0}^* S(x_0,t), \\ S(0,t) = 0. \end{cases} \quad (3.265)$$

The initial condition of survival probability $s(x_0,t)$ is

$$S(x_0,0) = \begin{cases} 1 & x_0 > 0, \\ 0 & x_0 \leq 0. \end{cases} \quad (3.266)$$

From the first equality in Eq. (3.264), one can obtain the relation between the PDF $f(x_0,t)$ of first-passage time and the survival probability:

$$f(x_0,t) = -\frac{\partial}{\partial t}S(x_0,t); \quad (3.267)$$

thus $f(x_0,t)$ also satisfies the backward Fokker-Planck equation (3.263), i.e.,

$$\frac{\partial f(x_0,t)}{\partial t} = \mathcal{L}_{x_0}^* f(x_0,t) \quad (3.268)$$

and the normalization $\int_0^\infty f(x_0, t)dt = 1$. The boundary condition for the PDF of first-passage time is $f(0, t) = \delta(t)$, since the particle has left the positive half-space at the initial position $x_0 = 0$.

Based on Eq. (3.268), the equation governing the moments of first-passage time can be obtained directly. Multiplying t and integrating over t on both sides of Eq. (3.268), we obtain the equation for the first moment of first-passage time $\langle t \rangle$:

$$\mathcal{L}_{x_0}^* \langle t \rangle = -1. \tag{3.269}$$

The boundary condition of Eq. (3.269) is the same as those of $S(x_0, t)$ (i.e., absorbing at $x = 0$), due to its relation with the survival probability $S(x_0, t)$:

$$\langle t \rangle = \int_0^\infty t f(x_0, t)dt = \int_0^\infty S(x_0, t)dt. \tag{3.270}$$

Similarly, define $\langle t^n \rangle$ as the nth moment of first-passage time, we obtain the recurrence relationship:

$$-n\langle t^{n-1} \rangle = \mathcal{L}_{x_0}^* \langle t^n \rangle, \tag{3.271}$$

which means all the moments of the first-passage time can be found through repeated integration. With the specified absorbing boundary condition, Eq. (3.269) has a unique solution [Gardiner (1983); Wang *et al.* (2019c)]

$$\langle t \rangle = \int_0^{x_0} W^{-1}(y) \int_y^\infty \frac{W(x)}{D(x)} dx dy, \tag{3.272}$$

where

$$W(x) = \exp\left[\int_0^x \frac{F(x')}{D(x')} dx'\right]. \tag{3.273}$$

The exact solutions of Eq. (3.269) in a bounded domain with different kinds of boundary conditions can be found in [Gardiner (1983)]. For example, if the particle moves in a bounded domain $[0, L]$ with reflecting boundary condition at $x = L$, then

$$\langle t \rangle = \int_0^{x_0} W^{-1}(y) \int_y^L \frac{W(x)}{D(x)} dx dy \tag{3.274}$$

with the same $W(x)$ as semi-infinite problem.

For the simplest Brownian motion with $F(x) \equiv 0$ and $D(x) \equiv D$ in Eq. (3.272), we find that the mean first exit time is infinite. Thus, although the one dimensional diffusion is recurrent, i.e., the particle surely reaches the origin, the mean first exit time it takes is infinite. If the particle is confined in a bounded domain $[0, L]$, the solution of Eq. (3.274) is

$$\langle t \rangle = \frac{x_0}{2D}(2L - x_0). \tag{3.275}$$

The Eq. (3.275) shows that the mean first exit time $\langle t \rangle$ increases monotonously with respect to x_0 and the size L of the domain. On the other hand, D is the diffusion coefficient, representing the strength of fluctuation. The mean first exit time $\langle t \rangle$ decreases with the increase of D, which is also expected.

Besides the mean first exit time $\langle t \rangle$, the more detailed properties of first-passage time can be obtained through the conditional probability density $p(x, t|x_0, 0)$. By using the (forward) Fokker-Planck equation (3.255) and ignoring the absorbing boundary, the solution for initial condition $p(x, 0|x_0, 0) = \delta(x - x_0)$ is given by Greens function (or heat kernel)

$$G(x, t|x_0, 0) = \frac{1}{\sqrt{4\pi Dt}} \exp\left[-\frac{(x - x_0)^2}{4Dt}\right]. \tag{3.276}$$

The absorbing boundary condition enforces $p(0, t|x_0, 0) = 0$ for all x_0 and t. The solution satisfying this boundary condition can be obtained by using the "image method" [Redner (2001)], i.e.,

$$p(x, t|x_0, 0) = G(x, t|x_0, 0) - G(x, t| - x_0, 0), \tag{3.277}$$

which clearly satisfies the differential equation, the initial condition and the boundary condition in domain $x > 0$. Then we obtain the survival probability

$$S(x_0, t) = \int_0^\infty p(x, t|x_0, 0)dx = \text{erf}\left(\frac{x}{\sqrt{4Dt}}\right), \tag{3.278}$$

where $\text{erf}(z)$ is the error function. For large t, the survival probability scales as

$$S(x_0, t) \simeq \frac{x}{\sqrt{\pi Dt}}. \tag{3.279}$$

Therefore, the asymptotic form of PDF $f(x_0, t)$ of first-passage time is

$$f(x_0, t) = -\frac{\partial}{\partial t}S(x_0, t) \simeq \frac{x_0}{\sqrt{4\pi Dt^3}} \propto t^{-3/2} \tag{3.280}$$

for large t. This result is consistent to the previous discussion of divergent mean first exit time.

3.5.2   First-Passage Properties for Coupled Langevin Equation

We have just discussed the first-passage problem of the uncoupled Langevin equation (3.253), which can display both normal and anomalous diffusion due to external force and multiplicative noise [Cherstvy *et al.* (2013); Cherstvy and Metzler (2014); Wang *et al.* (2019c); Leibovich and Barkai (2019)]. In spite of this, the characteristic time scale of this system is still finite. Nowadays, more and more physical and biological systems display anomalous diffusion with a divergent characteristic time scale. Therefore, we investigate the coupled Langevin system here:

$$\frac{d}{ds}x(s) = F(x(s)) + \sqrt{2D(x(s))}\xi(s),$$

$$\frac{d}{ds}t(s) = \eta(s).$$

(3.281)

In order to derive a general result of first-passage properties, we assume $t(s)$ here is an arbitrary subordinator with characteristic function $\Phi(\lambda)$, i.e., [Applebaum (2009)]

$$\langle e^{-\lambda t(s)} \rangle = e^{-s\Phi(\lambda)}. \tag{3.282}$$

Similar to the previous case of uncoupled Langevin equation, we start from the conditional probability density $p(x,t|x_0,0)$, which satisfies the backward Fokker-Planck equation now [Risken (1989); Carmi and Barkai (2011); Cairoli and Baule (2015a); Deng *et al.* (2017)]:

$$\frac{\partial p(x,t|x_0,0)}{\partial t} = \int_0^t K(t-t')\mathcal{L}_{x_0}^* p(x,t'|x_0,0)dt'. \tag{3.283}$$

Compared with the uncoupled one Eq. (3.263), the kernel $K(t)$ comes from the subordinator $t(s)$. The relationship between them is embodied by the identity

$$\hat{K}(\lambda) = \frac{\lambda}{\Phi(\lambda)}, \tag{3.284}$$

where $\hat{K}(\lambda)$ is the Laplace transform $(t \to \lambda)$ of kernel $K(t)$. Equation (3.283) can recover Eq. (3.263) when the subordinator is removed, that is, $t(s) \equiv s$ resulting in $\Phi(\lambda) = \lambda$ and $K(t) = \delta(t)$.

Therefore, the only difference between Eqs. (3.263) and (3.283) is the kernel $K(t)$. For convenience, we treat Eq. (3.283) in Laplace λ space and obtain

$$\lambda\hat{p}(x,\lambda|x_0,0) - \delta(x-x_0) = \hat{K}(\lambda)\mathcal{L}_{x_0}^*\hat{p}(x,\lambda|x_0,0). \tag{3.285}$$

Integrating it over x, one arrives at the equation of survival probability in Laplace space:

$$\lambda \hat{S}(x_0, \lambda) - 1 = \hat{K}(\lambda)\mathcal{L}^*_{x_0}\hat{S}(x_0, \lambda). \tag{3.286}$$

Then by virtue of the relationship between first-passage time and survival probability in Laplace space, i.e.,

$$\hat{f}(x_0, \lambda) = 1 - \lambda \hat{S}(x_0, \lambda), \tag{3.287}$$

we obtain the equation for first-passage time

$$\lambda \hat{f}(x_0, \lambda) = \hat{K}(\lambda)\mathcal{L}^*_{x_0}\hat{f}(x_0, \lambda), \tag{3.288}$$

which is

$$\frac{\partial}{\partial t}f(x_0, t) = \int_0^t K(t - t')\mathcal{L}^*_{x_0}f(x_0, t')dt' \tag{3.289}$$

in t space. The only difference between Eqs. (3.268) and (3.289) is still the kernel $K(t)$.

On the other hand, using the relationship $\hat{K}(\lambda) = \lambda/\Phi(\lambda)$, Eq. (3.288) can be reduced to

$$\Phi(\lambda)\hat{f}(x_0, \lambda) = \mathcal{L}^*_{x_0}\hat{f}(x_0, \lambda). \tag{3.290}$$

Then using the identity for the first moment of first-passage time

$$\langle t \rangle = -\left.\frac{\partial}{\partial \lambda}\hat{f}(x_0, \lambda)\right|_{\lambda=0} \tag{3.291}$$

and $\Phi(\lambda = 0) = 0$, $\hat{f}(x_0, 0) = 1$ (from Eq. (3.287)), we obtain the equation for mean first exit time

$$\mathcal{L}^*_{x_0}\langle t \rangle = -\Phi'(0). \tag{3.292}$$

Comparing Eq. (3.292) with Eq. (3.269), we find an interesting phenomenon that the mean first exit time for coupled Langevin equation is just a constant $\Phi'(0)$ multiple of the uncoupled one. This result has been also derived through the subordinator approach in [Deng *et al.* (2017)]. Based on this finding, we can make a direct determination of whether mean first exit time is finite or infinite. For example, if the subordinator $t(s)$ is the Lévy α-stable process with $0 < \alpha < 1$, that is,

$$\Phi(\lambda) = \lambda^\alpha, \tag{3.293}$$

making $\Phi'(0)$ diverge, then we obtain an infinite mean first exit time. Otherwise, if it is the α-dependent subordinator with $1 < \alpha < 2$ implying [Wang *et al.* (2019b)]

$$\Phi(\lambda) = \mu_1\lambda - \mu_\alpha\lambda^\alpha, \tag{3.294}$$

or the tempered Lévy α-stable process with $0 < \alpha < 1$ and truncation parameter μ meaning [Meerschaert and Sikorskii (2011); Gajda and Magdziarz (2010, 2011); Magdziarz *et al.* (2007); Kumar and Vellaisamy (2015); Molina-Garcia *et al.* (2018)]

$$\Phi(\lambda) = (\lambda + \mu)^\alpha - \mu^\alpha, \tag{3.295}$$

then we have a finite mean first exit time.

A more general version has been shown in [Deng *et al.* (2017)], where the Gaussian white noise ξ in Eqs. (3.253) and (3.281) is extended to the Lévy noise in n dimensions which is defined through the Lévy-Khintchine formula [Applebaum (2009)]. Now for simplicity, we present the main result in one dimension. Denote this Lévy process as $L(t)$, i.e.,

$$\frac{d}{dt}L(t) = \xi(t). \tag{3.296}$$

It has a specific form with its characteristic function [Applebaum (2009)], i.e., for all $t \geq 0$,

$$\langle e^{ikL(t)} \rangle = e^{t\phi(k)}, \tag{3.297}$$

where

$$\phi(k) = iak - \frac{1}{2}bk^2 + \int_{\mathbb{R}\setminus\{0\}} \left[e^{iky} - 1 - iky\chi_{\{|y|<1\}} \right] \nu(dy). \tag{3.298}$$

The first two terms of $\phi(k)$ represent the Brownian part of the Lévy process $L(t)$, while the last term of integration is a pure jump process in a Lévy process. $\xi(t)$ recovers the Gaussian white noise when the constant a and the measure ν in Eq. (3.298) is zero. It is well-known that different noise contributes to different spatial operator in the macroscopic equations. For Gaussian white noise together with the external force $f(x)$, the operator in forward Fokker-Planck equation (3.255) is $\mathcal{L}_x$ in Eq. (3.256). For the Lévy noise, the new operator, denoted as $\mathcal{A}_x$, is defined as the inverse Fourier transform of $\phi(k)$, i.e.,

$$\mathcal{A}_x p(x) = \frac{1}{2\pi} \int_{-\infty}^{\infty} e^{-ikx} \phi(k)\tilde{p}(k)dk, \tag{3.299}$$

where $\tilde{p}(k)$ is the Fourier transform of $p(x)$. Further considering the form of $\phi(k)$ in Eq. (3.298), we have

$$\mathcal{A}_x p(x) = -\frac{\partial}{\partial x}F(x)p(x) + a\frac{\partial}{\partial x}p(x) + \frac{b}{2}\frac{\partial^2}{\partial x^2}p(x)$$
$$+ \int_{\mathbb{R}\setminus\{0\}} \left[p(x+y) - p(x) - y\frac{\partial}{\partial x}p(x)\chi_{\{|y|<1\}} \right] \nu(dy) \tag{3.300}$$

being the pseudo-differential operator. In this case, the backward Fokker-Planck equation for the coupled Langevin equation becomes

$$\frac{\partial p(x,t|x_0,0)}{\partial t} = \int_0^t K(t-t')\mathcal{A}^*_{x_0}p(x,t'|x_0,0)dt', \qquad (3.301)$$

where $\mathcal{A}^*_x$ is the conjugate operator of $\mathcal{A}_x$ in Eq. (3.300). Since the derivations of the equations for PDF of first-passage time and mean first exit time are independent of the spatial operator, we immediately obtain the equations

$$\frac{\partial f(x_0,t)}{\partial t} = \int_0^t K(t-t')\mathcal{A}^*_{x_0}f(x_0,t')dt' \qquad (3.302)$$

and

$$\mathcal{A}^*_{x_0}\langle t \rangle = -\Phi'(0) \qquad (3.303)$$

as the analogues of Eqs. (3.289) and (3.292).

3.5.3 *Hitting Probability and Escape Probability*

We have investigated the first-passage properties of a particle which moves in the positive half-space or a bounded domain $[0,L]$. For the latter case, people sometimes are also interested in which end the particle exits, leading to the concept of a hitting probability. Denote $P_0(x_0)$ and $P_L(x_0)$ as the hitting probability of the particle exiting this domain from left boundary $x = 0$ and right boundary $x = L$, respectively, given that it starts at position x_0. Then

$$P_0(x_0) + P_L(x_0) = 1. \qquad (3.304)$$

Since the hitting probability only concerns about which end the particle exits rather than how long it exits, we can define an auxiliary quantity

$$P(x,x_0) = \int_0^\infty p(x,t|x_0,0)dt, \qquad (3.305)$$

by integrating the conditional probability density $p(x,t|x_0,0)$ over time t. The quantity $P(x,x_0)$ can be interpreted as the mean time that the trajectory spends at x before exiting the domain given the initial position x_0 [Naeh *et al.* (1990); Karlin and Taylor (1981); Deng *et al.* (2017)].

Let us consider the model in Eq. (3.281) and the backward Fokker-Planck equation for condition probability density $p(x,t|x_0,0)$ in Eq. (3.283). The Laplace form of this equation for $x \notin (0,L)$ is

$$\lambda \hat{p}(x,\lambda|x_0,0) = \hat{K}(\lambda)\mathcal{L}^*_{x_0}\hat{p}(x,\lambda|x_0,0). \qquad (3.306)$$

Since

$$\hat{p}(x,0|x_0,0) = \int_0^\infty p(x,t|x_0,0)dt \equiv P(x,x_0), \qquad (3.307)$$

the equation governing $P(x,x_0)$ will be obtained by taking $\lambda = 0$ in Eq. (3.306). Further considering that $\hat{K}(\lambda) = \lambda/\Phi(\lambda)$ and $\Phi(0) = 0$, we finally obtain

$$\mathcal{L}_{x_0}^* P(x,x_0) = 0, \qquad (3.308)$$

which is independent of the specific subordinator $t(s)$. Since $P_0(x_0) = P(0,x_0)$, we obtain the equation for hitting probability as

$$\mathcal{L}_{x_0}^* P_0(x_0) = 0. \qquad (3.309)$$

The boundary condition for this equation is

$$P_0(x_0)|_{x_0=0} = 1, \quad P_0(x_0)|_{x_0=L} = 0. \qquad (3.310)$$

We can find the key element in Eq. (3.309) is the spatial operator $\mathcal{L}_{x_0}^*$, which is closely related to the Gaussian white noise in Eq. (3.281). Through the previous derivations of first-passage problem, we find this operator will be replaced by the new one $\mathcal{A}_{x_0}^*$ in Eq. (3.301) when $\xi(t)$ in Eq. (3.281) becomes a general Lévy noise.

Another important difference between Brownian motion and a general Lévy process is the continuity of the trajectories. More precisely, the Brownian trajectories are continuous while the path of the α-stable Lévy process is discontinuous. Therefore, the particle may exit the domain and reach any position beyond the domain, by jumping one of the boundaries $x = 0$ and $x = L$. In this case, people are interested in another quantity, escape probability, in the exit phenomenon. This quantity describes the probability of a particle starting at position x_0, first escaping a domain D and landing in a subset E of D^c (the complement of D), which is denoted as $P_E(x_0)$ [see Fig. 3.7].

Now, the corresponding equation for $P_E(x_0)$ is just to replace the spatial operator in Eq. (3.309) by $\mathcal{A}_{x_0}^*$, i.e.,

$$\mathcal{A}_{x_0}^* P_E(x_0) = 0. \qquad (3.311)$$

The corresponding boundary condition for this equation is

$$P_E(x_0)|_{x_0 \in E} = 1, \quad P_E(x_0)|_{x_0 \in D^c \backslash E} = 0. \qquad (3.312)$$

Some specific analytic solution and numerical simulations of this equation can be found in [Deng *et al.* (2017)].

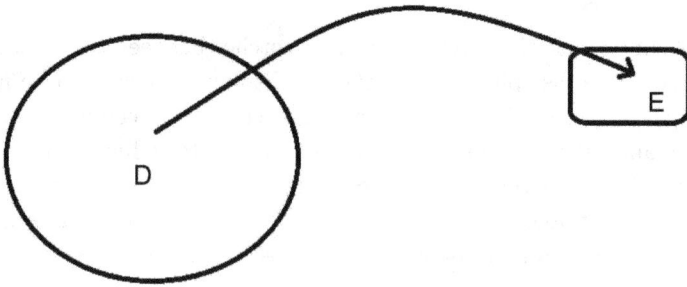

Fig. 3.7: Sketch map of the escape probability.

3.5.4 *Intermittent Search Strategy*

One of the most important rules of first-passage time is to quantify the efficacy of the diffusive transport of a particle to reach a target [Redner (2001); Gardiner (1983)]. Apart from the first-passage time, there are still some different quantities used to assess the efficiency of search strategies, such as the energy necessary for reaching the first prey and the number of preys collected in a given time [Bénichou *et al.* (2011)]. From the kinetic point of view, the first-passage time of reaching a target can be easily described and solved. Some parameters might need to be determined to minimize the mean first-passage time.

Searching a target is a natural demand in the real world. At the same time, many physical or biological problems can be regarded as the search processes, describing how a searcher finds a target located in an unknown position. At the macroscopic scale, it is exemplified as animals searching for food or a shelter [Bell (1991)]. At the microscopic scale, one can cite the localization by a protein of a specific DNA sequence or the active transport of vesicles in cells [Bénichou *et al.* (2011)]. In these examples, intermittent search strategies have been proved to play a crucial role in optimizing the search time of randomly hidden targets [Bénichou *et al.* (2005a); Lomholt *et al.* (2008)]. This kind of search behavior could be extended to broader research domains such as the theory of stochastic processes [Bartumeus *et al.* (2002)], applied mathematics [Stone (1975)], and molecular biology [Coppey *et al.* (2004)]; and it also motivates some new interesting research topics [Xu and Deng (2018a,b)].

For the intermittent search process, it switches between two phases — a slow phase and a fast phase. The searcher displays a slow reactive motion in the first phase, during which the target can be detected. The latter fast phase aims at relocating into unvisited regions to reduce oversampling, during which the searcher is unable to detect the target. We will show some examples of how to evaluate the first-passage time of the intermittent search process at the macroscopic and microscopic scales.

3.5.4.1 *Macroscopic scale*

Anyone who has ever lost keys knows that an intermittent behavior containing local scanning phases and relocating phases is often adopted instinctively. Actually, numerous studies of foraging behavior of a broad range of animal species show that such intermittent behavior is commonly observed and the duration time of search and displacement phases vary widely [O'Brien *et al.* (1990); Bell (1991); Kramer and McLaughlin (2015)]. Now we give an explicit description of the intermittent searching trajectories. It is assumed that the searcher alternates its movements with two distinct phases:

(1) A scanning phase, named phase 1, during which the sensory organs of the searcher explore its immediate vicinity. This phase is modeled as a slow diffusive movement (Brownian motion with diffusion coefficient D). The target is found when this movement reaches the target location for the first time. A minimal mean time spent in this phase is assumed to be $\tau_1^{\min}$, since focusing and finding the target require a minimum time.

(2) A motion phase, named phase 2, during which the searcher moves fast and is unable to detect targets. The relocating movements in this phase are characterized by a ballistic motion (Lévy walk with velocity V) [O'Brien *et al.* (1990)]. In the case of animals, two successive ballistic phases are usually correlated in the angles. Here, it is assumed that the phase 2 is always in the same direction for a one-dimensional problem.

Moreover, it is assumed the searcher randomly switches from phase 1(2) to phase 2(1) with a fixed rate per unit time, $z_1(z_2)$, that is, with no temporal memory. It leads to exponentially distributed phase durations, in agreement with numerous experimental studies [Pierce-Shimomura *et al.* (1999); Hill *et al.* (2000); Fujiwara *et al.* (2002); Li *et al.* (2008)], the mean duration of phase i being $\tau_i = 1/z_i$.

For this one-dimensional problem, the chosen geometry is a single target at $x = 0$ on a segment of size L with boundary conditions. The geometry can be regarded as the case of regularly spaced targets with L being the typical distance between targets. The state of the searcher can be described by two elements, the position x and the phase i ($i = 1$ for slow detection phase and $i = 2$ for ballistic relocation phase). Let $p_i(t, x)$ be the survival probability when the searcher starts from position x and in phase i, the target has not been found at time t. Then $p_i(t, x), i = 1, 2$ satisfy the backward Fokker-Planck equation [Gardiner (1983); Redner (2001)]:

$$\frac{\partial p_1(t, x)}{\partial t} = D \frac{\partial^2 p_1(t, x)}{\partial x^2} + \frac{1}{\tau_1}[p_2(t, x) - p_1(t, x)],$$

$$\frac{\partial p_2(t, x)}{\partial t} = -V \frac{\partial p_2(t, x)}{\partial x} + \frac{1}{\tau_2}[p_1(t, x) - p_2(t, x)]. \tag{3.313}$$

Equation (3.313) can be obtained through the similar procedure of deriving Eq. (3.263) and the Chapman-Kolmogorov equation with two phases:

$$p(x, i, t|x_0, i_0, 0) = \sum_{i=1}^{2} \int_{-\infty}^{\infty} p(x, i, t|x', i', t')p(x', i', t'|x_0, i_0, 0)dx', \tag{3.314}$$

where $p(x, i, t|x_0, i_0, 0)$ represents the conditional probability density of position x and state i at time t given that the initial position is x_0 and initial phase is i_0 at $t = 0$. Then denote $t_i(x)$ as the mean first-passage time at the target for the searcher starting from x in phase i. By virtue of the relationship between mean first-passage time and the survival probability

$$t_i(x) = -\int_0^\infty t \frac{\partial p_i(t, x)}{\partial t} dt = \int_0^\infty p_i(t, x)dt, \tag{3.315}$$

we obtain the equations for $t_i(x)$ as

$$D \frac{d^2 t_1(x)}{dx^2} + \frac{1}{\tau_1}[t_2(x) - t_1(x)] = -1,$$

$$-V \frac{dt_2}{dx} + \frac{1}{\tau_2}[t_1(x) - t_2(x)] = -1. \tag{3.316}$$

These differential equations have to be closed with boundary conditions. Since the target at $x = 0$ can be found only in phase 1 and the periodic boundary conditions have been specified, the boundary condition is

$$t_1(0) = t_1(L) = 0, \quad t_2(0) = t_2(L). \tag{3.317}$$

The mean search time $\langle t \rangle$ is defined as the average of $t_1(x)$ over the initial position x of the searcher, which is uniformly distributed over the domain $[0, L]$, as the searcher initially does not know where the target is located. The result is [Bénichou *et al.* (2011, 2005a,b)]:

$$\langle t \rangle = (\tau_1 + \tau_2) \left(\frac{L}{2} \frac{(e^{\alpha+\beta} - 1)\sqrt{1 + 4r} + (1 + 2r)(e^\beta - e^\alpha)}{\sqrt{1 + 4r}(e^\beta - 1)(e^\alpha - 1)\tau_2 V} - \frac{1}{r} - 1 \right) \tag{3.318}$$

with

$$r = \frac{\tau_2^2 V^2}{D\tau_1}, \tag{3.319}$$

$$\alpha = \frac{L}{2} \left(-\frac{1}{\tau_2 V} + \sqrt{\frac{1}{\tau_2^2 V^2} + \frac{4}{D\tau_1}} \right), \tag{3.320}$$

$$\beta = -\frac{L}{2} \left(\frac{1}{\tau_2 V} + \sqrt{\frac{1}{\tau_2^2 V^2} + \frac{4}{D\tau_1}} \right). \tag{3.321}$$

If we consider the large L limit for the case of rare hidden targets in a large domain, we have the asymptotic form:

$$\langle t \rangle \simeq \frac{L(\tau_1 + \tau_2)(D\tau_1 + 2\tau_2^2 V^2)}{2\tau_2 V \sqrt{D\tau_1}\sqrt{D\tau_1 + 4\tau_2^2 V^2}}. \tag{3.322}$$

Note that the mean first-passage time $\langle t \rangle$ scales as L for intermittence search strategy while the mean detection time is $t_{\text{diff}} = L^2/12D$ which is much larger for large L. Furthermore, the mean search time $\langle t \rangle$ can be minimized for the specific option of τ_1 and τ_2:

$$\tau_1^{\text{opt}} = \tau_1^{\text{min}} \tag{3.323}$$

and τ_2^{opt} satisfies the relation [Bénichou *et al.* (2005b)]

$$\tau_1^3 + 6\frac{\tau_1^2 \tau_2^2}{\tau} - 8\frac{\tau_2^5}{\tau^2} = 0, \tag{3.324}$$

where $\tau_1 = \tau_1^{\text{min}}$ and $\tau = D/V^2$ is an extra characteristic time. Besides, as the extension of the exponentially distributed relocation times, it has been shown that the search process with Lévy distributed relocation times significantly outperforms better [Lomholt *et al.* (2008)], since the resulting Lévy walks reduce oversampling and thus further optimize the intermittent search strategy in the critical situation of rare targets.

3.5.4.2 *Microscopic scale*

The typical example for a search process at microscopic scale is that a single protein binds a specific target sequence of base pairs (target site) on a long DNA molecule. The best known theoretical model for this facilitated diffusion of proteins searching for DNA targets was originally developed by Berg, Winter, and von Hippel (BHW) [Berg *et al.* (1981); Winter and von Hippel (1981); Berg and von Hippel (1985)], which assumes that the protein's trajectory is an intermittent search process. It randomly switches between two distinct phases of motion, 3D diffusion in solution, and 1D diffusion along DNA (sliding). There are no correlations between the two transport phases so that the main reason of speeding up the search process is the effective reduction in the dimensionality of the protein motion.

More precisely, the BHW says that a single protein searches for a single binding site on a long DNA strand of N base pairs, the length of each of them being b. Suppose that there are R rounds labeled $i = 1, \cdots, R$ on a given search. In the ith round the protein spends a time $T_{3,i}$ diffusion in the cytosol followed by a period $T_{1,i}$ sliding along the DNA. Then the total search time is

$$T = \sum_{i=1}^{R}(T_{3,i} + T_{1,i}), \tag{3.325}$$

and the mean search time is

$$\tau = r(\tau_3 + \tau_1), \tag{3.326}$$

where r is the mean number of rounds and τ_3 and τ_1 are the mean durations of each phase of 3D and 1D diffusion. Let n be the mean number of sites scanned during each 1D diffusion with $n \ll N$. If the binding site of DNA following the 3D phase is distributed uniformly along the DNA, then the probability of finding the specific promoter site is $p = n/N$. It follows that the probability of finding the site after R rounds is $(1-p)^{R-1}p$. Hence, the mean number of rounds is

$$r = \frac{1}{p} = \frac{N}{n}. \tag{3.327}$$

Assuming that the 1D sliding occurs via normal diffusion with diffusion coefficient D_1, we have

$$\tau = \frac{N}{n}(\tau_1 + \tau_3), \quad n = \frac{2\sqrt{D_1\tau_1}}{b}. \tag{3.328}$$

Since τ_3 mainly depends on the cellular environment and is thus unlikely to vary significantly between proteins, it is reasonable to minimize the mean search time τ with respect to τ_1 while τ_3 is kept fixed. So we set $d\tau/d\tau_1 = 0$ and obtain the optimal search strategy $\tau_1 = \tau_3$ and the optimal search time

$$\tau^{\text{opt}} = 2\tau_3 \frac{N}{n} = Nb\sqrt{\frac{\tau_3}{D_1}}. \tag{3.329}$$

Comparing it with the result ($\tau_{3D} = N\tau_3$) of pure 3D diffusion by setting $\tau_1 = 0$ and $n = 1$, the intermittent search strategy is faster by a factor $n/2$.

A more complicated analysis is needed in order to take into consideration the boundary conditions. For example, the DNA is treated as a finite track of length $l = L + M$ with reflecting boundary conditions at $x = -M$ and $x = +L$ and a pointlike target at $x = 0$. Let $p_3(t)$ be the PDF that the protein in solution at time $t = 0$ binds to the DNA at time t at a random position. Then

$$p_3(t) = z_3 e^{-z_3 t}, \tag{3.330}$$

where $\tau_3 = 1/z_3$ is the mean time spent in solution. Let $p_1(x,t)$ be the condition probability density in which the protein disassociates from the DNA at time t without finding the target, given that it is at position x along the DNA at time $t = 0$:

$$p_1(x,t) = z_1 e^{-z_1 t} P(x,t), \tag{3.331}$$

where $\tau_1 = 1/z_1$ is the mean time of sliding phase, and $P(x,t)$ is the survival probability that the protein starts at x has not found the target at time t. Therefore,

$$f(x,t) = -\frac{dP(x,t)}{dt} \tag{3.332}$$

is the PDF of first-passage time associated with diffusion along the DNA strand. Let $q_1(x,t)$ be the PDF that the protein starting at x finds the target at time t:

$$q_1(x,t) = e^{-z_1 t} f(x,t). \tag{3.333}$$

Suppose that in a given trial, a protein starting at x executes $n - 1$ excursions before finding the target with $t_1, \cdots, t_n$ the residence times on DNA and $\tau_1, \cdots, \tau_{n-1}$ the excursion times. Based on the quantities defined previously, the probability density for such a sequence of events with

$$t = \sum_{i=1}^{n} t_i + \sum_{i=1}^{n-1} \tau_i \tag{3.334}$$

is

$$P_n(x, \{t_i, \tau_i\}) = q_1(t_n) p_3(\tau_{n-1}) p_1(t_{n-1}) \cdots p_1(t_2) p_3(\tau_1) p_1(x, t_1), \tag{3.335}$$

where

$$p_1(t) = \langle p_1(x,t) \rangle, \quad q_1(t) = \langle q_1(x,t) \rangle,$$

$$\langle g(x,t) \rangle \equiv \frac{1}{L+M} \int_{-M}^{L} g(x,t)dx,$$

(3.336)

for arbitrary function g. Let $F(x,t)$ be the PDF of first-passage time of finding the target. It can be calculated by summing over all possible numbers of excursions and intervals of time. Thus we have

$$F(t) = \langle F(x,t) \rangle$$

$$= \sum_{n=1}^{\infty} \int_0^{\infty} \cdots \int_0^{\infty} dt_1 \cdots dt_n d\tau_1 \cdots d\tau_{n-1} \delta \left(\sum_{i=1}^{n} t_i + \sum_{i=1}^{n-1} \tau_i - t \right)$$

$$\times q_1(t_n) \prod_{i=1}^{n-1} p_3(\tau_i) \prod_{i=1}^{n-1} p_1(t_i).$$

(3.337)

Taking advantage of the technique of Laplace transform, we obtain

$$\hat{F}(\lambda) = \hat{f}(z_1 + \lambda) \left[1 - \frac{1 - \hat{f}(z_1 + \lambda)}{(1 + \lambda/z_1)(1 + \lambda/z_3)} \right]^{-1},$$

(3.338)

where $\hat{f}(\lambda) = \int_0^{\infty} e^{-\lambda t} \langle f(x,t) \rangle dt$. Further, the mean first-passage time is

$$\tau = -\left. \frac{d\hat{F}(\lambda)}{d\lambda} \right|_{\lambda=0} = \frac{1 - \hat{f}(z_1)}{\hat{f}(z_1)} (z_1^{-1} + z_3^{-1}).$$

(3.339)

The quantity $\hat{f}(z_1)$ is the Laplace transform of the averaged $f(x,t)$ with respect to x. If $x < 0$ ($x > 0$), then $f(x,t)$ represents the PDF of first-passage time for a 1D Brownian particle on the interval $[-M,0]$ ($[0,L]$) with a reflecting boundary condition at $x = -M$ ($x = L$) and an absorbing boundary condition at $x = 0$. Based on the discussions of first-passage time in Sec. 3.5, we obtain

$$\hat{f}(\lambda) = \frac{1}{L+M} \sqrt{\frac{D_1}{z_1}} \left[\tanh(L\sqrt{z_1/D_1}) + \tanh(M\sqrt{z_1/D_1}) \right].$$

(3.340)

Substituting Eq. (3.340) into Eq. (3.339) yields

$$\tau = \left[\frac{(L+M)/\sqrt{\tau_1 D_1}}{\tanh(L/\sqrt{\tau_1 D_1}) + \tanh(M\sqrt{\tau_1/D_1})} - 1 \right] (\tau_1 + \tau_3),$$

(3.341)

which recovers the original result of [Berg *et al.* (1981)]. It also recovers Eq. (3.328) when $L/\sqrt{\tau_1 D_1}, M/\sqrt{\tau_1 D_1} \gg 1$.

Chapter 4

Algorithms for the Models Governing Functional Distribution

As described in the previous sections, there are many microscopic models to describe anomalous diffusion, including continuous time random walks, Langevin type equation, Lévy processes, and fractional Brownian motion, etc. Macroscopically, fractional partial differential equations (PDEs) derived from the microscopic models are the most popular and effective models for describing anomalous diffusion. With the deep research of anomalous diffusion, the fractional PDEs governing the functional distribution of particles' trajectories are also developed [Turgeman et al. (2009); Wu et al. (2016)] and the fractional PDEs are named as the fractional Feynman-Kac equations, which cover the forward and the backward fractional Feynman-Kac equations. Usually, the functional is defined as $A = \int_0^t U(x(\tau))d\tau$, where $U(x)$ is a prescribed function depending on specific applications [Agmon (1984); Carmi et al. (2010)], and $x(\tau)$ denotes a trajectory of particle. It is easy to find that fractional substantial derivative in the fractional Feynman-Kac equation is a non-local time-space coupled operator and the equation is with the complex parameters, which bring about many challenges on regularity and numerical analyses and naturally lead to the attention of scholars for the development of numerical methods for fractional Feynman-Kac equations. So far, there have been some works on solving fractional Feynman-Kac equation numerically [Chen and Deng (2018); Deng et al. (2015); Deng and Zhang (2017); Deng et al. (2018a); Nie et al. (2019)]. Among them, [Chen and Deng (2015)] provides numerical approximation about fractional substantial derivative; [Deng et al. (2015)] solves forward and backward fractional Feynman-Kac equations numerically by finite difference and finite element methods; [Deng and Zhang (2017)] provides the H^1 error estimate for the backward fractional Feynman-Kac equation with the assumptions $U(x) > 0$ and $Re(\rho) > 0$, where $Re(\rho)$ means the real part of ρ; [Deng *et al.* (2018a)] provides an efficient time-stepping method to solve the forward fractional Feynman-Kac equation and performs error analysis in the measure norm.

In this chapter, we give some numerical schemes for solving the fractional Feynman-Kac equations. In Sec. 4.1, we introduce the numerical methods for solving the backward fractional Feynman-Kac equations, in which we first provide the $L1$ and convolution quadrature methods to approximate the fractional substantial derivative, and then we give the equivalent form of the backward fractional Feynman-Kac equation for reaction and diffusion processes, at last, we provide the finite difference and finite element methods to solve the backward fractional Feynman-Kac equation. In Sec. 4.2, we introduce the time discretization for the forward fractional Feynman-Kac equation and give the complete error analysis. Throughout, C denotes a generic positive constant, whose value may differ at different occurrences.

4.1 Numerical Schemes for Backward Fractional Feynman-Kac Equations

In this section, we first give some basic approximations to the (tempered) fractional substantial derivatives, which will be used to develop the numerical schemes for the backward fractional Feynman-Kac equations. And then we provide the equivalent form of backward fractional Feynman-Kac equations. Finally we apply these approximations to solve backward fractional Feynman-Kac equations with different physical backgrounds and the detailed theoretical analyses are also presented.

4.1.1 *Basic Approximations to the Fractional Substantial Derivatives*

In this subsection, we provide some approximation formulae to (tempered) fractional substantial derivatives which will be used to construct numerical scheme in the subsequent subsections. At first, some notations are introduced as follows. Let time step size $\tau = T/N$, $N \in \mathbb{N}^+$, $t_i = i\tau$, $i = 0, 1, \ldots, N$ and $0 = t_0 < t_1 < \cdots < t_N = T$, where T is the fixed final time.

For any $\alpha > 0$, $\mu \geq 0$, the tempered Riemann-Liouville fractional integral [Li *et al.* (2019)] is defined to be

$$
{}_0I_t^{\alpha,\mu}v(t) = e^{-\mu t}\,{}_0I_t^{\alpha}\left(e^{\mu t}v(t)\right) = \frac{1}{\Gamma(\alpha)}\int_0^t (t-s)^{\alpha-1}e^{-\mu(t-s)}v(s)ds, \quad (4.1)
$$

where ${}_0I_t^{\alpha}$ denotes the Riemann-Liouville fractional integral [Podlubny (1999)], i.e.,

$$
{}_0I_t^{\alpha}v(t) = \frac{1}{\Gamma(\alpha)}\int_0^t (t-s)^{\alpha-1}v(s)ds. \quad (4.2)
$$

And for $n-1 < \alpha < n$, $n \in \mathbb{N}^+$, $\mu \geq 0$, the tempered Riemann-Liouville fractional derivative and tempered Caputo fractional derivative [Li *et al.* (2019)] are respectively defined by

$$
\begin{aligned}
D_t^{\alpha,\mu}v(t) &= e^{-\mu t}D_t^{\alpha}\left(e^{\mu t}v(t)\right)\\
&= \frac{e^{-\mu t}}{\Gamma(n-\alpha)}\frac{d^n}{dt^n}\int_0^t \frac{e^{\mu s}v(s)}{(t-s)^{\alpha-n+1}}ds
\end{aligned} \quad (4.3)
$$

and

$$
\begin{aligned}
{}^C D_t^{\alpha,\mu}v(t) &= e^{-\mu t}\,{}^C D_t^{\alpha}\left(e^{\mu t}v(t)\right)\\
&= \frac{e^{-\mu t}}{\Gamma(n-\alpha)}\int_0^t (t-s)^{n-\alpha-1}\frac{d^n\left(e^{\mu s}v(s)\right)}{ds^n}ds,
\end{aligned} \quad (4.4)
$$

where D_t^{α} and ${}^C D_t^{\alpha}$ denote the Riemann-Liouville fractional derivative and Caputo fractional derivative [Podlubny (1999)] respectively defined by

$$
\begin{aligned}
D_t^{\alpha}v(t) &= \frac{1}{\Gamma(n-\alpha)}\frac{d^n}{dt^n}\int_0^t \frac{v(s)}{(t-s)^{\alpha-n+1}}ds,\\
{}^C D_t^{\alpha}v(t) &= \frac{1}{\Gamma(n-\alpha)}\int_0^t (t-s)^{n-\alpha-1}\frac{d^n v(s)}{ds^n}ds.
\end{aligned} \quad (4.5)
$$

Then we present the definitions of Riemann-Liouville fractional substantial derivative and Caputo fractional substantial derivative with $0 < \alpha < 1$ [Li *et al.* (2019)],

$$
\begin{aligned}
\mathcal{D}_t^{\alpha}v(x,t) &= \frac{1}{\Gamma(1-\alpha)}\left(\frac{\partial}{\partial t}+\rho U(x)\right)\int_0^t \frac{e^{-(t-s)\rho U(x)}}{(t-s)^{\alpha}}v(x,s)ds\\
&= \frac{1}{\Gamma(1-\alpha)}e^{-t\rho U(x)}\frac{\partial}{\partial t}\int_0^t \frac{e^{s\rho U(x)}}{(t-s)^{\alpha}}v(x,s)ds,
\end{aligned} \quad (4.6)
$$

$$
{}^C\mathcal{D}_t^{\alpha}v(x,t) = \frac{1}{\Gamma(1-\alpha)}e^{-t\rho U(x)}\int_0^t \frac{1}{(t-s)^{\alpha}}\frac{\partial}{\partial s}\left(e^{s\rho U(x)}v(x,s)\right)ds,
$$

where ρ is a complex number and $U(x)$ is a prescribed function. By Eq. (4.5), the Riemann-Liouville fractional substantial derivative and Caputo fractional substantial derivative can be rewritten as

$$\mathcal{D}_t^\alpha v(x,t) = e^{-t\rho U(x)} D_t^\alpha \left(e^{t\rho U(x)} v(x,t) \right), \tag{4.7}$$

and

$$^C\mathcal{D}_t^\alpha v(x,t) = e^{-t\rho U(x)} \, ^C D_t^\alpha \left(e^{t\rho U(x)} v(x,t) \right). \tag{4.8}$$

For the detailed definitions and properties of the fractional integrals and derivatives, one can refer to Appendix A.4.

From the above definitions of the fractional substantial derivatives, one can note that there are many challenges to approximate them since they are non-local, singular and time-space coupled operators. The two popular discretization schemes to them are: $L1$ method and convolution quadrature method; see [Chen and Deng (2014); Deng and Zhang (2017); Nie *et al.* (2019)] for more details. Now, we introduce these approximations.

4.1.1.1   *Approximating fractional substantial derivatives by L1 method*

Let's start with a simple idea: we use product integral rule to approximate it [Zhang and Deng (2017)]. By the definition of Caputo fractional substantial derivative with $0 < \alpha < 1$, we rewrite it as

$$^C\mathcal{D}_t^\alpha v(x,t)|_{t=t_n} = \frac{e^{-t_n\rho U(x)}}{\Gamma(1-\alpha)} \sum_{j=0}^{n-1} \int_{t_j}^{t_{j+1}} \frac{1}{(t_n-s)^\alpha} \frac{\partial}{\partial s}(e^{s\rho U(x)}v(x,s))ds. \tag{4.9}$$

Using

$$\bar{\partial}_s(e^{s\rho U(x)}v(x,s)) = \frac{e^{t_{j+1}\rho U(x)}v(x,t_{j+1}) - e^{t_j\rho U(x)}v(x,t_j)}{\tau} \tag{4.10}$$

to approximate $\frac{\partial}{\partial s}(e^{s\rho U(x)}v(x,s))$ when $s \in (t_j, t_{j+1})$, one can get the following approximation, i.e., L1 scheme

$$^C\mathcal{D}_t^\alpha v(x,t)|_{t=t_n} \approx \frac{e^{-t_n\rho U(x)}}{\Gamma(1-\alpha)} \sum_{j=0}^{n-1} \int_{t_j}^{t_{j+1}} \frac{1}{(t_n-s)^\alpha} \bar{\partial}_s(e^{s\rho U(x)}v(x,s))ds$$

$$= \sum_{j=0}^{n} w_{n-j}^\alpha(x)v(x,t_j), \tag{4.11}$$

where

$$w_0^\alpha(x) = \frac{1}{\Gamma(1-\alpha)} \int_{t_{n-1}}^{t_n} (t_n - s)^{-\alpha} \tau^{-1} ds,$$

$$w_n^\alpha(x) = -\frac{1}{\Gamma(1-\alpha)} e^{-\rho U(x) t_n} \int_{t_0}^{t_1} (t_n - s)^{-\alpha} \tau^{-1} ds, \tag{4.12}$$

and

$$w_j^\alpha(x) = \frac{e^{-\rho U(x) t_j}}{\Gamma(1-\alpha)} \left(\int_{t_{n-j-1}}^{t_{n-j}} (t_n - s)^{-\alpha} \tau^{-1} ds \right.$$
$$\left. - \int_{t_{n-j}}^{t_{n-j+1}} (t_n - s)^{-\alpha} \tau^{-1} ds \right), \quad j = 1, \cdots, n-1. \tag{4.13}$$

4.1.1.2   *Approximating fractional substantial derivatives by convolution quadrature*

In recent years, convolution quadrature introduced in [Lubich (1988a,b, 2004)] has been widely used to approximate time fractional derivatives [Jin *et al.* (2016); Lubich *et al.* (1996)]. Here, we first provide backward difference formulas of order k [Lubich (1988a,b, 2004); Podlubny (1999)], i.e.,

$$\delta_{\tau,k}(\zeta) = \frac{1}{\tau} \sum_{i=1}^{k} \frac{1}{i} (1 - \zeta)^i, \quad k = 1, 2, \cdots, 6. \tag{4.14}$$

Then we discuss the approximations of Riemann-Liouville fractional derivative with $0 < \alpha < 1$. Denote $\hat{v}(\lambda)$ as the Laplace transform of $v(t)$. Using the fact $\widehat{D_t^\alpha v}(\lambda) = \lambda^\alpha \hat{v}(\lambda)$, $\alpha \in (0,1)$ [Podlubny (1999)], one has

$$D_t^\alpha v(t)|_{t=t_n} \approx \frac{1}{\tau^\alpha} \sum_{j=0}^{n} d_{n-j}^{k,\alpha} v(t_j), \tag{4.15}$$

where

$$\sum_{j=0}^{\infty} d_j^{k,\alpha} \zeta^j = (\tau \delta_{\tau,k}(\zeta))^\alpha. \tag{4.16}$$

And the weights $d_i^{1,\alpha}$ have following properties.

Lemma 4.1 ([Chen *et al.* (2009); Podlubny (1999)]). *For $0 < \alpha < 1$, the weights $d_i^{1,\alpha}$ defined in Eq. (4.15) satisfy*

$$d_0^{1,\alpha} = 1; \quad d_i^{1,\alpha} < 0, \ i \geq 1; \quad \sum_{i=0}^{n-1} d_i^{1,\alpha} > 0; \quad \sum_{i=0}^{\infty} d_i^{1,\alpha} = 0; \tag{4.17}$$

and

$$\frac{1}{n^\alpha \Gamma(1-\alpha)} < \sum_{i=0}^{n-1} d_i^{1,\alpha} = -\sum_{i=n}^{\infty} d_i^{1,\alpha} \le \frac{1}{n^\alpha} \qquad \text{for } n \ge 1. \qquad (4.18)$$

Proof. Taylor's expansion gives

$$d_0^{1,\alpha} = 1, \quad d_i^{1,\alpha} = \left(1 - \frac{\alpha+1}{i}\right) d_{i-1}^{1,\alpha}, \quad i = 1, 2, \cdots. \qquad (4.19)$$

Due to $0 < \alpha < 1$, one has

$$d_i^{1,\alpha} < 0, \quad i \ge 1. \qquad (4.20)$$

Using the definitions of $d_i^{1,\alpha}$ and taking $\zeta = 1$ in Eq. (4.16) lead to

$$\sum_{i=0}^{\infty} d_i^{1,\alpha} = 0, \qquad (4.21)$$

which implies $\sum_{i=0}^{n-1} d_i^{1,\alpha} > 0$.

Denote $v_n = -n^\alpha \sum_{i=n}^{\infty} d_i^{1,\alpha} = n^\alpha \sum_{i=0}^{n-1} d_i^{1,\alpha}$, $n \ge 1$. The facts $\sum_{i=0}^{\infty} d_i^{1,\alpha} = 0$ and $\sum_{i=0}^{n} d_i^{1,\alpha} = \sum_{i=0}^{n} \binom{i-\alpha-1}{i} = \binom{n-\alpha}{n}$ give

$$-\sum_{i=n}^{\infty} d_i^{1,\alpha} = \sum_{i=0}^{n-1} d_i^{1,\alpha} = \binom{n-1-\alpha}{n-1}, \qquad (4.22)$$

which leads to

$$\frac{-\sum_{i=n}^{\infty} d_i^{1,\alpha}}{-\sum_{i=n+1}^{\infty} d_i^{1,\alpha}} = \binom{n-1-\alpha}{n-1} \Big/ \binom{n-\alpha}{n} = \frac{n}{n-\alpha} = 1 + \frac{\alpha}{n-\alpha}. \qquad (4.23)$$

For $x \in (-1, 1)$, it holds

$$(1+x)^\alpha = 1 + \alpha x + \frac{\alpha(\alpha-1)}{2!} x^2 + \frac{\alpha(\alpha-1)(\alpha-2)}{3!} x^3 + \cdots. \qquad (4.24)$$

The fact $0 < \alpha < 1$ yields

$$\frac{(n+1)^\alpha}{n^\alpha} = \left(1 + \frac{1}{n}\right)^\alpha = 1 + \alpha\frac{1}{n} + \frac{\alpha(\alpha-1)}{2!}\left(\frac{1}{n}\right)^2$$
$$+ \frac{\alpha(\alpha-1)(\alpha-2)}{3!}\left(\frac{1}{n}\right)^3 + \cdots < 1 + \frac{\alpha}{n}. \qquad (4.25)$$

Thus

$$\frac{-\sum_{i=n}^{\infty} d_i^{1,\alpha}}{-\sum_{i=n+1}^{\infty} d_i^{1,\alpha}} > \frac{(n+1)^\alpha}{n^\alpha}. \qquad (4.26)$$

Furthermore, there is $v_{n+1} < v_n$, i.e.,

$$\sum_{i=0}^{n} d_i^{1,\alpha} < \frac{n^\alpha}{(n+1)^\alpha} \sum_{i=0}^{n-1} d_i^{1,\alpha} \quad \text{for } n \geq 1. \tag{4.27}$$

Since

$$\frac{t^{-\alpha}}{\Gamma(1-\alpha)} = D_t^\alpha (t-0)^0 = \lim_{\tau \to 0, n\tau = t} \tau^{-\alpha} \sum_{i=0}^{n} d_i^{1,\alpha}, \tag{4.28}$$

for $t = 1$, we obtain

$$\frac{1}{n^\alpha \Gamma(1-\alpha)} < \sum_{i=0}^{n-1} d_i^{1,\alpha} = -\sum_{i=n}^{\infty} d_i^{1,\alpha} \quad \text{for} \quad n \geq 1. \tag{4.29}$$

Next we use mathematical induction to prove

$$\sum_{i=0}^{n-1} d_i^{1,\alpha} = -\sum_{i=n}^{\infty} d_i^{1,\alpha} \leq \frac{1}{n^\alpha} \quad \text{for } n \geq 1. \tag{4.30}$$

It is easy to see that Eq. (4.30) holds for $n = 1, 2$. Supposing that

$$\sum_{i=0}^{s-1} d_i^{1,\alpha} = -\sum_{i=s}^{\infty} d_i^{1,\alpha} \leq \frac{1}{s^\alpha}, \quad s = 1, 2, \cdots, n-1, \tag{4.31}$$

and using Eq. (4.27), we have

$$\sum_{i=0}^{n-1} d_i^{1,\alpha} < \frac{(n-1)^\alpha}{n^\alpha} \sum_{i=0}^{n-2} d_i^{1,\alpha} \leq \frac{(n-1)^\alpha}{n^\alpha} \frac{1}{(n-1)^\alpha} = \frac{1}{n^\alpha} \quad \text{for } n \geq 2. \tag{4.32}$$

Thus the desired results are obtained. $\square$

Lemma 4.2 ([Deng *et al.* (2015)]). *Let $R \geq 0$, $\epsilon^j \geq 0$, $j = 0, 1, \cdots, N$ and ϵ^j satisfy*

$$\epsilon^n \leq -\sum_{j=1}^{n-1} d_j^{1,\alpha} \epsilon^{n-j} + R, \quad n \geq 1. \tag{4.33}$$

Then we have the estimates:

(1) when $0 < \alpha < 1$,

$$\epsilon^n \leq \left(\sum_{j=0}^{n-1} d_j^{1,\alpha} \right)^{-1} R \leq n^\alpha \Gamma(1-\alpha) R; \tag{4.34}$$

(2) *when* $\alpha \to 1$,

$$\epsilon^n \leq nR. \tag{4.35}$$

Proof. (1) Case $0 < \alpha < 1$: we prove the following estimate by the mathematical induction,

$$\epsilon^n \leq \left(\sum_{j=0}^{n-1} d_j^{1,\alpha} \right)^{-1} R. \tag{4.36}$$

Taking $n = 1$ in Eq. (4.33), the desired estimate for $n = 1$ holds obviously. Supposing that

$$\epsilon^s \leq \left(\sum_{j=0}^{s-1} d_j^{1,\alpha} \right)^{-1} R, \quad s = 1, 2, \cdots, n-1, \tag{4.37}$$

thus from Eq. (4.33), one has

$$
\begin{aligned}
\epsilon^n &\leq -\sum_{j=1}^{n-1} d_j^{1,\alpha} \epsilon^{n-j} + R \leq -\sum_{j=1}^{n-1} d_j^{1,\alpha} \left(\sum_{i=0}^{n-j-1} d_i^{1,\alpha} \right)^{-1} R + R \\
&\leq -\sum_{j=1}^{n-1} d_j^{1,\alpha} \left(\sum_{i=0}^{n-1} d_i^{1,\alpha} \right)^{-1} R + R \\
&\leq \left(1 - \sum_{j=0}^{n-1} d_j^{1,\alpha} \right) \left(\sum_{i=0}^{n-1} d_i^{1,\alpha} \right)^{-1} R + R \leq \left(\sum_{i=0}^{n-1} d_i^{1,\alpha} \right)^{-1} R.
\end{aligned} \tag{4.38}
$$

Lemma 4.1 gives

$$\epsilon^n \leq \left(\sum_{j=0}^{n-1} d_j^{1,\alpha} \right)^{-1} R \leq n^\alpha \Gamma(1-\alpha) R. \tag{4.39}$$

(2) Now we discuss the case $\alpha \to 1$. Since $\Gamma(1-\alpha) \to \infty$ as $\alpha \to 1$ in the estimate (4.34), we need to provide an estimate of the other form. We prove the following estimate by the mathematical induction:

$$\epsilon^n \leq nR. \tag{4.40}$$

Eq. (4.33) holds obviously for $n = 1$. Suppose that

$$\epsilon^s \leq sR, \quad s = 1, 2, \cdots, n-1. \tag{4.41}$$

Thus according to Eq. (4.33), we have

$$\epsilon^n \leq -\sum_{j=1}^{n-1} d_j^{1,\alpha} \epsilon^{n-j} + R \leq -\sum_{j=1}^{n-1} d_j^{1,\alpha}(n-j)R + R$$

$$\leq -\sum_{j=1}^{n-1} d_j^{1,\alpha}(n-1)R + R \leq (n-1)R + R = nR.$$

(4.42)

$\square$

Now we begin to introduce to how to approximate the fractional substantial derivative by convolution quadrature. The fact $^C D_t^\alpha v(t) = D_t^\alpha v(t) - \frac{v(0)}{t^\alpha \Gamma(1-\alpha)}$ for $0 < \alpha < 1$ and Eq. (4.15) give

$$^C D_t^\alpha v(t)|_{t=t_n} \approx \frac{1}{\tau^\alpha} \sum_{j=0}^{n} d_{n-j}^{k,\alpha} v(t_j) - \frac{v(0)}{t_n^\alpha \Gamma(1-\alpha)}.$$

(4.43)

According to Eq. (4.5), the Caputo fractional substantial derivative can be approximated as

$$^C \mathcal{D}_t^\alpha v(x,t)|_{t=t_n} \approx \frac{1}{\tau^\alpha} \sum_{j=0}^{n} w_{n-j}^{k,\alpha}(x)v(x,t_j) - \frac{e^{-\rho U(x)t_n} v(x,0)}{t_n^\alpha \Gamma(1-\alpha)},$$

(4.44)

where

$$w_j^{k,\alpha}(x) = e^{-\rho U(x)t_j} d_j^{k,\alpha},$$

(4.45)

and $d_j^{k,\alpha}$ are defined in Eq. (4.16). On the other hand, Eq. (4.15) and the fact $^C \mathcal{D}_t^\alpha v(x,t) = D_t^\alpha(v(x,t) - e^{t\rho U(x)}v(x,0))$ for $0 < \alpha < 1$ yield

$$^C \mathcal{D}_t^\alpha v(x,t)|_{t=t_n} \approx \frac{1}{\tau^\alpha} \sum_{j=0}^{n} w_{n-j}^{k,\alpha}(x)(v(x,t_j) - e^{-t_j \rho U(x)} v(x,0)),$$

(4.46)

where $w_j^{k,\alpha}(x)$ are defined in Eq. (4.45).

Besides, using the fact $\widehat{\mathcal{D}_t^\alpha v}(\lambda) = (\lambda + \rho U(x))^\alpha \hat{v}(\lambda)$ for $0 < \alpha < 1$ to get the approximation is also a good way. Thus the Riemann-Liouville fractional substantial derivative can be approximated as

$$\mathcal{D}_t^\alpha v(x,t)|_{t=t_n} \approx \frac{1}{\tau^\alpha} \sum_{j=0}^{n} w_{n-j}^{k,\alpha}(x)v(x,t_j), \ 0 < \alpha < 1,$$

(4.47)

and the approximation of Caputo fractional substantial derivative is

$$^C \mathcal{D}_t^\alpha v(x,t)|_{t=t_n} \approx \frac{1}{\tau^\alpha} \sum_{j=0}^{n} w_{n-j}^{k,\alpha}(x)v(x,t_j)$$

$$- \frac{e^{-\rho U(x)t_n} v(x,0)}{t_n^\alpha \Gamma(1-\alpha)}, \ 0 < \alpha < 1,$$

(4.48)

where

$$\sum_{j=0}^{\infty} w_j^{k,\alpha}(x)\zeta^j = \tau^\alpha(\delta_{\tau,k}(\zeta) + \rho U(x))^\alpha. \qquad (4.49)$$

Obviously, the weights $w_j^{k,\alpha}(x)$ used in Eq. (4.48) are not equal to the ones in Eq. (4.44), but both of the formulas can approximate $^C\mathcal{D}_t^\alpha$ effectively.

4.1.2 *Equivalent Form of Backward Fractional Feynman-Kac Equation for Reaction and Diffusion Processes*

So far, there have been many numerical methods to solve backward fractional Feynman-Kac equations, the main idea of which is to get equivalent form of the equations by separating the fractional substantial derivative from diffusion operator, being helpful in constructing numerical scheme and doing a complete theoretical analysis. Here, we provide the derivation of equivalent form of backward fractional Feynman-Kac equation for reaction and diffusion processes [Hou and Deng (2018)],

$$\frac{\partial}{\partial t} G_{x_0}(\rho, t) = K\mathcal{D}_t^{1-\alpha,\mu,r} \mathfrak{L} G_{x_0}(\rho, t)$$
$$+ (\mu^\alpha \mathcal{D}_t^{1-\alpha,\mu,r} - \mu) \left(G_{x_0}(\rho, t) - e^{-\rho U(x_0)t} e^{r(x_0)t} \right) \quad (4.50)$$
$$+ (r(x_0) - \rho U(x_0)) G_{x_0}(\rho, t)$$

in a finite domain Ω, $0 < \alpha < 1$, $0 < t \leq T$ with initial and boundary conditions

$$G_{x_0}(\rho, 0) = G_0(x_0, \rho), \qquad x_0 \in \Omega, \qquad (4.51)$$

and

$$G_{x_0}(\rho, t) = 0, \qquad x_0 \in \mathbb{D}, \qquad 0 < t \leq T. \qquad (4.52)$$

Here, $G_{x_0}(\rho, t)$ is the solution of Eq. (4.50) and it is the Fourier transform of $G_{x_0}(A, t)$, being the probability density function of the functional A at time t with the initial position x_0, where A is the functional associated with some specified function $U(x_0)$, i.e., $A = \int_0^t U(x_0(s))ds$ and $x_0(t)$ is a trajectory of particle starting at x_0, and ρ is the Fourier pair of A; K denotes a positive constant and $\mu \geq 0$; $\mathfrak{L}$ means a diffusion operator, such as Laplace operator, (tempered) fractional Laplacian operator and so on; $\mathbb{D}$ is a specific boundary/domain depending on $\mathfrak{L}$; $\mathcal{D}_t^{1-\alpha,\mu,r}$ is defined by

$$\mathcal{D}_t^{\alpha,\mu,r} G_{x_0}(\rho, t) = \frac{1}{\Gamma(1-\alpha)} \left(\frac{\partial}{\partial t} + \mu - r(x_0) + \rho U(x_0) \right)$$
$$\times \int_0^t \frac{e^{-(t-s)(\mu-r(x_0)+\rho U(x_0))}}{(t-s)^\alpha} G_{x_0}(\rho, s)ds, \qquad (4.53)$$

where $r(x_0)$ stands for the reaction rate and satisfies $\sup_{x_0 \in \bar{\Omega}} r(x_0) \leq 0$. And it is easy to see that when $r(x_0) = 0$ and $\mu = 0$, we have $\mathcal{D}_t^{1-\alpha,\mu,r} = \mathcal{D}_t^{1-\alpha}$.

Simple calculations and the definition of the Riemann-Liouville fractional derivative lead to

$$\mathcal{D}_t^{1-\alpha,\mu,r} G_{x_0}(\rho,t) = e^{-t(\mu-r(x_0)+\rho U(x_0))}$$
$$\times D_t^{1-\alpha}\left(e^{t(\mu-r(x_0)+\rho U(x_0))} G_{x_0}(\rho,t)\right). \tag{4.54}$$

Under the assumption that $G_{x_0}(\rho,t)$ is sufficiently regular, multiplying $e^{t(\mu-r(x_0)+\rho U(x_0))}$ and applying $_0 I_t^{1-\alpha}$ on both sides of Eq. (4.50), then it becomes

$$^C D_t^\alpha \left(e^{t(\mu-r(x_0)+\rho U(x_0))} G_{x_0}(\rho,t)\right)$$
$$= K e^{t(\mu-r(x_0)+\rho U(x_0))} \mathfrak{L} G_{x_0}(\rho,t) + \mu \, _0 I_t^{1-\alpha} e^{\mu t} \tag{4.55}$$
$$+ \mu^\alpha e^{(\mu-r(x_0)+\rho U(x_0))t} G_{x_0}(\rho,t) - \mu^\alpha e^{\mu t}.$$

Furthermore, multiplying $e^{-t(\mu-r(x_0)+\rho U(x_0))}$ on both sides of Eq. (4.55) leads to the equivalent form

$$^C\mathcal{D}_t^{\alpha,\mu,r} G_{x_0}(\rho,t) = K \mathfrak{L} G_{x_0}(\rho,t) + \mu e^{-t(\mu-r(x_0)+\rho U(x_0))} \, _0 I_t^{1-\alpha} e^{\mu t}$$
$$+ \mu^\alpha G_{x_0}(\rho,t) - \mu^\alpha e^{t(r(x_0)-\rho U(x_0))}, \tag{4.56}$$

where

$$^C\mathcal{D}_t^{\alpha,\mu,r} G_{x_0}(\rho,t)$$
$$= e^{-t(\mu-r(x_0)+\rho U(x_0))} \, ^C D_t^\alpha \left(e^{t(\mu-r(x_0)+\rho U(x_0))} G_{x_0}(\rho,t)\right). \tag{4.57}$$

Similar to the above derivation of the equivalent form for Eq. (4.50), we can also get the equivalent forms for the backward fractional Feynman-Kac equations under different physical backgrounds [Chen and Deng (2015, 2018); Deng et al. (2015); Deng and Zhang (2017)].

4.1.3 Finite Difference Method for Backward Fractional Feynman-Kac Equation

There are many discussions about finite difference scheme for backward fractional Feynman-Kac equations and the related theoretical analyses [Chen and Deng (2018); Deng et al. (2015); Deng and Zhang (2017); Nie et al. (2019); Zhang and Deng (2017)]. In this subsection, we use finite difference method to discretize backward fractional Feynman-Kac equation presented in [Carmi and Barkai (2011); Carmi *et al.* (2010); Turgeman *et al.* (2009)],

$$\frac{\partial}{\partial t}G_{x_0}(\rho,t) = -K\mathcal{D}_t^{1-\alpha}\mathfrak{L}G_{x_0}(\rho,t)$$

$$- \rho U(x_0)G_{x_0}(\rho,t), \qquad x_0 \in \Omega,\ 0 < t \leq T \tag{4.58}$$

with given initial condition and boundary condition, where $\Omega = (0,l)$ and $\mathfrak{L}$ is a diffusion operator, such as negative Laplace operator; T is a fixed terminal time; $0 < \alpha < 1$; $K > 0$ is a constant; ρ is a complex number; $U(x_0)$ is a prescribed function; the definition of $G_{x_0}(\rho,t)$ is described in Eq. (4.50); $\mathcal{D}_t^{1-\alpha}$ is Riemann-Liouville fractional substantial derivative defined in Eq. (4.6).

4.1.3.1 *Preliminaries and numerical scheme*

For convenience, we take $\mathfrak{L} = -\Delta$ in Eq. (4.58), i.e.,

$$\begin{cases} \dfrac{\partial}{\partial t}G_{x_0}(\rho,t) = K\,\mathcal{D}_t^{1-\alpha}\Delta G_{x_0}(\rho,t) \\[2mm] \qquad\qquad - \rho U(x_0)G_{x_0}(\rho,t), & x_0 \in \Omega,\ 0 < t \leq T, \\[2mm] G_{x_0}(\rho,0) = G_0(x_0,\rho), & x_0 \in \Omega, \\[2mm] G_{x_0}(\rho,t) = 0, & x_0 \in \partial\Omega,\ 0 < t \leq T. \end{cases} \tag{4.59}$$

According to the skill introduced in Sec. 4.1.2, we can reformulate Eq. (4.59) with the Caputo fractional substantial derivative, i.e.,

$$\begin{cases} {}^C\!\mathcal{D}_t^{\alpha}G_{x_0}(\rho,t) = K\Delta G_{x_0}(\rho,t), & x_0 \in \Omega,\ 0 < t \leq T, \\[2mm] G_{x_0}(\rho,0) = G_0(x_0,\rho), & x_0 \in \Omega, \\[2mm] G_{x_0}(\rho,t) = 0, & x_0 \in \partial\Omega,\ 0 < t \leq T. \end{cases} \tag{4.60}$$

Remark 4.1. If we take $\mathfrak{L} = (-\Delta)^s$ or $\mathfrak{L} = -(\Delta + \mu)^s$, $s \in (0,1)$, the boundary condition of Eq. (4.59) should be modified [Deng *et al.* (2018b)], i.e.,

$$\begin{cases} \dfrac{\partial}{\partial t}G_{x_0}(\rho,t) = -K\mathcal{D}_t^{1-\alpha}\mathfrak{L}G_{x_0}(\rho,t) \\[2mm] \qquad\qquad - \rho U(x_0)G_{x_0}(\rho,t), & x_0 \in \Omega,\ 0 < t \leq T, \\[2mm] G_{x_0}(\rho,0) = G_0(x_0,\rho), & x_0 \in \Omega, \\[2mm] G_{x_0}(\rho,t) = 0, & x_0 \in \Omega^c,\ 0 < t \leq T, \end{cases} \tag{4.61}$$

where Ω^c means the complementary set of Ω with respect to $\mathbb{R}^n$ and the definitions of $(-\Delta)^s$ and $-(\Delta + \mu)^s$ can refer to (A.40) and (A.41).

Let the mesh size $h = l/M$ and time step size $\tau = T/N$, N, $M \in \mathbb{N}^+$. Denote mesh points $x_{0,i} = ih$ for $i = 0, 1, \cdots, M$, and $t_n = n\tau$, $n = 0, 1, \cdots, N$. Introduce $G(x_0, t) = G_{x_0}(\rho, t)$, G_i^n as the numerical approximation to $G(x_{0,i}, t_n)$ and $U_i = U(x_{0,i})$. Here we use second order central difference formula for Laplace operator, i.e.,

$$\Delta G(x_0, t)|_{(x_{0,i}, t)} = \frac{G(x_{0,i-1}, t) - 2G(x_{0,i}, t) + G(x_{0,i+1}, t)}{h^2} + \mathcal{O}(h^2),$$
(4.62)

and apply Eq. (4.44) to discretize Caputo fractional substantial derivative, that is

$$
\begin{aligned}
&{}^C\mathcal{D}_t^\alpha G(x_0, t)|_{(x_0, t_n)} \\
&= \frac{1}{\tau^\alpha} \sum_{j=0}^n w_{n-j}^{k,\alpha}(x_0) \left(G(x_0, t_j) - e^{-\rho U(x_0) t_j} G(x_0, 0) \right) + \mathcal{O}(\tau^k)
\end{aligned}
$$
(4.63)

for $k = 1, 2, 3, 4, 5$.

Remark 4.2. As for taking $\mathcal{L} = (-\Delta)^s$ or $\mathcal{L} = -(\Delta + \mu)^s$, $s \in (0, 1)$, the detailed discretization of $\mathcal{L}$ can be referred to Appendix A.5.

Thus we have

$$
\begin{aligned}
&\frac{1}{\tau^\alpha} \sum_{j=0}^n w_{i,n-j}^{k,\alpha} \left(G(x_{0,i}, t_j) - e^{-\rho U_i t_j} G(x_{0,i}, 0) \right) \\
&= K \frac{G(x_{0,i-1}, t_n) - 2G(x_{0,i}, t_n) + G(x_{0,i+1}, t_n)}{h^2} + r_i^n,
\end{aligned}
$$
(4.64)

where

$$|r_i^n| \le C_P(\tau^k + h^2) \quad and \quad w_{i,j}^{k,\alpha} = w_j^{k,\alpha}(x_{0,i}).$$
(4.65)

Here C_P is a constant independent of τ and h. Multiplying τ^α on both sides of Eq. (4.64) yields

$$
\begin{aligned}
&\sum_{j=1}^n w_{i,n-j}^{k,\alpha} \left(G(x_{0,i}, t_j) - e^{-t_j \rho U_i} G(x_{0,i}, 0) \right) \\
&= K\tau^\alpha \frac{G(x_{0,i-1}, t_n) - 2G(x_{0,i}, t_n) + G(x_{0,i+1}, t_n)}{h^2} + R_i^n,
\end{aligned}
$$
(4.66)

where

$$|R_i^n| \le C_P \tau^\alpha (\tau^k + h^2), \quad k = 1, 2, 3, 4, 5.$$
(4.67)

Thus the numerical scheme can be written as

$$G_i^n - \frac{K\tau^\alpha}{h^2}(G_{i+1}^n - 2G_i^n + G_{i-1}^n)$$

$$= \sum_{j=0}^{n-1} w_{i,j}^{k,\alpha} e^{-t_{n-j}\rho U_i} G_i^0 - \sum_{j=1}^{n-1} w_{i,j}^{k,\alpha} G_i^{n-j}, \quad n \geq 1 \tag{4.68}$$

with $i = 1, 2, \cdots, M - 1$ and $k = 1, 2, 3, 4, 5$. It's worthwhile to noting that the second term on the right-hand side of Eq. (4.68) automatically vanishes when $n = 1$.

4.1.3.2   *Stability and convergence for the first order scheme*

In this subsection, we mainly discuss the unconditional stability and convergence of the first order scheme in discrete l^2 norm and l^∞ norm under the assumptions that ρ is a real number and $0 \leq \rho U_i \leq \eta$ for some fixed constant η, i.e., taking $k = 1$ in Eq. (4.68), we have

$$G_i^n - \frac{K\tau^\alpha}{h^2}(G_{i+1}^n - 2G_i^n + G_{i-1}^n)$$

$$= \sum_{j=0}^{n-1} w_{i,j}^{1,\alpha} e^{-t_{n-j}\rho U_i} G_i^0 - \sum_{j=1}^{n-1} w_{i,j}^{1,\alpha} G_i^{n-j}, \quad n \geq 1. \tag{4.69}$$

Here we first introduce some grid functions. Define $\mathbf{u}^n = \{u_i^n | 0 \leq i \leq M, \ n \geq 0\}$ and $\mathbf{v}^n = \{v_i^n | 0 \leq i \leq M, \ n \geq 0\}$, $M \in \mathbb{N}^+$. Introduce the discrete l^2 scalar product and discrete l^2 norm in one-dimensional domain as

$$(\mathbf{u}^n, \mathbf{v}^n) = h \sum_{i=1}^{M-1} u_i^n v_i^n, \quad \|\mathbf{u}^n\| = (\mathbf{u}^n, \mathbf{u}^n)^{1/2}, \tag{4.70}$$

and l^∞ norm as

$$\|\mathbf{u}^n\|_\infty = \max_{0 \leq i \leq M} |u_i^n|. \tag{4.71}$$

Denote

$$(u_i^n)_x = (u_{i+1}^n - u_i^n)/h, \quad (u_i^n)_{\bar{x}} = (u_i^n - u_{i-1}^n)/h;$$

$$(\mathbf{u}^n, \mathbf{v}^n] = h \sum_{i=1}^{M} u_i^n v_i^n, \quad \|\mathbf{u}^n\| = (\mathbf{u}^n, \mathbf{u}^n]^{1/2}; \tag{4.72}$$

in particular, if $\mathbf{u}_0^n = 0$ and $\mathbf{u}_M^n = 0$, there hold

$$(\mathbf{u}^n, ((\mathbf{v}^n)_{\bar{x}})_x) = -((\mathbf{u}^n)_{\bar{x}}, (\mathbf{v}^n)_{\bar{x}}] \text{ and } \|\mathbf{u}^n\|^2 \leq \frac{l^2}{8}\|(\mathbf{u}^n)_{\bar{x}}\|^2, \tag{4.73}$$

where $(\mathbf{u}^n)_x = ((\mathbf{u}_0^n)_x, (\mathbf{u}_1^n)_x, \cdots, (\mathbf{u}_{M-1}^n)_x, 0)$, $(\mathbf{u}^n)_{\bar{x}} = (0, (\mathbf{u}_1^n)_{\bar{x}}, \cdots,$ $(\mathbf{u}_{M-1}^n)_{\bar{x}}, (\mathbf{u}_M^n)_{\bar{x}})$ and l means length of domain Ω.

Then we begin to prove the stability of our scheme (4.69).

Theorem 4.1. *When $0 \leq \rho U_i \leq \eta$, the difference scheme (4.69) is unconditionally stable.*

Proof. Let $\ddot{G}_i^n$ be the approximate solution of G_i^n. Taking $\epsilon_i^n = \ddot{G}_i^n - G_i^n, i = 1, 2, \cdots, M - 1$ and according to Eq. (4.69), one has the following perturbation equation

$$\epsilon_i^n - \frac{K\tau^\alpha}{h^2}(\epsilon_{i+1}^n - 2\epsilon_i^n + \epsilon_{i-1}^n) = \sum_{j=0}^{n-1} w_{i,j}^{1,\alpha} e^{-t_{n-j}\rho U_i}\epsilon_i^0 - \sum_{j=1}^{n-1} w_{i,j}^{1,\alpha}\epsilon_i^{n-j}. \quad (4.74)$$

The fact $w_{i,j}^{1,\alpha} = w_j^{1,\alpha}(x_{0,i})$ and Eq. (4.45) give

$$\epsilon_i^n - \frac{K\tau^\alpha}{h^2}(\epsilon_{i+1}^n - 2\epsilon_i^n + \epsilon_{i-1}^n) = \sum_{j=0}^{n-1} d_j^{1,\alpha} e^{-t_n\rho U_i}\epsilon_i^0 - \sum_{j=1}^{n-1} d_j^{1,\alpha} e^{-t_j\rho U_i}\epsilon_i^{n-j}$$

$$(4.75)$$

with $\epsilon_0^n = \epsilon_M^n = 0$. Multiplying $h\epsilon_i^n$ and summing i from 1 to $M - 1$, one has

$$h\sum_{i=1}^{M-1} (\epsilon_i^n)^2 - h\frac{K\tau^\alpha}{h^2} \sum_{i=1}^{M-1} (\epsilon_{i+1}^n - 2\epsilon_i^n + \epsilon_{i-1}^n)\epsilon_i^n$$

$$= h\sum_{i=1}^{M-1}\sum_{j=0}^{n-1} d_j^{1,\alpha} e^{-t_n\rho U_i}\epsilon_i^0\epsilon_i^n - h\sum_{i=1}^{M-1}\sum_{j=1}^{n-1} d_j^{1,\alpha} e^{-t_j\rho U_i}\epsilon_i^{n-j}\epsilon_i^n. \quad (4.76)$$

Simple calculations and the definition of $\|\cdot\|$ lead to

$$h\sum_{i=1}^{M-1} (\epsilon_i^n)^2 = \|\epsilon^n\|^2 \quad (4.77)$$

and

$$-h\frac{K\tau^\alpha}{h^2} \sum_{i=1}^{M-1} (\epsilon_{i+1}^n - 2\epsilon_i^n + \epsilon_{i-1}^n)\epsilon_i^n$$

$$= -K\tau^\alpha h \sum_{i=1}^{M-1} ((\epsilon_i^n)_{\bar{x}})_x\epsilon_i^n = -K\tau^\alpha(\epsilon^n, ((\epsilon^n)_{\bar{x}})_x] \quad (4.78)$$

$$= K\tau^\alpha((\epsilon^n)_{\bar{x}}, (\epsilon^n)_{\bar{x}}] = K\tau^\alpha\|(\epsilon^n)_{\bar{x}}]\|^2 \geq \frac{8K\tau^\alpha}{l^2}\|\epsilon^n\|^2 \geq 0.$$

Using $e^{-n\tau\rho U_i} \in [e^{-\eta T}, 1]$, Eq. (4.45) and Lemma 4.1, we obtain

$$d_j^{1,\alpha} \leq w_{i,j}^{1,\alpha} = e^{-j\tau\rho U_i}d_j^{1,\alpha} \leq e^{-\eta T}d_j^{1,\alpha} < 0, \quad j \geq 1, \quad (4.79)$$

and

$$0 < e^{-t_n \rho U_i} \sum_{j=0}^{n-1} d_j^{1,\alpha} \leq \sum_{j=0}^{n-1} d_j^{1,\alpha}, \quad n \geq 1. \tag{4.80}$$

Thus there hold

$$h \sum_{i=1}^{M-1} \sum_{j=0}^{n-1} d_j^{1,\alpha} e^{-t_n \rho U_i} \epsilon_i^0 \epsilon_i^n \leq h \sum_{i=1}^{M-1} \sum_{j=0}^{n-1} d_j^{1,\alpha} e^{-t_n \rho U_i} \frac{(\epsilon_i^0)^2 + (\epsilon_i^n)^2}{2}$$

$$\leq \frac{1}{2} \sum_{j=0}^{n-1} d_j^{1,\alpha} (\|\epsilon^0\|^2 + \|\epsilon^n\|^2), \tag{4.81}$$

and

$$-h \sum_{i=1}^{M-1} \sum_{j=1}^{n-1} d_j^{1,\alpha} e^{-t_j \rho U_i} \epsilon_i^{n-j} \epsilon_i^n \leq -h \sum_{i=1}^{M-1} \sum_{j=1}^{n-1} d_j^{1,\alpha} e^{-t_j \rho U_i} \frac{(\epsilon_i^{n-j})^2 + (\epsilon_i^n)^2}{2}$$

$$\leq -\frac{1}{2} \sum_{j=1}^{n-1} d_j^{1,\alpha} (\|\epsilon^{n-j}\|^2 + \|\epsilon^n\|^2). \tag{4.82}$$

Hence

$$\|\epsilon^n\|^2 \leq \frac{1}{2} \sum_{j=0}^{n-1} d_j^{1,\alpha} (\|\epsilon^0\|^2 + \|\epsilon^n\|^2) - \frac{1}{2} \sum_{j=1}^{n-1} d_j^{1,\alpha} (\|\epsilon^{n-j}\|^2 + \|\epsilon^n\|^2)$$

$$\leq \left(1 + \frac{1}{2} \sum_{j=1}^{n-1} d_j^{1,\alpha}\right) \|\epsilon^0\|^2 - \frac{1}{2} \sum_{j=1}^{n-1} d_j^{1,\alpha} \|\epsilon^{n-j}\|^2. \tag{4.83}$$

Next, we prove that $\|\epsilon^n\|^2 \leq \|\epsilon^0\|^2$ by mathematical induction. For $n = 1$, Eq. (4.83) holds obviously. Supposing

$$\|\epsilon^s\|^2 \leq \|\epsilon^0\|^2 \quad \text{for} \quad s = 1, 2, \cdots, n-1, \tag{4.84}$$

and using Eq. (4.83), one has

$$\|\epsilon^n\|^2 \leq \left(1 + \frac{1}{2} \sum_{i=1}^{n-1} d_i^{1,\alpha}\right) \|\epsilon^0\|^2 - \frac{1}{2} \sum_{i=1}^{n-1} d_i^{1,\alpha} \|\epsilon^{n-i}\|^2$$

$$\leq \left(1 + \frac{1}{2} \sum_{i=1}^{n-1} d_i^{1,\alpha}\right) \|\epsilon^0\|^2 - \frac{1}{2} \sum_{i=1}^{n-1} d_i^{1,\alpha} \|\epsilon^0\|^2 = \|\epsilon^0\|^2. \tag{4.85}$$

So the proof is complete.   □

In the following, we provide the error analysis in l^2 norm.

Theorem 4.2. *Let G_i^n be approximate solution of $G(x_{0,i}, t_n)$ computed by the difference scheme (4.69) with assumption $0 \leq \rho U_i \leq \eta$. Then*

$$\|G(x_{0,i}, t_n) - G_i^n\| \leq C_P \Gamma(1-\alpha) l^{1/2} T^\alpha (\tau + h^2), \quad 0 < \alpha < 1, \quad (4.86)$$

where C_P is defined in Eq. (4.65) and $(x_{0,i}, t_n) \in (0,l) \times (0,T]$, $i = 1, 2, \cdots, M-1$; $n = 1, 2, \cdots, N$.

Proof. Denote $\varepsilon_i^n = G(x_{0,i}, t_n) - G_i^n$. Subtracting Eq. (4.66) from Eq. (4.69) and using $\varepsilon_i^0 = 0$ lead to

$$\varepsilon_i^n - \frac{K\tau^\alpha}{h^2} \left(\varepsilon_{i+1}^n - 2\varepsilon_i^n + \varepsilon_{i-1}^n \right) = - \sum_{j=1}^{n-1} e^{-j\tau \rho U_i} d_j^{1,\alpha} \varepsilon_i^{n-j} + R_i^n, \quad n \geq 1, \quad (4.87)$$

where R_i^n is defined by Eq. (4.67). Multiplying $h\varepsilon_i^n$ and summing up for i from 1 to $M-1$, it yields

$$h \sum_{i=1}^{M-1} (\varepsilon_i^n)^2 - \tau^\alpha K h \sum_{i=1}^{M-1} \frac{\varepsilon_{i+1}^n - 2\varepsilon_i^n + \varepsilon_{i-1}^n}{h^2} \varepsilon_i^n$$

$$= -h \sum_{i=1}^{M-1} \sum_{j=1}^{n-1} e^{-\rho U_i j\tau} d_j^{1,\alpha} \varepsilon_i^{n-j} \varepsilon_i^n + h \sum_{i=1}^{M-1} R_i^n \varepsilon_i^n. \quad (4.88)$$

According to the proof of Theorem 4.1, one has

$$h \sum_{i=1}^{M-1} (\varepsilon_i^n)^2 = \|\varepsilon^n\|^2;$$

$$-\tau^\alpha K h \sum_{i=1}^{M-1} \frac{\varepsilon_{i+1}^n - 2\varepsilon_i^n + \varepsilon_{i-1}^n}{h^2} \varepsilon_i^n \geq 0; \quad (4.89)$$

$$-h \sum_{i=1}^{M-1} \sum_{j=1}^{n-1} e^{-\rho U_i j\tau} d_j^{1,\alpha} \varepsilon_i^{n-j} \varepsilon_i^n \leq -\frac{1}{2} \sum_{j=1}^{n-1} d_j^{1,\alpha} \left(\|\varepsilon^{n-j}\|^2 + \|\varepsilon^n\|^2 \right).$$

From Lemma 4.1, we have

$$h \sum_{i=1}^{M-1} |R_i^n \varepsilon_i^n| = \tau^\alpha h \sum_{i=1}^{M-1} |r_i^n \varepsilon_i^n|$$

$$\leq \tau^\alpha h \sum_{i=1}^{M-1} \left[\frac{\tau^\alpha}{2 \sum_{j=0}^{n-1} d_j^{1,\alpha}} (r_i^n)^2 + \frac{\sum_{j=0}^{n-1} d_j^{1,\alpha}}{2\tau^\alpha} (\varepsilon_i^n)^2 \right]$$

$$= \frac{\tau^{2\alpha} h}{2 \sum_{j=0}^{n-1} d_j^{1,\alpha}} \sum_{i=1}^{M-1} (r_i^n)^2 + \frac{1}{2} \sum_{j=0}^{n-1} d_j^{1,\alpha} \|\varepsilon^n\|^2$$

$$\leq \frac{\tau^{2\alpha} n^\alpha \Gamma(1-\alpha)}{2} h(M-1) C_P^2 \left(\tau + h^2\right)^2 + \frac{1}{2} \sum_{j=0}^{n-1} d_j^{1,\alpha} \|\varepsilon^n\|^2$$

$$\leq \frac{T^\alpha \Gamma(1-\alpha) l C_P^2}{2} \tau^\alpha \left(\tau + h^2\right)^2 + \frac{1}{2} \sum_{j=0}^{n-1} d_j^{1,\alpha} \|\varepsilon^n\|^2$$

$$= \frac{C_1}{2} \tau^\alpha \left(\tau + h^2\right)^2 + \frac{1}{2} \sum_{j=0}^{n-1} d_j^{1,\alpha} \|\varepsilon^n\|^2,$$

$$(4.90)$$

where

$$C_1 = l C_P^2 \Gamma(1-\alpha) T^\alpha. \tag{4.91}$$

Thus

$$\|\varepsilon^n\|^2 \leq -\frac{1}{2} \sum_{j=1}^{n-1} d_j^{1,\alpha} \left(\|\varepsilon^{n-j}\|^2 + \|\varepsilon^n\|^2 \right)$$

$$+ \frac{C_1}{2} \tau^\alpha \left(\tau + h^2\right)^2 + \frac{1}{2} \sum_{j=0}^{n-1} d_j^{1,\alpha} \|\varepsilon^n\|^2 \tag{4.92}$$

$$= -\frac{1}{2} \sum_{j=1}^{n-1} d_j^{1,\alpha} \|\varepsilon^{n-j}\|^2 + \frac{C_1}{2} \tau^\alpha \left(\tau + h^2\right)^2 + \frac{1}{2} \|\varepsilon^n\|^2,$$

that is

$$\|\varepsilon^n\|^2 \leq -\sum_{j=1}^{n-1} d_j^{1,\alpha} \|\varepsilon^{n-j}\|^2 + C_1 \tau^\alpha \left(\tau + h^2\right)^2. \tag{4.93}$$

By Lemma 4.2, we obtain

$$\|\varepsilon^n\|^2 \leq n^\alpha \Gamma(1-\alpha) C_1 \tau^\alpha \left(\tau + h^2\right)^2 = l C_P^2 \Gamma(1-\alpha) T^\alpha \Gamma(1-\alpha) T^\alpha \left(\tau + h^2\right)^2.$$

$$(4.94)$$

Therefore,

$$\|G(x_{0,i}, t_n) - G_i^n\| = \|\varepsilon_i^n\| \leq C_P \Gamma(1-\alpha) l^{1/2} T^\alpha \left(\tau + h^2\right). \tag{4.95}$$

$\square$

Remark 4.3. The proof for l^∞ stability of the difference scheme (4.69) is similar to the one of Theorem 4.2.

Remark 4.4. When we take $\mathcal{L}$ as other positive diffusion operators, such as $(-\Delta)^s$ or $-(\Delta + \mu)^s$, $s \in (0,1)$, the l^2 and l^∞ stabilities can be got similarly.

Finally we present the convergence results in l^∞ norm.

Theorem 4.3. *Let G_i^n be the approximate solution of $G(x_{0,i}, t_n)$ computed by the difference scheme (4.69) with the assumption $0 \le \rho U_i \le \eta$. Then the error estimates are*

$$\|G(x_{0,i}, t_n) - G_i^n\|_\infty \le C_P \Gamma(1 - \alpha) T^\alpha (\tau + h^2) \quad \text{for } 0 < \alpha < 1; \quad (4.96)$$

and

$$\|G(x_{0,i}, t_n) - G_i^n\|_\infty \le C_P T \tau^{\alpha-1}(\tau + h^2) \quad \text{for } \alpha \to 1, \quad (4.97)$$

where C_P is defined by Eq. (4.65) and $(x_{0,i}, t_n) \in (0, l) \times (0, T]$, $i = 1, 2, \cdots, M - 1$; $n = 1, 2, \cdots, N$.

Proof. Let $G(x_{0,i}, t_n)$ be the exact solution of Eq. (4.59) at the mesh point $(x_{0,i}, t_n)$ and denote $\varepsilon_i^n = G(x_{0,i}, t_n) - G_i^n$, $\varepsilon^n = [\varepsilon_0^n, \varepsilon_1^n, \cdots, \varepsilon_M^n]$. Subtracting Eq. (4.69) from Eq. (4.66) and using $\varepsilon_i^0 = 0$, we obtain

$$\varepsilon_i^n - \frac{K\tau^\alpha}{h^2}\left(\varepsilon_{i+1}^n - 2\varepsilon_i^n + \varepsilon_{i-1}^n\right) = -\sum_{k=1}^{n-1} w_{i,k}^{1,\alpha} \varepsilon_i^{n-k} + R_i^n, \quad n \ge 1. \quad (4.98)$$

Assume that

$$|\varepsilon_{i_0}^n| = \|\varepsilon^n\|_\infty = \max_{0 \le i \le M} |\varepsilon_i^n|, \quad (4.99)$$

and $R_{max} = \max\limits_{0 \le i \le M,\, 0 \le n \le N} |R_i^n|$. Then the following estimates hold

$$
\begin{aligned}
\|\varepsilon^n\|_\infty = |\varepsilon_{i_0}^n| &\le |\varepsilon_{i_0}^n| - \frac{K\tau^\alpha}{h^2}\left(|\varepsilon_{i_0+1}^n| - 2|\varepsilon_{i_0}^n| + |\varepsilon_{i_0-1}^n|\right) \\
&= \left(1 + 2\frac{K\tau^\alpha}{h^2}\right)|\varepsilon_{i_0}^n| - \frac{K\tau^\alpha}{h^2}\left(|\varepsilon_{i_0+1}^n| + |\varepsilon_{i_0-1}^n|\right) \\
&\le \left|\varepsilon_{i_0}^n - \frac{K\tau^\alpha}{h^2}\left(\varepsilon_{i_0+1}^n - 2\varepsilon_{i_0}^n + \varepsilon_{i_0-1}^n\right)\right| \\
&= \left|-\sum_{j=1}^{n-1} w_{i_0,j}^{1,\alpha} \varepsilon_{i_0}^{n-j} + R_{i_0}^n\right| \\
&\le -\sum_{j=1}^{n-1} w_{i_0,j}^{1,\alpha} \|\varepsilon^{n-j}\|_\infty + R_{max}.
\end{aligned}
\quad (4.100)
$$

Thus

$$\|\varepsilon^n\|_\infty \leq -\sum_{j=1}^{n-1} d_j^{1,\alpha} \|\varepsilon^{n-j}\|_\infty + R_{max}. \tag{4.101}$$

Hence, Lemma 4.2 leads to

$$\|G(x_{0,i}, t_n) - G_i^n\|_\infty \leq C_P \Gamma(1 - \alpha) T^\alpha (\tau + h^2) \quad \text{for } 0 < \alpha < 1; \tag{4.102}$$

and

$$\|G(x_{0,i}, t_n) - G_i^n\|_\infty \leq C_P T \tau^{\alpha-1}(\tau + h^2) \quad \text{for } \alpha \to 1. \tag{4.103}$$

□

Remark 4.5. When ρ is a complex number, i.e., $\rho = a + b\mathbf{i}$, and similar to the proof of Theorem 4.3 but with $0 \leq aU_i \leq \eta$, the same results on numerical stability and convergence can be obtained.

Remark 4.6. When we take $\mathfrak{L}$ as other positive diffusion operators, such as $(-\Delta)^s$ or $-(\Delta + \mu)^s$, $s \in (0, 1)$, the l^∞ error can also be got similarly.

4.1.3.3 *Numerical experiments*

Consider the backward fractional Feynman-Kac equation (4.59) with a finite domain $\Omega = (0, 1)$ and a finite time $T = 1$. Here we choose $K = 0.5$, $U(x_0) = x_0$, $\rho = 1 + \mathbf{i}$, the forcing function

$$f(x_0, t) = \frac{\Gamma(4 + \alpha)}{\Gamma(4)} e^{-\rho x_0 t} t^3 \sin(\pi x_0) - K e^{-\rho x_0 t}(t^{3+\alpha} + 1)(\rho^2 t^2 \sin(\pi x_0)$$
$$- 2\pi \rho t \cos(\pi x_0) - \pi^2 \sin(\pi x_0)), \tag{4.104}$$

and the initial condition $G_0(x_0, \rho) = \sin(\pi x_0)$. Thus Eq. (4.59) has the exact solution

$$G_{x_0}(\rho, t) = e^{-\rho x_0 t}(t^{3+\alpha} + 1) \sin(\pi x_0). \tag{4.105}$$

Tables 4.1 and 4.2 show that the algorithms with $k = 1$ and 4 have the global truncation errors $\mathcal{O}(\tau + h^2)$ and $\mathcal{O}(\tau^4 + h^2)$ at time $T = 1$ respectively.

Table 4.1: The maximum errors and convergence orders for Eq. (4.68), i.e., $k = 1$, when $U(x_0) = x_0$, $\rho = 1 + \mathbf{i}$, $K = 0.5$, $\tau = h^2$.

h	$\alpha = 0.1$		$\alpha = 0.5$		$\alpha = 0.9$	
	Error	Rate	Error	Rate	Error	Rate
1/10	1.156E-02		1.195e-02		1.356E-02	
1/20	2.918E-03	1.9864	3.017E-03	1.9861	3.422E-03	1.9867
1/40	7.312E-04	1.9966	7.562E-04	1.9962	8.587E-04	1.9946
1/80	1.829E-04	1.9991	1.893E-04	1.9981	2.148E-04	1.9990

Table 4.2: The maximum errors and convergence orders for Eq. (4.68) with $k = 4$, when $U(x_0) = x_0$, $\rho = 1 + \mathbf{i}$, $K = 0.5$, $h = \tau^2$.

τ	$\alpha = 0.1$		$\alpha = 0.5$		$\alpha = 0.9$	
	Error	Rate	Error	Rate	Error	Rate
1/10	1.156E-04		1.099E-04		1.049E-04	
1/20	7.228E-06	3.9998	6.869E-06	3.9998	6.484E-06	4.0154
1/40	4.517E-07	4.0000	4.293E-07	4.0003	4.041E-07	4.0042
1/80	2.778E-08	4.0233	2.657E-08	4.0143	2.551E-08	3.9855

4.1.4 Finite Element Scheme for Backward Fractional Feynman-Kac Equation

Besides finite difference scheme, finite element scheme is also an effective numerical scheme for backward fractional Feynman-Kac equation. In this subsection, we discuss the application of finite element on backward time tempered fractional Feynman-Kac equation [Wu *et al.* (2016)],

$$
\begin{cases}
\dfrac{\partial}{\partial t} G_{x_0}(\rho, t) = K \mathcal{D}_t^{1-\alpha, \mu} \Delta G_{x_0}(\rho, t) - \rho U(x_0) G_{x_0}(\rho, t) \\
\qquad + (\mu^\alpha \mathcal{D}_t^{1-\alpha,\mu} - \mu)(G_{x_0}(\rho,t) - e^{-t\rho U(x_0)}), \ x_0 \in \Omega, \ 0 < t \le T, \\
G_{x_0}(\rho, 0) = G_0(x_0, \rho), \qquad\qquad\qquad\qquad\quad x_0 \in \Omega, \\
G_{x_0}(\rho,t)|_{x_0=a} = \psi_l(\rho,t), \quad G_{x_0}(\rho,t)|_{x_0=b} = \psi_r(\rho,t), \qquad 0 < t \le T,
\end{cases}
$$
$$(4.106)$$

where $\Omega = (a,b)$; K is a positive constant; $0 < \alpha < 1$; $\mu \ge 0$; Δ means the Laplace operator; ρ is a complex number; $U(x_0)$ is a prescribed function; $\psi_l(\rho,t)$ and $\psi_r(\rho,t)$ are two given functions; the meaning of $G_{x_0}(\rho,t)$ can refer to Eq. (4.50); $\mathcal{D}_t^{\alpha,\mu}$ denotes the tempered Riemann-Liouville fractional substantial derivative and the definition is

$$
\mathcal{D}_t^{\alpha,\mu} G_{x_0}(\rho, t) = \frac{1}{\Gamma(1-\alpha)} \left(\mu + \rho U(x_0) + \frac{\partial}{\partial t} \right)
$$
$$
\times \int_0^t \frac{e^{-(t-s)(\mu+\rho U(x_0))}}{(t-s)^\alpha} G_{x_0}(\rho, s) ds.
$$

(4.107)

Moreover, if $\mu = 0$, Eq. (4.106) reduces to the standard backward fractional Feynman-Kac equation (4.59).

4.1.4.1 *Preliminaries and numerical scheme*

For convenience, we can get the equivalent form of Eq. (4.106) by the skill introduced in Sec. 4.1.2, i.e.,

$$
\begin{cases}
{}^C\mathcal{D}_t^{\alpha,\mu} G_{x_0}(\rho, t) - \mu^\alpha G_{x_0}(\rho, t) \\
\qquad = K\Delta G_{x_0}(\rho, t) + f(x_0, \rho, t), \quad x_0 \in \Omega,\ 0 < t \le T, \\
G_{x_0}(\rho, 0) = G_0(x_0, \rho), \qquad\qquad\qquad\qquad x_0 \in \Omega, \\
G_{x_0}(\rho, t)|_{x_0=a} = \psi_l(\rho, t), \quad G_{x_0}(\rho, t)|_{x_0=b} = \psi_r(\rho, t), \qquad 0 < t \le T,
\end{cases}
$$

(4.108)

where

$$
{}^C\mathcal{D}_t^{\alpha,\mu} G_{x_0}(\rho, t) = e^{-t(\mu+\rho U(x_0))}\ {}^C D_t^\alpha \left(e^{t(\mu+\rho U(x_0))} G_{x_0}(\rho, t) \right),
$$

(4.109)

and

$$
f(x_0, \rho, t) = -\mu^\alpha e^{-t\rho U(x_0)} + \mu e^{-t(\mu+\rho U(x_0))}\ {}_0 I_t^{1-\alpha}(e^{\mu t}).
$$

(4.110)

To get an effective numerical scheme for Eq. (4.106), we first homogenize the initial condition and the boundary condition. Introduce $W(x_0, \rho, t)$ as

$$
G_{x_0}(\rho, t) = W(x_0, \rho, t) + G_0(x_0, \rho)e^{-t(\mu+\rho U(x_0))}
$$
$$
+ \frac{x_0 - a}{b - a}(\psi_r(\rho, t)e^{t(\mu+\rho U(b))} - G_0(b, \rho))e^{-t(\mu+\rho U(x_0))}
$$
$$
+ \frac{b - x_0}{b - a}(\psi_l(\rho, t)e^{t(\mu+\rho U(a))} - G_0(a, \rho))e^{-t(\mu+\rho U(x_0))}.
$$

(4.111)

Thus $W(x_0, \rho, t)$ satisfies

$$
W(x_0, \rho, 0) = W(a, \rho, t) = W(b, \rho, t) = 0.
$$

(4.112)

Without loss of generality, we assume that $G_0(x_0, \rho) = \psi_l(\rho, t) = \psi_r(\rho, t) = 0$.

Let the time step size $\tau = T/N$ and $t_n = n\tau$, $n = 1, 2, \cdots, N$, $N \in \mathbb{N}^+$. Set the mesh size $h = \frac{(b-a)}{M}$ and $x_j = a + jh$, $j = 0, 1, \cdots, M$, $M \in \mathbb{N}^+$.

Introduce $\bar{v}$ as the conjugate of function v. Denote $L^p(\Omega)$ as the Lebesgue space and $H^k(\Omega)$ as the Sobolev space, i.e.,

$$\int_\Omega |u|^p dx \leq C, \text{ if } u \in L^p(\Omega);$$
$$D^\alpha u \in L^2(\Omega), \text{ if } u \in H^k(\Omega), \quad (4.113)$$

where D is a weak derivative; see Appendix A.4 for more details; $\alpha \leq k$. Denote $H_0^k(\Omega)$ as the closure of $C_0^\infty(\Omega)$ in $H^k(\Omega)$. The norms of $L^2(\Omega)$ and $H^1(\Omega)$ can be defined by

$$\|u\|_{L^2(\Omega)} = \left(\int_\Omega |u|^2 dx\right)^{1/2}, \quad \|u\|_{H^1(\Omega)} = \|u\|_{L^2(\Omega)} + |u|_{H^1(\Omega)}, \quad (4.114)$$

where $|u|_{H^1(\Omega)}$ is the semi-norm of $H^1(\Omega)$ defined as

$$|u|_{H^1(\Omega)} = \|Du\|_{L^2(\Omega)}. \quad (4.115)$$

As usual, we find the weak solution $G_{x_0}(\cdot,\cdot) \in H_0^1(\Omega)$ of Eq. (4.108) such that for any $v \in H_0^1(\Omega)$, there holds

$$\left({}^C D_t^{\alpha,\mu} G_{x_0}(\rho,t), v\right) - (\mu^\alpha G_{x_0}(\rho,t), v) + K(\nabla G_{x_0}(\rho,t), \nabla v) = (f, v), \quad (4.116)$$

where $(u, v) = \int_\Omega u\bar{v} dx$ denotes the complex inner product in $L^2(\Omega)$. Introduce the finite element space

$$\mathbb{S}_h = \{v \in H_0^1 \cap C(\bar{\Omega}) : v|_{I_m} \in P_q(I_m)\}, \quad (4.117)$$

where $P_q(I_m)$ denotes the space of polynomials of degree no greater than q, $q \in \mathbb{N}^+$ on $I_m = (x_{m-1}, x_m)$, $m = 1, 2, \ldots, M$.

As for the temporal discretization, using $G_0(x_0, \rho) = 0$ and the relationship between Riemann-Liouville fractional derivative and Caputo fractional derivative, we have

$$
{}^C D_t^{\alpha,\mu} G_{x_0}(\rho,t) = e^{-t(\mu+\rho U(x_0))} {}^C D_t^\alpha \left(e^{t(\mu+\rho U(x_0))} G_{x_0}(\rho,t)\right)
$$
$$
= e^{-t(\mu+\rho U(x_0))} D_t^\alpha \left(e^{t(\mu+\rho U(x_0))} G_{x_0}(\rho,t)\right). \quad (4.118)
$$

Similar to the approximation of fractional substantial derivative, using Eq. (4.15) with $k = 1$ leads to

$$
{}^C D_t^{\alpha,\mu} G_{x_0}(\rho,t)|_{t=t_n} = \frac{1}{\tau^\alpha} \sum_{j=0}^n d_j^{1,\alpha} e^{-t_j(\mu+\rho U(x_0))} G_{x_0}(\rho, t_{n-j}) + \mathcal{O}(\tau)
$$
$$
= \frac{1}{\tau^\alpha} \sum_{j=0}^n d_j^{1,\alpha,\mu} e^{-t_j \rho U(x_0)} G_{x_0}(\rho, t_{n-j}) + \mathcal{O}(\tau),
$$

$$\quad (4.119)$$

where

$$d_j^{1,\alpha,\mu} = d_j^{1,\alpha} e^{-t_j \mu}. \tag{4.120}$$

Thus the fully discrete finite element scheme can be described as: for $n = 1, 2, \cdots, N$, find $G_h^n \in \mathbb{S}_h$ such that

$$\left(\frac{1}{\tau^\alpha} \sum_{j=0}^{n} d_j^{1,\alpha,\mu} e^{-t_j \rho U(x_0)} G_h^{n-j}, v \right)$$

$$- (\mu^\alpha G_h^n, v) + K(\nabla G_h^n, \nabla v) = (f^n, v) \quad \forall v \in \mathbb{S}_h \tag{4.121}$$

with $G_h^0 = R_h G_0(x_0, \rho) = 0$, where $f^n = f(x_0, \rho, t_n)$ and $R_h : H_0^1 \to \mathbb{S}_h$ is Ritz projection defined as

$$(\nabla R_h u, \nabla v_h) = (\nabla u, \nabla v_h) \quad \forall v_h \in \mathbb{S}_h. \tag{4.122}$$

For convenience, by the fact $e^{-\alpha\mu\tau} = 1 - \alpha\mu\tau + \mathcal{O}(\tau^2)$, we introduce

$$d_0^{(\alpha,\mu)} = d_0^{1,\alpha,\mu} - e^{-\alpha\mu\tau}(\mu\tau)^\alpha, \quad d_j^{(\alpha,\mu)} = d_j^{1,\alpha,\mu}, \quad j = 1, 2, \cdots, N. \tag{4.123}$$

Thus Eq. (4.121) can be rewritten as

$$\left(\frac{1}{\tau^\alpha} \sum_{j=0}^{n} d_j^{(\alpha,\mu)} e^{-t_j \rho U(x_0)} G_h^{n-j}, v \right) + K(\nabla G_h^n, \nabla v) = (f^n, v) \quad \forall v \in \mathbb{S}_h. \tag{4.124}$$

4.1.4.2 *Stability and error analysis*

In this subsection, we discuss the stability and error analysis with the assumption $Re(\rho U(x_0)) > 0$, where $Re(\lambda)$ means the real part of λ. Firstly, we introduce some lemmas about $d_i^{1,\alpha,\mu}$ and $d_i^{(\alpha,\mu)}$, $i = 0, 1, \cdots, N$.

Lemma 4.3. *For $0 < \alpha < 1$, the weights $d_i^{1,\alpha,\mu}$ defined in Eq. (4.120) satisfy*

$$d_0^{1,\alpha,\mu} = 1; \quad d_i^{1,\alpha,\mu} < 0, \ i \geq 1; \quad \sum_{i=0}^{n-1} d_i^{1,\alpha,\mu} > (1 - e^{-\mu\tau})^\alpha, \ n \geq 0; \tag{4.125}$$

and

$$d_i^{1,\alpha,\mu} = e^{-\mu\tau} \left(1 - \frac{\alpha+1}{i} \right) d_{i-1}^{1,\alpha,\mu}, \quad i \geq 1. \tag{4.126}$$

Proof. This lemma can be proved by the properties of $d_i^{1,\alpha}$ and the definitions of $d_i^{1,\alpha,\mu}$, that is, Lemma 4.1 and Eq. (4.120). $\qquad\square$

Lemma 4.4. *For any* $L \in \mathbb{N}^+$, $0 < \alpha < 1$, *and* $\mathbf{u} = (\mathbf{u}_0, \mathbf{u}_1, \cdots, \mathbf{u}_L)^T$, *it holds*

$$\sum_{n=0}^{L} \sum_{j=0}^{n} d_j^{(\alpha,\mu)} \mathbf{u}_{n-j} \mathbf{u}_n \geq 0. \tag{4.127}$$

Here $\mathbf{u}^T$ *denotes the transpose of* $\mathbf{u}$.

Proof. By the definitions of $d_j^{(\alpha,\mu)}$ and Lemma 4.3, one has

$$d_0^{(\alpha,\mu)} = 1 - e^{-\alpha\mu\tau}(\mu\tau)^\alpha = (e^{-\alpha\mu\tau})((e^{\mu\tau})^\alpha - (\mu\tau)^\alpha) > 0;$$

$$d_j^{(\alpha,\mu)} \leq 0 \quad for \ j \geq 1;$$

$$\sum_{j=0}^{n} d_j^{(\alpha,\mu)} > (1 - e^{-\mu\tau})^\alpha - e^{-\alpha\mu\tau}(\mu\tau)^\alpha$$

$$= e^{-\alpha\mu\tau}((e^{\mu\tau} - 1)^\alpha - (\mu\tau)^\alpha) > 0, \tag{4.128}$$

where the inequality $e^x > 1 + x$, $x > 0$ has been used.

Define a lower triangular Toeplitz matrix $\mathbf{B} = (b_{i,j})_{L+1,L+1}$ with $b_{i,j} = d_{i-j}^{(\alpha,\mu)}$, $j = 0, \cdots, i$. Then matrix $\mathbf{B}$ is row and column diagonally-dominant with $b_{k,k} > 0$. Using the Gerschgorin theorem [Isaacson and Keller (1996)], we obtain

$$\sum_{n=0}^{L} \sum_{j=0}^{n} d_j^{(\alpha,\mu)} \mathbf{u}_{n-j} \mathbf{u}_n = \mathbf{u}^T \mathbf{B} \mathbf{u} = \frac{1}{2} \mathbf{u}^T (\mathbf{B} + \mathbf{B}^T) \mathbf{u} \geq 0. \tag{4.129}$$

The proof is complete. $\qquad\square$

Next we provide the stability and convergence analysis for numerical scheme (4.124) with the assumption $Re(\rho U(x_0)) > 0$.

Theorem 4.4. *The finite element scheme* (4.124) *is unconditionally stable. Assuming that* $G_{x_0}(\rho, \cdot) \in H_0^1(\Omega) \cap H^r(\Omega)$, *then the approximation solution* G_h^n *satisfies*

$$\|\mathbf{E}\| \leq C(\tau + h^{\min(q,r-1)}), \tag{4.130}$$

where $\mathbf{E} = (\|G_{x_0}(\rho, t_1) - G_h^1\|_{H^1(\Omega)}, \cdots, \|G_{x_0}(\rho, t_N) - G_h^N\|_{H^1(\Omega)})$ *with* $G_{x_0}(\rho, t_n)$ *satisfying Eq.* (4.116) *and* $\|\cdot\|$ *means* l^2 *norm defined in Eq.* (4.70).

Proof. Taking $v = G_h^n$ in Eq. (4.124) yields

$$\frac{1}{\tau^\alpha} \sum_{j=0}^{n} d_j^{(\alpha,\mu)} \left(e^{-t_j \rho U(x_0)} G_h^{n-j}, G_h^n \right) + K \left(\nabla G_h^n, \nabla G_h^n \right) = (f^n, G_h^n). \tag{4.131}$$

Noting that $Re(\rho U(x_0)) \geq 0$, we have

$$\left|\left(e^{-t_j \rho U(x_0)} G_h^{n-j}, G_h^n\right)\right| \leq \|G_h^{n-j}\|_{L^2(\Omega)} \|G_h^n\|_{L^2(\Omega)}. \quad (4.132)$$

Therefore

$$\frac{1}{\tau^\alpha} \sum_{j=0}^n d_j^{(\alpha,\mu)} \left\|G_h^{n-j}\right\|_{L^2(\Omega)} \|G_h^n\|_{L^2(\Omega)}$$
$$+ K |G_h^n|_{H^1(\Omega)}^2 \leq \epsilon \|f^n\|_{L^2(\Omega)}^2 + \frac{1}{4\epsilon} \|G_h^n\|_{L^2(\Omega)}^2. \quad (4.133)$$

Summing up Eq. (4.133) for n from 1 to L, then adding $\tau^{1-\alpha} d_0^{(\alpha,\mu)} \|G_h^0\|_{L^2(\Omega)}^2$ on both sides of the obtained result, and using Lemma 4.4, we have

$$K \sum_{n=1}^L \tau |G_h^n|_{H^1(\Omega)}^2 \leq \tau \sum_{n=1}^L \left(\epsilon \|f^n\|_{L^2(\Omega)}^2 + \frac{(b-a)^2}{4\epsilon\pi^2} |G_h^n|_{H^1(\Omega)}^2\right)$$
$$+ \tau^{1-\alpha} d_0^{(\alpha,\mu)} \|G_h^0\|_{L^2(\Omega)}^2. \quad (4.134)$$

Choosing $\epsilon = \frac{(b-a)^2}{2\pi^2 K}$, we have

$$K \sum_{n=1}^L \tau |G_h^n|_{H^1(\Omega)}^2 \leq \frac{\tau(b-a)^2}{\pi^2 K} \sum_{n=1}^L \|f^n\|_{L^2(\Omega)}^2 + 2\tau^{1-\alpha} d_0^{(\alpha,\mu)} \|G_h^0\|_{L^2(\Omega)}^2. \quad (4.135)$$

So, the finite element scheme (4.124) is unconditionally stable.

Now, we provide the error estimate. Decompose the error into two terms, i.e., $G_{x_0}(\rho, t_n) - G_h^n = \eta^n + \theta^n$ with $\eta(t) = G_{x_0}(\rho, t) - R_h G_{x_0}(\rho, t)$, $\eta^n = \eta(t_n)$ and $\theta^n = R_h G_{x_0}(\rho, t_n) - G_h^n$. Then for any $v \in \mathbb{S}_h$, one has

$$\left(\frac{1}{\tau^\alpha} \sum_{j=0}^n d_j^{(\alpha,\mu)} e^{-t_j \rho U(x_0)} \theta^{n-j}, v\right) + K(\nabla \theta^n, \nabla v) = (r_1^n + r_2^n, v), \quad (4.136)$$

where r_1^n satisfying $\|r_1^n\|_{L^2(\Omega)} \leq C_2 \tau$ with $1 \leq n \leq N$ is the time directional error term introduced in Eq. (4.119) and the space projection error

$$\|r_2^n\|_{L^2(\Omega)}$$

$$= \left\| -\frac{1}{\tau^\alpha} \sum_{j=0}^n d_j^{(\alpha,\mu)} e^{-t_j \rho U(x_0)} \eta^{n-j} \right\|_{L^2(\Omega)}$$

$$\leq \frac{1}{\tau^\alpha} \left\| e^{-t_n(\mu+\rho U(x_0))} \sum_{j=0}^n (Q_j^\alpha - Q_{j-1}^\alpha) e^{t_{n-j}(\mu+\rho U(x_0))} \eta^{n-j} \right\|_{L^2(\Omega)}$$

$$+ e^{-\alpha\mu\tau} \mu^\alpha \|\eta^n\|_{L^2(\Omega)}$$

$$\leq \frac{1}{\tau^\alpha} \left\| e^{-t_n(\mu+\rho U(x_0))} \sum_{j=0}^{n-1} Q_{n-1-j}^\alpha \int_{t_j}^{t_{j+1}} \frac{\partial}{\partial t} \left(e^{t(\mu+\rho U(x_0))} \eta(t) \right) dt \right\|_{L^2(\Omega)}$$

$$+ e^{-\alpha\mu\tau} \mu^\alpha \|\eta^n\|_{L^2(\Omega)}$$

$$\leq \frac{1}{\tau^\alpha} \sum_{j=0}^{n-1} Q_{n-1-j}^\alpha \int_{t_j}^{t_{j+1}} \left\| \left(\mu + \rho U(x_0) + \frac{\partial}{\partial t} \right) \eta(t) \right\|_{L^2(\Omega)} dt$$

$$+ e^{-\alpha\mu\tau} \mu^\alpha \|\eta^n\|_{L^2(\Omega)}$$

$$\leq \sum_{j=0}^{n-1} \frac{Q_{n-1-j}^\alpha}{\tau^{\alpha-1}} \left(\left(\mu + \rho \sup_{x_0 \in \Omega} U(x_0) \right) \|\eta(t)\|_{L^\infty([0,T];L^2(\Omega))} \right.$$

$$\left. + \left\| \frac{\partial \eta(t)}{\partial t} \right\|_{L^\infty([0,T];L^2(\Omega))} \right) + \mu^\alpha \|\eta^n\|_{L^2(\Omega)}.$$

$$(4.137)$$

Here $Q_{-1}^\alpha = 0$, $Q_j^\alpha = \sum_{i=0}^j d_i^{1,\alpha} > 0$, and $\eta^0 = 0$ have been used in the third step. Due to

$$(1-\zeta)^\alpha = \sum_{j=0}^\infty d_j^{1,\alpha} \zeta^j = Q_0^\alpha + \sum_{j=1}^\infty (Q_j^\alpha - Q_{j-1}^\alpha)\zeta^j = (1-\zeta) \sum_{j=0}^\infty Q_j^\alpha \zeta^j, \quad (4.138)$$

then Q_j^α are the coefficients of the power series of the function $(1-\zeta)^{\alpha-1}$. The Stirling formula [Lubich *et al.* (1996)] leads to

$$\sum_{j=0}^{n-1} (-1)^j \binom{\alpha-1}{j} = (-1)^{n-1} \binom{\alpha-2}{n-1} = \frac{n^{1-\alpha}}{\Gamma(2-\alpha)} + \mathcal{O}(n^{-\alpha}). \quad (4.139)$$

Hence

$$\tau^{1-\alpha} \sum_{j=0}^{n-1} Q_{n-1-j}^\alpha = \frac{t_n^{1-\alpha}}{\Gamma(2-\alpha)} + \tau\mathcal{O}\left(t_n^{-\alpha}\right) \leq CT^{1-\alpha}. \quad (4.140)$$

Taking $v = \theta^n$ in Eq. (4.136) and using projection properties give

$$K \sum_{n=1}^{N} \tau \, |\theta^n|^2_{H^1(\Omega)} \leq \frac{\tau(b-a)^2}{\pi^2 K} \sum_{n=1}^{N} \|r_1^n + r_2^n\|^2_{L^2(\Omega)} + 2\tau^{1-\alpha} d_0^{(\alpha,\mu)} \|\theta^0\|^2_{L^2(\Omega)}$$

$$\leq C \frac{2(b-a)^2}{\pi^2 K} T \left((b-a)\tau^2 + h^{2\min(q+1,r)} \right).$$
(4.141)

Triangle inequality leads to

$$\|\mathbf{E}\|^2 \leq 2 \sum_{n=1}^{N} \tau \, |\theta^n|^2_{H^1(\Omega)} + 2 \sum_{n=1}^{N} \tau \, |\eta^n|^2_{H^1(\Omega)}$$

$$\leq C \frac{4(b-a)^2}{\pi^2 K^2} T \left((b-a)\tau^2 + h^{2\min(q+1,r)} \right)$$
(4.142)

$$+ 2C\|G_{x_0}(\rho,t)\|^2_{L^\infty([0,T],H^r(\Omega))} T h^{2\min(q,r-1)}$$

$$\leq C(\tau + h^{\min(q,r-1)}).$$

$\square$

4.1.4.3 *Numerical experiments*

In this subsection, we present several examples to verify the theoretical results. Here, we choose $\Omega = (0,1)$, $K = 1$ and $U(x_0) = x_0$.

Example 4.1. In this example, we add a source term in Eq. (4.108), such that its exact solution is

$$G_{x_0}(\rho,t) = (t^2 + 1)e^{-(\mu+\rho x_0)t}(\sin(x_0) - x_0 \sin(1)).$$
(4.143)

The corresponding source term and the initial and boundary conditions can be derived from the exact solution.

Set

$$G_{x_0}(\rho,t) = W(x_0,\rho,t) + e^{-(\mu+\rho x_0)t}(\sin(x_0) - x_0 \sin(1)).$$
(4.144)

Then $W(x_0,\rho,t)$ solves

$$e^{-(\mu+\rho x_0)t} \, {}^C D_t^\alpha \left(e^{(\mu+\rho x_0)t} W(x_0,\rho,t) \right) - \mu^\alpha W(x_0,\rho,t)$$

$$= K\Delta W(x_0,\rho,t) + g(x_0,\rho,t),$$
(4.145)

where

$$g(x_0,\rho,t) = e^{-(\mu+\rho x_0)t} (\sin(x_0) - x_0 \sin(1)) \left(\frac{2}{\Gamma(3-\alpha)} t^{2-\alpha} - \mu^\alpha t^2 \right)$$

$$+ Kt^2 e^{-(\mu+\rho x_0)t} \left[(\rho^2 t^2 - 1) \sin(x_0) - 2\rho t \cos(x_0) \right.$$

$$\left. + \left(2\rho t - \rho^2 t^2 x_0 \right) \sin(1) \right].$$
(4.146)

We choose $q = 1$ for the finite element discretization. We take $\mu = 3$, $T = 1$, $(\alpha, \rho) = (0.3, 1 + \mathbf{i})$, $(0.5, 5)$, $(0.8, 10\mathbf{i})$ and $\tau = 1/2^4, 1/2^5, 1/2^6$ respectively. Table 4.3 shows the errors $\|\mathbf{E}\|$ (defined in Theorem 4.4) and convergence rates which well confirm the theoretical analysis.

Table 4.3: Numerical results of the finite element scheme for Example 4.1 with $\mu = 3$ and $T = 1$.

		h	$1/2^4$	$1/2^5$	$1/2^6$
$\alpha = 0.3, \rho = 1+\mathbf{i}$	$\tau = h$		3.134E-04	1.556E-04	7.748E-05
		Rate		1.0106	1.0055
	$\tau = h^2$		3.084E-04	1.541E-04	7.704E-05
		Rate		1.0008	1.0002
$\alpha = 0.5, \rho = 5$	$\tau = h$		2.962E-04	1.470E-04	7.315E-05
		Rate		1.0106	1.0069
	$\tau = h^2$		2.882E-04	1.445E-04	7.228E-05
		Rate		0.9963	0.9991
$\alpha = 0.8, \rho = 10\mathbf{i}$	$\tau = h$		2.89E-03	1.42E-03	7.02E-04
		Rate		1.0248	1.0151
	$\tau = h^2$		2.76E-03	1.38E-03	6.93E-04
		Rate		0.9947	0.9987

Example 4.2. In this example, we take

$$G_0(x_0, \rho) = 0, \quad \psi_l(\rho, t) = t, \quad \psi_r(\rho, t) = e^{-t} - 1. \tag{4.147}$$

Then one can let

$$G_{x_0}(\rho, t) = W(x_0, \rho, t) + \left[(e^{-t} - 1)e^{(\mu + \rho)t} x_0 + t e^{\mu t}(1 - x_0) \right] e^{-(\mu + \rho x_0)t} \tag{4.148}$$

and

$g(x_0, \rho, t)$

$$= - \left[x_0 \, {}^C\!D_t^\alpha \left((e^{-t} - 1) e^{(\mu + \rho)t} \right) + (1 - x_0) \, {}^C\!D_t^\alpha \left(t e^{\mu t} \right) \right] e^{-(\mu + \rho x_0)t}$$

$$+ \mu^\alpha \left((e^{-t} - 1) x_0 e^{\rho t} + t(1 - x_0) - 1 \right) e^{-\rho x_0 t} + \mu e^{-(\mu + \rho x_0)t} \, {}_0 I_t^{1-\alpha} \left(e^{\mu t} \right)$$

$$+ K \left[(e^{-t} - 1) \left(\rho^2 t^2 x_0 - 2\rho t \right) e^{\rho t} + t^2 \left(-\rho^2 t x_0 + \rho^2 t + 2\rho \right) \right] e^{-\rho t x_0}. \tag{4.149}$$

Here we take $\rho = 5\mathbf{i}$ and $T = 0.5$ and use the $|\cdot|_{H^1(\Omega)}$ to measure the error. Since the exact solution is unknown, the numerical solutions obtained at $\tau = h = 1/2^{12}$ have been regarded as the exact solution. Table 4.4 shows the error and convergence rates when $(\alpha, \mu) = (0.3, 0)$, $(0.5, 3)$, $(0.8, 5)$ and $\tau = 1/2^4, 1/2^5, 1/2^6$.

Table 4.4: Numerical results of the finite element scheme for Example 4.2 with $\rho = 5\mathbf{i}$ and $T = 0.5$.

	h	$1/2^4$	$1/2^5$	$1/2^6$
	$\tau = h$	6.817E-02	3.409E-02	1.705E-02
		Rate	0.9997	1.0000
$\alpha = 0.3, \mu = 0$	$\tau = h^2$	6.811E-02	3.407E-02	1.704E-02
		Rate	0.9993	0.9998
	$\tau = h$	7.210E-02	3.605E-02	1.802E-02
		Rate	1.0001	1.0002
$\alpha = 0.5, \mu = 3$	$\tau = h^2$	7.164E-02	3.583E-02	1.792E-02
		Rate	0.9994	0.9998
	$\tau = h$	8.02E-02	4.05E-02	2.03E-02
		Rate	0.9879	0.9931
$\alpha = 0.8, \mu = 5$	$\tau = h^2$	7.54E-02	3.77E-02	1.88E-02
		Rate	1.0008	1.0002

4.2 Numerical Schemes for Forward Fractional Feynman-Kac Equation

Due to the non-commutativity of the Riemann-Liouville fractional substantial derivative and the diffusion operators, say, the Laplace operator, i.e. $-\Delta \mathcal{D}_t^{1-\alpha} \neq -\mathcal{D}_t^{1-\alpha}\Delta$, we can't separate the operators $-\Delta$ from $\mathcal{D}_t^{1-\alpha}$ to get the equivalent form of forward fractional Feynman-Kac equation, which causes many troubles on construction of numerical scheme and doing error analysis. As we all know, there are relatively less research works on solving the forward fractional Feynman-Kac equation numerically. In this section, we introduce an efficient time-stepping method to solve the forward fractional Feynman-Kac equation and provide error analysis in the measure norm [Deng *et al.* (2018a)].

For illustrating the main ideas of the scheme used to solve the forward fractional Feynman-Kac equation, here we first introduce the time discretization of the traditional time fractional diffusion equation [Metzler and Klafter (2000b)], i.e.,

$$
\begin{cases}
\dfrac{\partial u(x,t)}{\partial t} - \Delta D_t^{1-\alpha} u(x,t) = 0, & x \in \Omega,\ 0 < t \le T, \\[2mm]
u(x,0) = u_0(x), & x \in \Omega, \\[2mm]
u(x,t) = 0, & x \in \partial\Omega,\ 0 < t \le T,
\end{cases} \tag{4.150}
$$

where $u(x,t)$ means the solution; $0 < \alpha < 1$; Δ denotes the Laplace operator; $D_t^{1-\alpha}$ denotes the Riemann-Liouville fractional derivative which is defined in Eq. (4.5). The idea of solving Eq. (4.150) is the same as the one for following forward time tempered fractional Feynman-Kac equation presented in [Deng *et al.* (2018a)],

$$
\begin{cases}
\mathcal{D}_t^{1,\mu} G(x,\rho,t) - (\mu^\alpha + \Delta)\mathcal{D}_t^{1-\alpha,\mu} G(x,\rho,t) \\[2mm]
\quad = -G(x,\rho,0)(\mu^\alpha \mathcal{D}_t^{1-\alpha,\mu} - \mu)e^{-\rho U(x)t}, & x \in \Omega,\ 0 < t \le T, \\[2mm]
G(x,\rho,0) = G_0(x), & x \in \Omega, \\[2mm]
G(x,\rho,t) = 0, & x \in \partial\Omega,\ 0 < t \le T,
\end{cases} \tag{4.151}
$$

where Δ denotes Laplace operator and $\mathcal{D}_t^{1,\mu} = \frac{\partial}{\partial t} + \mu + \rho U(x)$ is the tempered substantial derivative; $0 < \alpha < 1$ and $\mu \ge 0$; Ω means a bounded domain and $\mathcal{D}_t^{\alpha,\mu}$ is tempered Riemann-Liouville fractional substantial derivative which is defined in Eq. (4.107). The solution $G(x,\rho,t)$, depending on the parameter ρ, represents the characteristic function of the joint probability density function $G(x,A,t)$ of finding a particle at position x and time t with functional value $\int_0^t U(x(\tau))d\tau = A$, i.e., $G(x,\rho,t) = \int_{\mathbb{R}} e^{\rho A} G(x,A,t)dA$.

4.2.1 Time-Stepping Scheme for Traditional Time Fractional Diffusion Equation

Here we use the Laplace transform and convolution quadrature techniques [Lubich (1988a,b)] for numerical analysis of Eq. (4.150). The idea is to consider the Laplace form of Eq. (4.150), namely

$$
(\lambda - \lambda^{1-\alpha}\Delta)\hat{u}(x,\lambda) = u_0(x). \tag{4.152}
$$

Simple calculations lead to

$$
\lambda^{1-\alpha}(\lambda^\alpha - \Delta)\hat{u}(x,\lambda) = u_0(x). \tag{4.153}
$$

Let $t_n = n\tau$, $n = 0,1,\cdots,N$ be a uniform partition of the time interval $[0,T]$, with time step size $\tau = T/N$, $N \in \mathbb{N}^+$. Denote $u_n(x)$ as the approximation of $u(x,t_n)$. By denoting $\zeta = e^{-\lambda\tau}$, we approximate λ, $\hat{u}(\cdot,\lambda)$ and

u_0 in Eq. (4.153) by $\frac{1-\zeta}{\tau}$, $\tau \sum_{n=1}^{\infty} u_n \zeta^n$ and ζu_0, respectively. This gives

$$\delta_{\tau,1}(\zeta)^{1-\alpha}(\delta_{\tau,1}(\zeta)^{\alpha} - \Delta) \sum_{n=1}^{\infty} u_n \zeta^n = \frac{\zeta}{\tau} u_0, \qquad (4.154)$$

where $\delta_{\tau,1}(\zeta)$ is defined in Eq. (4.14). Introduce $\bar{\partial}_{\tau}^{\alpha}$ as an approximation of the Riemann-Liouville fractional derivative by

$$\bar{\partial}_{\tau}^{\alpha} u_n = \frac{1}{\tau^{\alpha}} \sum_{j=1}^{n} d_{n-j}^{1,\alpha} u_j, \qquad (4.155)$$

where $d_j^{1,\alpha}$ is defined in Eq. (4.16). Straightforward calculation leads to

$$\delta_{\tau,1}(\zeta)^{\alpha} \sum_{n=1}^{\infty} u_n \zeta^n = \frac{1}{\tau^{\alpha}} \left(\sum_{j=0}^{\infty} d_j^{1,\alpha} \zeta^j \right) \sum_{n=1}^{\infty} u_n \zeta^n = \sum_{n=1}^{\infty} (\bar{\partial}_{\tau}^{\alpha} u_n) \zeta^n. \quad (4.156)$$

Consequently, expanding Eq. (4.154) into a power series of ζ and considering the coefficients of the power series on both sides, we can obtain the following time-stepping scheme

$$\bar{\partial}_{\tau}^{1-\alpha}(\bar{\partial}_{\tau}^{\alpha} - \Delta) u_n = \begin{cases} \tau^{-1} u_0 & \text{if } n = 1, \\ 0 & \text{if } n \geq 2. \end{cases} \qquad (4.157)$$

The product rule leads to

$$\bar{\partial}_{\tau}^{1-\alpha} \bar{\partial}_{\tau}^{\alpha} u_n = \begin{cases} \tau^{-1} u_1 & \text{if } n = 1, \\ \tau^{-1}(u_n - u_{n-1}) & \text{if } n \geq 2, \end{cases} \qquad (4.158)$$

and the last equation reduces to

$$\begin{aligned} \tau^{-1} u_1 - \bar{\partial}_{\tau}^{1-\alpha} \Delta u_1 &= \tau^{-1} u_0, & \text{if } n = 1, \\ \frac{u_n - u_{n-1}}{\tau} - \bar{\partial}_{\tau}^{1-\alpha} \Delta u_n &= 0, & \text{if } n \geq 2. \end{aligned} \qquad (4.159)$$

Thus the time-stepping scheme of Eq. (4.150) can be rewritten as

$$\frac{u_n - u_{n-1}}{\tau} - \bar{\partial}_{\tau}^{1-\alpha} \Delta u_n = 0, \quad n = 1, 2, \cdots. \qquad (4.160)$$

4.2.2   Time-Stepping Scheme for Time Tempered Fractional Feynman-Kac Equation

In this subsection, we apply the method introduced in Sec. 4.2.1 to the time tempered fractional Feynman-Kac equation (4.151) and obtain regularity estimation about solution $G(x, \rho, t)$.

4.2.2.1 Derivation of the time-stepping scheme

Firstly we introduce some notations and function spaces. For $\kappa > 0$ and $\pi/2 < \theta < \pi$, we define sectors Σ_θ, $\Sigma_{\theta,\kappa}$ and $\Sigma_{\theta,\kappa}^\tau$ in the complex plane $\mathbb{C}$ as

$$\Sigma_\theta = \{\lambda \in \mathbb{C} \setminus \{0\}, |\arg(\lambda)| \leq \theta\},$$
$$\Sigma_{\theta,\kappa} = \{\lambda \in \mathbb{C} : |\lambda| \geq \kappa, |\arg(\lambda)| \leq \theta\},$$
$$\Sigma_{\theta,\kappa}^\tau = \{\lambda \in \mathbb{C} : |\lambda| \geq \kappa, |\arg(\lambda)| \leq \theta, Re(\lambda) \leq \kappa + 1, Im(\lambda) \leq \frac{\pi}{\tau}\},$$

$$(4.161)$$

and the contours $\Gamma_{\theta,\kappa}$ and $\Gamma_{\theta,\kappa}^\tau$ are defined by

$$\Gamma_{\theta,\kappa} = \{\lambda \in \mathbb{C} : |\lambda| = \kappa, |\arg(\lambda)| \leq \theta\} \cup \{\lambda \in \mathbb{C} : \lambda = re^{\pm i\theta}, r \geq \kappa\},$$
$$\Gamma_{\theta,\kappa}^\tau = \{\lambda \in \mathbb{C} : |\lambda| = \kappa, |\arg(\lambda)| \leq \theta\}$$
$$\cup \{\lambda \in \mathbb{C} : \lambda = re^{\pm i\theta}, \frac{\pi}{\tau \sin(\theta)} \geq r \geq \kappa\}.$$

$$(4.162)$$

Here $Re(\lambda)$ and $Im(\lambda)$ mean the real and imaginary parts of λ; $\mathbf{i}$ denotes the imaginary unit and $\mathbf{i}^2 = -1$. Denote $C(\bar{\Omega})$ as the space of continuous function whose norm is

$$\|u\|_{C(\bar{\Omega})} = \sup_{x \in \bar{\Omega}} |u(x)| \quad \text{for } u \in C(\bar{\Omega}).$$ $$(4.163)$$

And the dual space of $C(\bar{\Omega})$ is the space of finite signed measures on Ω, i.e., $\mathbb{M}(\Omega)$, whose norm is

$$\|\phi\|_{\mathbb{M}(\Omega)} = \sup_{f \in C(\bar{\Omega}), \|f\|_{C(\bar{\Omega})} \leq 1} |(f, \phi)| \quad \text{for } \phi \in \mathbb{M}(\Omega).$$ $$(4.164)$$

Then we present the Laplace transform of Eq. (4.151), namely

$$(\lambda + \mu + \rho U(x))\hat{G}(x, \rho, \lambda) - G_0(x)$$
$$- (\mu^\alpha + \Delta)(\lambda + \mu + \rho U(x))^{1-\alpha}\hat{G}(x, \rho, \lambda) \qquad (4.165)$$
$$= -G_0(x)(\mu^\alpha(\lambda + \mu + \rho U(x))^{1-\alpha} - \mu)(\lambda + \rho U(x))^{-1}.$$

Introducing the notations

$$\eta(x, \lambda) = (\lambda + \mu + \rho U(x))^\alpha - \mu^\alpha, \quad \beta(x, \lambda) = \lambda + \mu + \rho U(x) \quad (4.166)$$

with the abbreviations

$$\eta(\lambda) = \eta(\cdot, \lambda), \quad \beta(\lambda) = \beta(\cdot, \lambda), \qquad (4.167)$$

we reformulate Eq. (4.165) as

$$(\eta(\lambda) - \Delta)\beta(\lambda)^{1-\alpha}\hat{G}(x, \rho, \lambda) = G_0(x)\beta(\lambda)^{1-\alpha}\frac{\eta(\lambda)}{\lambda + \rho U(x)}. \qquad (4.168)$$

Thus

$$\hat{G}(x, \rho, \lambda) = \beta(\lambda)^{\alpha-1} (\eta(\lambda) - \Delta)^{-1} \left(\beta(\lambda)^{1-\alpha} \frac{G_0(x)\eta(\lambda)}{\lambda + \rho U(x)} \right). \qquad (4.169)$$

Since the non-commutativity between $(\eta(\lambda) - \Delta)^{-1}$ and $\beta(\lambda)^{1-\alpha}$, the two terms $\beta(\lambda)^{\alpha-1}$ and $\beta(\lambda)^{1-\alpha}$ in the expression above cannot be canceled. Using the inverse Laplace transform leads to

$$G(x, \rho, t) = \frac{1}{2\pi i} \int_{\kappa+1+i\mathbb{R}} e^{\lambda t} \beta(\lambda)^{\alpha-1} (\eta(\lambda) - \Delta)^{-1} \left(\beta(\lambda)^{1-\alpha} \frac{G_0(x)\eta(\lambda)}{\lambda + \rho U(x)} \right) d\lambda. \qquad (4.170)$$

From Proposition 4.1 below, we see that the integrand in Eq. (4.170) is an $\mathrm{M}(\Omega)$-valued analytic function for $\lambda \in \Sigma_{\theta,\kappa}^\tau$ defined in Eq. (4.161). Consequently, deforming the integration path from $\kappa + 1 + i\mathbb{R}$ to $\Gamma_{\theta,\kappa}$, then we have

$$G(x, \rho, t) = \frac{1}{2\pi i} \int_{\Gamma_{\theta,\kappa}} e^{t\lambda} \beta(\lambda)^{\alpha-1} (\eta(\lambda) - \Delta)^{-1} \left(\beta(\lambda)^{1-\alpha} \frac{G_0(x)\eta(\lambda)}{\lambda + \rho U(x)} \right) d\lambda. \qquad (4.171)$$

This integral representation will be used to estimate the error of the numerical solutions.

Proposition 4.1. *By choosing $\theta \in \left(\frac{\pi}{2}, \pi \right)$ sufficiently close to $\frac{\pi}{2}$ and $\kappa > 0$ sufficiently large (depending on the value $\mu + |\rho| \|U(x)\|_{C(\overline{\Omega})}$), we have the following results:*

(1) For all $x \in \Omega$ and $\lambda \in \Sigma_{\theta,\kappa}$, we have $\beta(\lambda) \in \Sigma_{\frac{3\pi}{4}, \frac{\kappa}{2}}$ and $\eta(\lambda) \in \Sigma_{\frac{3\pi}{4}, \frac{\kappa^\alpha}{2}}$, and

$$C|\lambda| \le |\beta(\lambda)| \le C|\lambda|, \quad C|\lambda|^\alpha \le |\eta(\lambda)| \le C|\lambda|^\alpha, \qquad (4.172)$$

where

$$\Sigma_{\theta,\kappa} = \{ \lambda \in \mathbb{C} : |\lambda| \ge \kappa, |\arg(\lambda)| \le \theta \}. \qquad (4.173)$$

So $\beta(\lambda)^{1-\alpha}$, $\beta(\lambda)^{\alpha-1}$, and $\eta(\lambda)$ are all $C(\overline{\Omega})$-valued analytic function of $\lambda \in \Sigma_{\theta,\kappa}$.

(2) The operator $(\eta(\lambda) - \Delta)^{-1} : \mathrm{M}(\Omega) \to \mathrm{M}(\Omega)$ is well-defined, bounded, and analytic with respect to $\lambda \in \Sigma_{\theta,\kappa}$, satisfying

$$\|\Delta(\eta(\lambda) - \Delta)^{-1}\|_{\mathrm{M}(\Omega) \to \mathrm{M}(\Omega)} \le C \quad \text{for all} \ \ \lambda \in \Sigma_{\theta,\kappa}, \qquad (4.174)$$

$$\|(\eta(\lambda) - \Delta)^{-1}\|_{\mathrm{M}(\Omega) \to \mathrm{M}(\Omega)} \le C|\lambda|^{-\alpha} \quad \text{for all} \ \ \lambda \in \Sigma_{\theta,\kappa}. \qquad (4.175)$$

(3) The contour integral (4.171) defines a solution of Eq. (4.151) under the given initial and boundary conditions, with the regularity $G(\cdot, t) \in \mathrm{M}(\Omega)$, $\mathcal{D}_t^{1,\mu} G(\cdot, \rho, t) \in \mathrm{M}(\Omega)$, $\mathcal{D}_t^{1-\alpha} G(\cdot, \rho, t) \in \mathrm{M}(\Omega)$, and $\Delta \mathcal{D}_t^{1-\alpha} G(\cdot, \rho, t) \in \mathrm{M}(\Omega)$ for $t \in (0, T]$. The solution given by Eq. (4.171) is called the mild solution of Eq. (4.151), with each term of Eq. (4.151) well-defined as a measure.

Before we provide the proof of Proposition 4.1, the following lemma is first given.

Lemma 4.5. *Let $L = \mu + |\rho| \|U(x)\|_{C(\bar{\Omega})}$. There exist positive constants $\theta_0 \in \left(\frac{\pi}{2}, \frac{5\pi}{8}\right)$, τ_0, and C_0 such that if $\theta \in \left(\frac{\pi}{2}, \theta_0\right)$ and $\tau \in (0, \tau_0]$, then*

$$- |\arg(\lambda)| - C_0 \tau \leq \arg\left(\frac{1 - e^{-\tau\lambda}}{\tau}\right) \leq |\arg(\lambda)| + C_0 \tau$$

$$\text{if } |\lambda| \neq 0, \quad |\arg(\lambda)| \leq \theta, \quad \text{and} \quad |Im(\lambda)| \leq \frac{\pi}{\tau} + L.$$
$$(4.176)$$

Proof. It is obvious that if $|\lambda| \neq 0$ and $\arg(\lambda) = 0$, then $\arg\left(\frac{1-e^{-\tau\lambda}}{\tau}\right) = 0$.

If $|\lambda| \neq 0$ and $\arg(\lambda) = \varphi \in (0, \theta]$ and $0 \leq Im(\lambda) \leq \pi/\tau + L$, then $\omega = \tau|\lambda| \sin(\varphi) \in (0, \pi + L\tau]$. Here we discuss the following two cases

Case 1: if $\omega \in (0, \pi]$, then $\arg\left(\frac{1-e^{-\tau\lambda}}{\tau}\right) \in [0, \pi)$;

Case 2: if $\omega \in (\pi, \pi + L\tau]$, then there exists a constant C_0 such that $\arg\left(\frac{1-e^{-\tau\lambda}}{\tau}\right) \in [-C_0\tau, 0)$.

For Case 2, we can get the conclusion easily.

In Case 1, if $\omega = \pi$, then $\arg(\frac{1-e^{-\tau\lambda}}{\tau}) = 0$ and Eq. (4.176) holds. If $\omega \in (0, \pi)$, then $\arg\left(\frac{1-e^{-\tau\lambda}}{\tau}\right) \in (0, \pi)$ and we prove $\arg\left(\frac{1-e^{-\tau\lambda}}{\tau}\right) \leq \varphi$ below.

Simple calculations lead to

$$
\begin{aligned}
\cot\left(\arg\left(\frac{1-e^{-\tau\lambda}}{\tau}\right)\right) &= \frac{1 - e^{-\tau|\lambda|\cos(\varphi)} \cos(\tau|\lambda| \sin(\varphi))}{e^{-\tau|\lambda|\cos(\varphi)} \sin(\tau|\lambda| \sin(\varphi))} \\
&= \frac{e^{\tau|\lambda|\cos(\varphi)} - \cos(\tau|\lambda| \sin(\varphi))}{\sin(\tau|\lambda| \sin(\varphi))} \\
&\geq \frac{1 + \tau|\lambda| \cos(\varphi) - \cos(\tau|\lambda| \sin(\varphi))}{\sin(\tau|\lambda| \sin(\varphi))} \\
&= \frac{1 + \omega \cot(\varphi) - \cos(\omega)}{\sin(\omega)},
\end{aligned}
$$
$$(4.177)$$

where we have used Taylor's expansion in the last inequality and set $\omega = \tau|\lambda|\sin(\varphi) \in (0, \pi)$. We shall prove $\cot\left(\arg\left(\frac{1-e^{-\tau\lambda}}{\tau}\right)\right) \geq \cot(\varphi)$ for $\omega \in (0, \pi)$, so that $0 \leq \arg\left(\frac{1-e^{-\tau\lambda}}{\tau}\right) \leq \varphi = \arg(\lambda)$. Finally, we consider the function

$$f(\omega) = 1 + \omega\cot(\varphi) - \cos(\omega) - \sin(\omega)\cot(\varphi), \quad \omega \in [0, \pi] \qquad (4.178)$$

with the fixed φ and variable ω (due to the change of $|\lambda|$). The derivative of f is

$$\begin{aligned} f'(\omega) &= \sin(\omega) + (1 - \cos(\omega))\cot(\varphi) \\ &= 2\sin\left(\frac{\omega}{2}\right)\cos\left(\frac{\omega}{2}\right) + 2\sin^2\left(\frac{\omega}{2}\right)\cot(\varphi) \\ &= 2\sin^2\left(\frac{\omega}{2}\right)\left(\cot\left(\frac{\omega}{2}\right) + \cot(\varphi)\right). \end{aligned} \qquad (4.179)$$

If $\varphi \in (0, \frac{\pi}{2}]$, then $f'(\omega) > 0$ for $\omega \in (0, \pi)$, which yields that the minimum value of f is achieved at $f(0) = 0$. If $\varphi \in (\frac{\pi}{2}, \theta]$, then $f'(\omega) > 0$ for $\omega \in (0, \pi - \varphi)$ and $f'(\omega) < 0$ for $\omega \in (\pi - \varphi, \pi]$, which means that the minimum value of f is achieved at either $f(0) = 0$ or $f(\pi) = 2 + \omega\cot(\varphi)$. In either case, the minimum value of f is achieved at one of the two end points, $\omega = 0$ and $\omega = \pi$ with

$$f(0) = 0 \quad \text{and} \quad f(\pi) = 2 + \pi\cot(\varphi). \qquad (4.180)$$

Choosing $\theta \in (\frac{\pi}{2}, \pi)$ sufficiently close to $\frac{\pi}{2}$, one has $f(\pi) \geq 0$. Consequently, $f(\omega) \geq 0$ for all $\omega \in (0, \pi)$. Thus $\cot\left(\arg\left(\frac{1-e^{-\tau\lambda}}{\tau}\right)\right) \geq \cot(\varphi)$ for all $\omega \in (0, \pi)$, which leads to $\arg\left(\frac{1-e^{-\tau\lambda}}{\tau}\right) \leq \varphi$.

Therefore, we have proved Eq. (4.176) in the case $\arg(\lambda) \in [0, \theta]$. The case $\arg(\lambda) \in [-\theta, 0)$ can be proved in the same way. $\square$

Next, we provide the proof of Proposition 4.1.

Proof. (1) Here, let $\theta \in (\frac{\pi}{2}, \theta_0)$ be a fixed angle. For all $\lambda \in \Sigma_{\theta,\kappa}$ and $x \in \Omega$, we have

$$\begin{aligned} |\arg(\beta(\lambda)) - \arg(\lambda)| &= |\arg(\lambda + \mu + \rho U(x)) - \arg(\lambda)| \\ &\leq \arcsin\left(\frac{|\mu + \rho U(x)|}{|\lambda|}\right) \\ &\leq \arcsin\left(\frac{|\mu| + |\rho|\|U(x)\|_{C(\bar{\Omega})}}{|\lambda|}\right). \end{aligned} \qquad (4.181)$$

When κ is large enough compared with $|\mu| + |\rho| \|U(x)\|_{C(\bar{\Omega})}$, the angle above is smaller than $\frac{\pi}{8}$ and

$$|\lambda + \mu + \rho U(x)| \geq \kappa - |\mu + \rho U(x)| \geq \frac{3\kappa}{4}. \tag{4.182}$$

Thus

$$\lambda + \mu + \rho U(x) \in \Sigma_{\theta + \frac{\pi}{8}, \frac{3\kappa}{4}} \quad \text{and} \quad (\lambda + \mu + \rho U(x))^\alpha \in \Sigma_{\alpha(\theta + \frac{\pi}{8}), (\frac{3\kappa}{4})^\alpha}, \tag{4.183}$$

which leads to $\beta(\lambda) \in \Sigma_{\frac{3\pi}{4}, \frac{\kappa}{2}}$. Analogously, one has

$$| \arg[(\lambda + \mu + \rho U(x))^\alpha - \mu^\alpha] - \arg[(\lambda + \mu + \rho U(x))^\alpha]|$$

$$\leq \arcsin\left(\frac{\mu^\alpha}{|\lambda + \mu + \rho U(x)|^\alpha}\right) \leq \arcsin\left(\frac{\mu^\alpha}{\left(\frac{3\kappa}{4}\right)^\alpha}\right). \tag{4.184}$$

When κ is large enough, the angle above is smaller than $\frac{3(1-\alpha)\pi}{4}$ and $\left(\frac{3\kappa}{4}\right)^\alpha - \mu^\alpha \geq \left(\frac{\kappa}{2}\right)^\alpha$. Consequently, we have

$$\eta(\lambda) = (\lambda + \mu + \rho U(x))^\alpha - \mu^\alpha \in \Sigma_{\alpha(\theta + \frac{\pi}{8}) + \frac{3(1-\alpha)\pi}{4}, (\frac{\kappa}{2})^\alpha} \subset \Sigma_{\frac{3\pi}{4}, \frac{\kappa^\alpha}{2}}. \tag{4.185}$$

Thus Eq. (4.172) is a consequence of the fact that $|\lambda|$ dominates μ and $U(x)$.

(2) Choose a fixed $x_0 \in \Omega$ and note $(\lambda + \mu + \rho U(x_0))^\alpha - \mu^\alpha \in \Sigma_{\frac{3\pi}{4}, \frac{\kappa^\alpha}{2}}$. Hence, the operator

$$((\lambda + \mu + \rho U(x_0))^\alpha - \mu^\alpha - \Delta)^{-1} : C(\bar{\Omega}) \to C(\bar{\Omega}) \cap H_0^1(\Omega) \tag{4.186}$$

is well defined, satisfying the following basic resolvent estimate:

$$\| ((\lambda + \mu + \rho U(x_0))^\alpha - \mu^\alpha - \Delta)^{-1} \|_{C(\bar{\Omega}) \to C(\bar{\Omega})} \leq$$
$$C |(\lambda + \mu + \rho U(x_0))^\alpha - \mu^\alpha|^{-1}, \tag{4.187}$$

which can be got by the analytic semigroup result [Lubich *et al.* (1996)] and resolvent estimate [Arendt *et al.* (2011)]. Reformulating the following equation

$$((\lambda + \mu + \rho U(x))^\alpha - \mu^\alpha - \Delta) \phi = f \tag{4.188}$$

as

$$((\lambda + \mu + \rho U(x_0))^\alpha - \mu^\alpha - \Delta) \phi$$
$$= f + ((\lambda + \mu + \rho U(x_0))^\alpha - (\lambda + \mu + \rho U(x))^\alpha) \phi, \tag{4.189}$$

and according to Eq. (4.187), we obtain

$$|(\lambda + \mu + \rho U(x_0))^\alpha - \mu^\alpha| \, \|\phi\|_{C(\bar{\Omega})}$$
$$= C \|f\|_{C(\bar{\Omega})} + C |(\lambda + \mu + \rho U(x_0))^\alpha - (\lambda + \mu + \rho U(x))^\alpha| \, \|\phi\|_{C(\bar{\Omega})}$$
$$\leq C \|f\|_{C(\bar{\Omega})} + C |U(x_0) - U(x)|^\alpha \|\phi\|_{C(\bar{\Omega})} \leq C \|f\|_{C(\bar{\Omega})} + C \|\phi\|_{C(\bar{\Omega})}. \tag{4.190}$$

Since $|\lambda| \geq \kappa$ and κ can be chosen to be large compared with μ and $|\rho| \|U(x)\|_{C(\bar{\Omega})}$, there holds

$$|(\lambda + \mu + \rho U(x_0))^\alpha - \mu^\alpha| \geq |\lambda|^\alpha - C\mu^\alpha - C|U(x_0)|^\alpha \geq \frac{1}{2}|\lambda|^\alpha. \quad (4.191)$$

The last two inequalities yield $\|\phi\|_{C(\bar{\Omega})} \leq C|\lambda|^{-\alpha}\|f\|_{C(\bar{\Omega})} + C|\lambda|^{-\alpha}\|\phi\|_{C(\bar{\Omega})}$. Again, when $|\kappa|$ is larger than some constant, $|\lambda|$ is sufficiently large so that the second term on the right-hand side can be absorbed by the left-hand side. Thus, we get the desired results. This proves the well-definedness and boundedness of the operator $(\eta(\lambda) - \Delta)^{-1} : C(\bar{\Omega}) \to C(\bar{\Omega})$ for $\lambda \in \Sigma_{\theta,\kappa}$ with

$$\|(\eta(\lambda) - \Delta)^{-1}\|_{C(\bar{\Omega}) \to C(\bar{\Omega})} \leq C|\lambda|^{-\alpha} \qquad \forall \lambda \in \Sigma_{\theta,\kappa}. \quad (4.192)$$

The duality between $\mathbb{M}(\Omega)$ and $C(\bar{\Omega})$ immediately implies the extended map $(\eta(\lambda) - \Delta)^{-1} : \mathbb{M}(\Omega) \to \mathbb{M}(\Omega)$ as well as the resolvent estimate (4.175). Besides, Eq. (4.175) gives

$$\|\Delta(\eta(\lambda) - \Delta)^{-1}\|_{\mathbb{M}(\Omega) \to \mathbb{M}(\Omega)} = \| -\mathbf{I} + \eta(\lambda)(\eta(\lambda) - \Delta)^{-1}\|_{\mathbb{M}(\Omega) \to \mathbb{M}(\Omega)}$$
$$\leq 1 + \|\eta(\lambda)(\eta(\lambda) - \Delta)^{-1}\|_{\mathbb{M}(\Omega) \to \mathbb{M}(\Omega)}$$
$$\leq 1 + C|\lambda|^\alpha \|(\eta(\lambda) - \Delta)^{-1}\|_{\mathbb{M}(\Omega) \to \mathbb{M}(\Omega)} \leq C,$$
$$(4.193)$$

where $\mathbf{I}$ means the identity operator.

(3) Note that

$$\|G(\cdot, \cdot, t)\|_{\mathbb{M}(\Omega)}$$
$$\leq C \int_{\Gamma_{\theta,\kappa}} e^{t|\lambda|\cos(\arg(\lambda))}$$
$$\times \left\| \beta(\lambda)^{\alpha-1}(\eta(\lambda) - \Delta)^{-1} \left(\beta(\lambda)^{1-\alpha} \frac{G_0(x)\eta(\lambda)}{\lambda + \rho U(x)} \right) \right\|_{\mathbb{M}(\Omega)} |d\lambda|$$
$$\leq C \int_{\Gamma_{\theta,\kappa}} e^{t|\lambda|\cos(\arg(\lambda))} \|\beta(\lambda)^{\alpha-1}\|_{C(\bar{\Omega})} \|(\eta(\lambda) - \Delta)^{-1}\|_{\mathbb{M}(\Omega) \to \mathbb{M}(\Omega)}$$
$$\times \left\| \frac{\beta(\lambda)^{1-\alpha}\eta(\lambda)}{\lambda + \rho U(x)} \right\|_{C(\bar{\Omega})} \|G_0(x)\|_{\mathbb{M}(\Omega)} |d\lambda|$$
$$\leq C \int_{\Gamma_{\theta,\kappa}} e^{t|\lambda|\cos(\arg(\lambda))} |\lambda|^{\alpha-1} |\lambda|^{-\alpha} |d\lambda|$$
$$\leq C \int_{\Gamma_{\theta,\kappa}} e^{t|\lambda|\cos(\arg(\lambda))} |\lambda|^{-1} |d\lambda|$$

$$\leq C \int_{\kappa}^{+\infty} e^{tr\cos(\theta)} r^{-1} dr + C \int_{-\theta}^{\theta} e^{t\kappa\cos(\varphi)} \kappa^{-1} \kappa d\varphi$$

$$\leq C \int_{\kappa t}^{+\infty} e^{s\cos(\theta)} s^{-1} ds + C \int_{-\theta}^{\theta} e^{t\kappa\cos(\varphi)} d\varphi$$

$$\leq C(1 + e^{\kappa T});$$

$$(4.194)$$

and analogously

$$\left\| \mathcal{D}_t^{1,\mu} G(\cdot,\cdot,t) \right\|_{M(\Omega)}$$

$$\leq C \int_{\Gamma_{\theta,\kappa}} e^{t|\lambda|\cos(\arg(\lambda))}$$

$$\times \left\| \beta(\lambda)^\alpha (\eta(\lambda) - \Delta)^{-1} \left(\beta(\lambda)^{1-\alpha} \frac{G_0(x)\eta(\lambda)}{\lambda + \rho U(x)} \right) \right\|_{M(\Omega)} |d\lambda|$$

$$\leq C \int_{\Gamma_{\theta,\kappa}} e^{t|\lambda|\cos(\arg(\lambda))} \left\| \beta(\lambda)^\alpha \right\|_{C(\bar\Omega)} \left\| (\eta(\lambda) - \Delta)^{-1} \right\|_{M(\Omega)\to M(\Omega)}$$

$$\times \left\| \frac{\beta(\lambda)^{1-\alpha}\eta(\lambda)}{\lambda + \rho U(x)} \right\|_{C(\bar\Omega)} \|G_0(x)\|_{M(\Omega)} |d\lambda|$$

$$\leq C \int_{\Gamma_{\theta,\kappa}} e^{t|\lambda|\cos(\arg(\lambda))} |\lambda|^\alpha |\lambda|^{-\alpha} |d\lambda|$$

$$\leq C \int_{\Gamma_{\theta,\kappa}} e^{t|\lambda|\cos(\arg(\lambda))} |d\lambda|$$

$$\leq C \int_{\kappa}^{+\infty} e^{tr\cos(\theta)} dr + C \int_{-\theta}^{\theta} e^{t\kappa\cos(\varphi)} \kappa d\varphi$$

$$\leq C t^{-1} \int_{\kappa t}^{+\infty} e^{s\cos(\theta)} ds + C \int_{-\theta}^{\theta} e^{t\kappa\cos(\varphi)} \kappa d\varphi$$

$$\leq C(t^{-1} + \kappa e^{\kappa T}).$$

$$(4.195)$$

Similarly, we also have

$$\|\Delta \mathcal{D}_t^{1-\alpha,\mu} G(\cdot,\cdot,t)\|_{M(\Omega)}$$

$$\leq C \int_{\Gamma_{\theta,\kappa}} e^{t|\lambda|\cos(\arg(\lambda))} \left\| \Delta(\eta(\lambda) - \Delta)^{-1} \left(\beta(\lambda)^{1-\alpha} \frac{G_0(x)\eta(\lambda)}{\lambda + \rho U(x)} \right) \right\|_{M(\Omega)} |d\lambda|$$

$$\leq C \int_{\Gamma_{\theta,\kappa}} e^{t|\lambda|\cos(\arg(\lambda))} \left\| \Delta(\eta(\lambda) - \Delta)^{-1} \right\|_{M(\Omega)\to M(\Omega)}$$

$$\times \left\| \frac{\beta(\lambda)^{1-\alpha}\eta(\lambda)}{\lambda + \rho U(x)} \right\|_{C(\bar\Omega)} \|G_0(x)\|_{M(\Omega)} |d\lambda|$$

$$\leq C \int_{\Gamma_{\theta,\kappa}} e^{t|\lambda|\cos(\arg(\lambda))} |d\lambda| \|G_0(x)\|_{\mathbb{M}(\Omega)}$$

$$\leq C \int_{\kappa}^{+\infty} e^{tr\cos(\theta)} dr + C \int_{-\theta}^{\theta} e^{t\kappa\cos(\varphi)} \kappa d\varphi$$

$$\leq C t^{-1} \int_{\kappa t}^{+\infty} e^{s\cos(\theta)} ds + C \int_{-\theta}^{\theta} e^{t\kappa\cos(\varphi)} \kappa d\varphi$$

$$\leq C(t^{-1} + \kappa e^{\kappa T}).$$

$$(4.196)$$

And the estimation $\|\mathcal{D}_t^{1-\alpha,\mu} G(\cdot,\cdot,t)\|_{\mathbb{M}(\Omega)} \leq C(t^{\alpha-1} + \kappa e^{\kappa T})$ can also be proved by the same way.

Applying the differential operators to the integral representation (4.171) implies that the solution $G(x,\rho,t)$ satisfies Eq. (4.151) with each term well-defined in $\mathbb{M}(\Omega)$. □

Next, we provide the derivation of the semi-discrete scheme. Eq. (4.155) and the definition of $\mathcal{D}_t^{1-\alpha,\mu}$ give that

$$\mathcal{D}_t^{1-\alpha,\mu} G(x,\rho,t)|_{t=t_n} \approx \bar{\mathcal{D}}_\tau^{1-\alpha,\mu} G(x,\rho,t_n)$$
$$= \frac{1}{\tau^{1-\alpha}} \sum_{j=1}^{n} d_{n-j}^{1,\alpha} e^{-t_{n-j}(\mu+\rho U(x))} G(x,\rho,t_j), \quad (4.197)$$

where $d_j^{1,\alpha}$ are defined in Eq. (4.15). Multiplying ζ^n on both sides and summing n from 1 to ∞ for the above equation, we obtain

$$\sum_{n=1}^{\infty} \bar{\mathcal{D}}_\tau^{1-\alpha,\mu} G(x,\rho,t_n)\zeta^n = \frac{1}{\tau^{1-\alpha}} \sum_{n=1}^{\infty} \sum_{j=1}^{n} d_{n-j}^{1,\alpha} e^{-t_{n-j}(\mu+\rho U(x))} G(x,\rho,t_j)\zeta^n$$
$$= \left(\frac{1 - e^{-\tau(\mu+\rho U(x))}\zeta}{\tau} \right)^{1-\alpha} \sum_{n=1}^{\infty} G(x,\rho,t_n)\zeta^n,$$

$$(4.198)$$

which follows by Eq. (4.16).

Denote $\eta_\tau(x,\lambda)$ and $\beta_\tau(x,\lambda)$ as the approximations of $\eta(x,\lambda)$ and $\beta(x,\lambda)$, respectively, i.e.,

$$\eta_\tau(x,\lambda) = \left(\frac{1 - e^{-\tau(\lambda+\mu+\rho U(x))}}{\tau} \right)^{\alpha} - \mu^{\alpha}, \quad \beta_\tau(x,\lambda) = \frac{1 - e^{-\tau(\lambda+\mu+\rho U(x))}}{\tau}$$

$$(4.199)$$

with the abbreviations

$$\eta_\tau(\lambda) = \eta_\tau(\cdot,\lambda), \quad \beta_\tau(\lambda) = \beta_\tau(\cdot,\lambda), \quad (4.200)$$

and choose $\frac{\tau e^{-\tau(\lambda+\rho U(x))}}{1-e^{-\tau(\lambda+\rho U(x))}}$ to be the approximation of $\frac{1}{\lambda+\rho U(x)}$. Let G_n be numerical approximation of $G(x,\rho,t_n)$. Then we start with approximating the problem on the Laplace transform side. In other words, we wish to construct the numerical solutions G_n, $n=1,2,\cdots$, satisfying the equation

$$(\eta_\tau(\lambda)-\Delta)\beta_\tau(\lambda)^{1-\alpha}\tau\sum_{n=1}^{\infty}G_n e^{-t_n\lambda}$$

$$=G_0\beta_\tau(\lambda)^{1-\alpha}\eta_\tau(\lambda)\frac{\tau e^{-\tau(\lambda+\rho U(x))}}{1-e^{-\tau(\lambda+\rho U(x))}}, \tag{4.201}$$

where $\tau\sum_{n=1}^{\infty}G_n e^{-t_n\lambda}$ can be seen as the approximation of the Laplace transform $\hat{G}(x,\rho,\lambda)$. To this end, we can construct G_n, $n=1,2,\cdots$, satisfying the following equation

$$\left(\left(\frac{1-e^{-\tau(\mu+\rho U(x))}\zeta}{\tau}\right)^{\alpha}-\mu^{\alpha}-\Delta\right)$$

$$\times\left(\frac{1-e^{-\tau(\mu+\rho U(x))}\zeta}{\tau}\right)^{1-\alpha}\sum_{n=1}^{\infty}G_n\zeta^n$$

$$=G_0\left(\frac{1-e^{-\tau(\mu+\rho U(x))}\zeta}{\tau}\right)^{1-\alpha} \tag{4.202}$$

$$\times\left(\left(\frac{1-e^{-\tau(\mu+\rho U(x))}\zeta}{\tau}\right)^{\alpha}-\mu^{\alpha}\right)\frac{e^{-\tau\rho U(x)}\zeta}{1-e^{-\tau\rho U(x)}\zeta}.$$

By simple calculations, we find that the last equation is equivalent to

$$\sum_{n=1}^{\infty}\left((\bar{\mathcal{D}}_\tau^{\alpha,\mu}-\mu^{\alpha}-\Delta)\bar{\mathcal{D}}_\tau^{1-\alpha,\mu}G_n\right)\zeta^n$$

$$=\sum_{n=1}^{\infty}\left(G_0\bar{\mathcal{D}}_\tau^{1-\alpha,\mu}(\bar{\mathcal{D}}_\tau^{\alpha,\mu}-\mu^{\alpha})e^{-\rho U(x)t_n}\right)\zeta^n. \tag{4.203}$$

Thus G_n, $n=1,2,\cdots$, are the solutions of

$$(\bar{\mathcal{D}}_\tau^{\alpha,\mu}-\mu^{\alpha}-\Delta)\bar{\mathcal{D}}_\tau^{1-\alpha,\mu}G_n=G_0\bar{\mathcal{D}}_\tau^{1-\alpha,\mu}(\bar{\mathcal{D}}_\tau^{\alpha,\mu}-\mu^{\alpha})e^{-\rho U(x)t_n}. \tag{4.204}$$

Similarly, it is straightforward to verify the following identity

$$\bar{\mathcal{D}}_\tau^{\alpha,\mu}\bar{\mathcal{D}}_\tau^{1-\alpha,\mu}G_n=\begin{cases}\dfrac{1}{\tau}G_1 & \text{if } n=1,\\[2mm]\bar{\mathcal{D}}_\tau^{1,\mu}G_n & \text{if } 2\leq n\leq N.\end{cases} \tag{4.205}$$

Thus Eq. (4.204) becomes

$$(\bar{\mathcal{D}}_\tau^{1,\mu} - (\mu^\alpha + \Delta)\bar{\mathcal{D}}_\tau^{1-\alpha,\mu})G_n$$
$$= G_0(\bar{\mathcal{D}}_\tau^{1,\mu} - \mu^\alpha\bar{\mathcal{D}}_\tau^{1-\alpha,\mu})e^{-\rho U(x)t_n}$$
$$= -G_0\left(\mu^\alpha\bar{\mathcal{D}}_\tau^{1-\alpha,\mu} - \frac{1-e^{-\mu\tau}}{\tau}\right)e^{-\rho U(x)t_n}, \qquad n = 1, 2, \cdots, N.$$

(4.206)

The scheme (4.204) can also be obtained by applying the implicit Euler scheme to the equation

$$\mathcal{D}_t^{1,\mu}G(x,\rho,t) - (\mu^\alpha + \Delta)\mathcal{D}_t^{1-\alpha,\mu}G(x,\rho,t)$$
$$= -G_0\left(\mu^\alpha\mathcal{D}_t^{1-\alpha,\mu} - \frac{1-e^{-\mu\tau}}{\tau}\right)e^{-\rho U(x)t},$$

(4.207)

which replaces a constant μ by $\frac{1-e^{-\mu\tau}}{\tau}$.

4.2.2.2 *Error estimates*

In this subsection, we estimate the error of the numerical solution given by Eq. (4.204). Applying Cauchy's integral formula, we have, for $\varrho_\kappa = e^{-\tau(\kappa+1)} \in (0,1)$,

$$G_n = \frac{1}{2\pi i}\int_{|\zeta|=\varrho_\kappa} \zeta^{-n-1}\sum_{i=1}^{\infty} G_i(x)\zeta^i d\zeta$$
$$= \frac{1}{2\pi i}\int_{\Gamma^\tau} e^{\lambda t_n}\left(\sum_{n=1}^{\infty} G_n(x)e^{-t_n\lambda}\right)\tau d\lambda,$$

(4.208)

where we take $\zeta = e^{-\lambda\tau}$ with the contour $\Gamma^\tau = \{\lambda = \kappa + 1 + iy : y \in \mathbb{R} \text{ and } |y| \leq \pi/\tau\}$. According to Eq. (4.201), it is easy to see

$$\sum_{n=1}^{\infty} G_n e^{-t_n\lambda} = \beta_\tau(\lambda)^{\alpha-1}\left(\eta_\tau(\lambda) - \Delta\right)^{-1}$$
$$\times \left(G_0(x)\beta_\tau(\lambda)^{1-\alpha}\frac{\eta_\tau(\lambda)e^{-\tau(\lambda+\rho U(x))}}{1 - e^{-\tau(\lambda+\rho U(x))}}\right),$$

(4.209)

which leads to

$$G_n = \frac{1}{2\pi i}\int_{\Gamma^\tau} e^{\lambda t_n}\beta_\tau(\lambda)^{\alpha-1}\left(\eta_\tau(\lambda) - \Delta\right)^{-1}$$
$$\times \left(G_0(x)\beta_\tau(\lambda)^{1-\alpha}\frac{\eta_\tau(\lambda)\tau e^{-\tau(\lambda+\rho U(x))}}{1 - e^{-\tau(\lambda+\rho U(x))}}\right)d\lambda$$
$$= \frac{1}{2\pi i}\int_{\Gamma^\tau_{\theta,\kappa}} e^{\lambda t_n}\beta_\tau(\lambda)^{\alpha-1}\left(\eta_\tau(\lambda) - \Delta\right)^{-1}$$
$$\times \left(G_0(x)\beta_\tau(\lambda)^{1-\alpha}\frac{\eta_\tau(\lambda)\tau e^{-\tau(\lambda+\rho U(x))}}{1 - e^{-\tau(\lambda+\rho U(x))}}\right)d\lambda,$$

(4.210)

where we have deformed the integration path (using Cauchy's theorem of complex analysis) from Γ^τ to $\Gamma^\tau_{\theta,\kappa}$. Such a deformation requires the integrand in Eq. (4.170) to be $\mathbb{M}(\Omega)$-valued analytic function for $\lambda \in \Sigma^\tau_{\theta,\kappa}$, which is a consequence of Proposition 4.2 below.

Proposition 4.2. *By choosing $\theta \in (\frac{\pi}{2}, \pi)$ sufficiently close to $\frac{\pi}{2}$ and $\kappa > 0$ sufficiently large (depending on $\mu + |\rho|\|U(x)\|_{C(\bar{\Omega})}$), there exists a positive constant τ_* (depending on θ and κ) such that the following estimates hold when $\tau \leq \tau_*$:*

(1) $\beta_\tau(\lambda)$, $\eta_\tau(\lambda) \in \Sigma_{\frac{3\pi}{4}}$ for $\lambda \in \Sigma^\tau_{\theta,\kappa}$, and

$$C|\lambda| \leq |\beta_\tau(\lambda)| \leq C|\lambda|, \quad C|\lambda|^\alpha \leq |\eta_\tau(\lambda)| \leq C|\lambda|^\alpha \quad \forall \lambda \in \Sigma^\tau_{\theta,\kappa}. \tag{4.211}$$

(2) The operator $(\eta_\tau(\lambda) - \Delta)^{-1}$ is bounded, and analytic in $\mathbb{M}(\Omega)$ for $\lambda \in \Sigma^\tau_{\theta,\kappa}$, satisfying

$$\|(\eta_\tau(\lambda) - \Delta)^{-1}\|_{\mathbb{M}(\Omega)\to\mathbb{M}(\Omega)} \leq C|\lambda|^{-\alpha} \quad \forall \lambda \in \Sigma^\tau_{\theta,\kappa}. \tag{4.212}$$

Proof. Introduce $\omega = \lambda + \mu + \rho U(x)$, with $\lambda \in \Sigma^\tau_{\theta,\kappa}$. For sufficiently small step size $\tau < \frac{\pi}{2\mu+2|\rho|\|U(x)\|_{C(\bar{\Omega})}}$, there holds

$$\tau|Im(\omega)| < \tau(|Im(\lambda)|+\mu+|\rho|\|U(x)\|_{C(\bar{\Omega})}) \leq \pi+\tau(\mu+|\rho|\|U(x)\|_{C(\bar{\Omega})}) < \frac{3}{2}\pi. \tag{4.213}$$

Hence, $1 - e^{-\tau\omega} = 0$ holds only when $\omega = 0$. In particular,

$$\text{if } \tau|\omega| \geq C, \text{ then } |1 - e^{-\tau\omega}| \geq C. \tag{4.214}$$

For $\lambda \in \Sigma^\tau_{\theta,\kappa}$, $\tau|Im(\lambda)| \leq \pi$ and $\tau|Re(\lambda)| \leq \tau(\kappa + 1) \leq \pi$ hold when $\tau \leq \frac{\pi}{\kappa+1}$. Thus

$$\tau|\lambda| \leq \tau|Im(\lambda)| + \tau|Re(\lambda)| \leq 2\pi,$$
$$\tau|\omega| \leq \tau|\lambda| + \tau(\mu + |\rho|\|U(x)\|_{C(\bar{\Omega})}) \leq \frac{5}{2}\pi. \tag{4.215}$$

By choosing $\kappa \geq 2(\mu + |\rho|\|U(x)\|_{C(\bar{\Omega})})$, Taylor's expansion yields, for $\lambda \in \Sigma^\tau_{\theta,\kappa}$,

$$|\beta_\tau(\lambda)| = \left|\frac{1-e^{-\tau w}}{\tau}\right| \leq C|\omega| \leq C\left(|\lambda| + \mu + |\rho|\|U(x)\|_{C(\bar{\Omega})}\right) \tag{4.216}$$
$$\leq C(|\lambda| + \kappa) \leq C|\lambda|,$$

where the fact $|\lambda| \geq \kappa$ for $\lambda \in \Sigma^\tau_{\theta,\kappa}$ is used. The inequality $|\beta_\tau(\lambda)| \leq C|\lambda|$ can be proved.

In the following, we consider two cases to prove $C|\lambda| \le |\beta_\tau(\lambda)|$ for $\lambda \in \Sigma_{\theta,\kappa}^\tau$. If $\tau|\omega|$ is smaller than some constant, by Taylor's expansion (with $|\mathcal{O}(\tau\omega)| < \frac{1}{2}$, due to the smallness of $\tau|\omega|$ assumed), we obtain

$$|\beta_\tau(\lambda)| = \left|\frac{1 - e^{-\tau\omega}}{\tau}\right| = |\omega(1 + \mathcal{O}(\tau\omega))| \ge \frac{1}{2}|\omega|$$

$$\ge \frac{1}{2}\left(|\lambda| - \mu - |\rho|\|U(x)\|_{C(\bar\Omega)}\right)$$

$$\ge \frac{1}{2}(|\lambda| - \kappa/2) \ge \frac{1}{4}|\lambda|,$$

$$(4.217)$$

where we have used $\kappa \ge 2(\mu + |\rho|\|U(x)\|_{C(\bar\Omega)})$ again and noted that $|\lambda| \ge \kappa$ for $\lambda \in \Sigma_{\theta,\kappa}^\tau$. If $\tau|\omega|$ is larger than the constant, then Eqs. (4.214) and (4.215) give

$$|\beta_\tau(\lambda)| = \left|\frac{1 - e^{-\tau\omega}}{\tau}\right| \ge \frac{C}{\tau} \ge C|\lambda|. \qquad (4.218)$$

In a word, under the conditions $\kappa \ge 2(\mu + |\rho|\|U(x)\|_{C(\bar\Omega)})$ and $\tau < \frac{\pi}{\kappa+1}$, we have proved that for some positive constants C_1 and C_2, there hold

$$C_1|\lambda| \le |\beta_\tau(\lambda)| = \left|\frac{1 - e^{-\tau(\lambda+\mu+\rho U(x))}}{\tau}\right| \le C_2|\lambda| \quad \forall \lambda \in \Sigma_{\theta,\kappa}^\tau. \qquad (4.219)$$

The last inequality further implies

$$C_1|\lambda|^\alpha - \mu^\alpha \le \left|\left(\frac{1 - e^{-\tau(\lambda+\mu+\rho U(x))}}{\tau}\right)^\alpha - \mu^\alpha\right| \le C_2|\lambda|^\alpha + \mu^\alpha. \qquad (4.220)$$

Choosing κ larger than some constant (depending on μ and $|\rho|\|U(x)\|_{C(\bar\Omega)}$), we have $\mu^\alpha \le \frac{C_1}{2}\kappa^\alpha \le \frac{C_1}{2}|\lambda|^\alpha$. Therefore one has

$$\frac{C_1}{2}|\lambda|^\alpha \le |\eta_\tau(\lambda)| = \left|\left(\frac{1 - e^{-\tau(\lambda+\mu+\rho U(x))}}{\tau}\right)^\alpha - \mu^\alpha\right| \le \left(\frac{C_1}{2} + C_2\right)|\lambda|^\alpha.$$

$$(4.221)$$

Thus the proof of Eq. (4.211) is complete. Next, we prove $\beta_\tau(\lambda)$, $\eta_\tau(\lambda) \in \Sigma_{\frac{3\pi}{4}}$ for $\lambda \in \Sigma_{\theta,\kappa}^\tau$. Lemma 4.5 gives

$$-|\arg(\lambda + \mu + \rho U(x))| - C_0\tau \le \arg\left(\frac{1 - e^{-\tau(\lambda+\mu+\rho U(x))}}{\tau}\right)$$

$$\le |\arg(\lambda + \mu + \rho U(x))| + C_0\tau,$$

$$(4.222)$$

which implies $-\frac{5\pi}{8} - C_0\tau \le \arg\left(\frac{1 - e^{-\tau(\lambda+\mu+\rho U(x))}}{\tau}\right) \le \frac{5\pi}{8} + C_0\tau$. Thus $\beta_\tau(\lambda) \in \Sigma_{\frac{3\pi}{4}}$ holds when the step size τ is smaller than some constant. Moreover, choosing κ large enough and using Eq. (4.219) imply

$$\left|\frac{1 - e^{-\tau(\lambda+\mu+\rho U(x))}}{\tau}\right| = |\beta_\tau(\lambda)| \ge C|\lambda| \ge C\kappa \quad \forall \lambda \in \Sigma_{\theta,\kappa}^\tau. \qquad (4.223)$$

The last two inequalities yield

$$\frac{1 - e^{-\tau(\lambda+\mu+\rho U(x))}}{\tau} \in \Sigma_{\frac{3\pi}{4}, C\kappa} \quad and \quad \left(\frac{1 - e^{-\tau(\lambda+\mu+\rho U(x))}}{\tau}\right)^{\alpha} \in \Sigma_{\frac{3\alpha\pi}{4}, C^{\alpha}\kappa^{\alpha}},$$

(4.224)

which further implies that (by choosing κ to be large enough and using the same argument for Eq. (4.185))

$$\eta_{\tau}(\lambda) = \left(\frac{1 - e^{-\tau(\lambda+\mu+\rho U(x))}}{\tau}\right)^{\alpha} - \mu^{\alpha} \in \Sigma_{\frac{3\pi}{4}, C\kappa^{\alpha}} \subset \Sigma_{\frac{3\pi}{4}}. \quad (4.225)$$

So the first conclusion of Proposition 4.2 has been proved and the second one can be got in the same way as the second conclusion of Proposition 4.1. □

Then we present the following convergence results.

Theorem 4.5. *There exists a positive constant* τ_* *(see Proposition 4.2) such that for* $\tau \le \tau_*$*, the solutions of Eqs. (4.151) and (4.206) satisfy the following error estimate:*

$$\|G(x, \rho, t_n) - G_n\|_{M(\Omega)} \le C_T \|G_0\|_{M(\Omega)} t_n^{-1} \tau, \qquad n = 1, 2, \cdots, N, \quad (4.226)$$

where the constant C_T *may grow exponentially with respect to* T *and the quantity* $\mu + |\rho| \|U(x)\|_{C(\bar{\Omega})}$.

Proof. Subtracting Eq. (4.210) from Eq. (4.171), one has

$$G(x, \rho, t_n) - G_n(x)$$
$$= \frac{1}{2\pi i} \int_{\Gamma_{\theta,\kappa} \backslash \Gamma_{\theta,\kappa}^{\tau}} e^{t_n \lambda} \beta(\lambda)^{\alpha-1} (\eta(\lambda) - \Delta)^{-1} \left(\beta(\lambda)^{1-\alpha} \frac{G_0 \eta(\lambda)}{\lambda + \rho U(x)}\right) d\lambda$$
$$+ \frac{1}{2\pi i} \int_{\Gamma_{\theta,\kappa}^{\tau}} e^{t_n \lambda} \left[\beta(\lambda)^{\alpha-1} (\eta(\lambda) - \Delta)^{-1} \left(\beta(\lambda)^{1-\alpha} \frac{G_0 \eta(\lambda)}{\lambda + \rho U(x)}\right)\right.$$
$$\left. - \beta_{\tau}(\lambda)^{\alpha-1} (\eta_{\tau}(\lambda) - \Delta)^{-1}\right.$$
$$\left. \times \left(\beta_{\tau}(\lambda)^{1-\alpha} \frac{G_0 \eta_{\tau}(\lambda) \tau e^{-\tau(\lambda+\rho U(x))}}{1 - e^{-\tau(\lambda+\rho U(x))}}\right)\right] d\lambda$$
$$=: J_1 + J_2.$$

(4.227)

Note that $|\lambda + \rho U(x)| \ge \frac{1}{2}|\lambda|$ on the contour $\Gamma_{\theta,\kappa}$, due to the largeness of κ compared with $\mu + |\rho| \|U(x)\|_{C(\bar{\Omega})}$. Introducing $|d\lambda|$ to be the arc length element on the contour $\Gamma_{\theta,\kappa}\backslash\Gamma_{\theta,\kappa}^{\tau}$, we obtain

$$\|J_1\|_{M(\Omega)} \le C\int_{\Gamma_{\theta,\kappa}\backslash\Gamma_{\theta,\kappa}^{\tau}} e^{t_n|\lambda|\cos(\theta)}$$

$$\times \left\|\beta(\lambda)^{\alpha-1}(\eta(\lambda)-\Delta)^{-1}\left(\beta(\lambda)^{1-\alpha}\frac{G_0\eta(\lambda)}{\lambda+\rho U(x)}\right)\right\|_{M(\Omega)} |d\lambda|,$$

$$(4.228)$$

where

$$\left\|\beta(\lambda)^{\alpha-1}(\eta(\lambda)-\Delta)^{-1}\left(\beta(\lambda)^{1-\alpha}\frac{G_0\eta(\lambda)}{\lambda+\rho U(x)}\right)\right\|_{M(\Omega)}$$

$$\le C\|\beta(\lambda)^{\alpha-1}\|_{C(\bar{\Omega})}\|(\eta(\lambda)-\Delta)^{-1}\|_{M(\Omega)\to M(\Omega)}\left\|\beta(\lambda)^{1-\alpha}\frac{G_0\eta(\lambda)}{\lambda+\rho U(x)}\right\|_{M(\Omega)}$$

$$\le C\|\beta(\lambda)\|_{C(\bar{\Omega})}^{\alpha-1}\|(\eta(\lambda)-\Delta)^{-1}\|_{M(\Omega)\to M(\Omega)}$$

$$\times \|\beta(\lambda)\|_{C(\bar{\Omega})}^{1-\alpha}\left\|\frac{\eta(\lambda)}{\lambda+\rho U(x)}\right\|_{C(\bar{\Omega})}\|G_0\|_{M(\Omega)}$$

$$\le C|\lambda|^{\alpha-1}|\lambda|^{-\alpha}|\lambda|^{1-\alpha}|\lambda|^{\alpha-1}\|G_0\|_{M(\Omega)}$$

$$\le C|\lambda|^{-1}\|G_0\|_{M(\Omega)}.$$

$$(4.229)$$

Thus, it holds

$$\|J_1\|_{M(\Omega)} \le C\|G_0\|_{M(\Omega)}\int_{\Gamma_{\theta,\kappa}\backslash\Gamma_{\theta,\kappa}^{\tau}} e^{t_n|\lambda|\cos(\theta)}|\lambda|^{-1}|d\lambda|$$

$$= C\|G_0\|_{M(\Omega)}\int_{\frac{\pi}{\tau\sin(\theta)}}^{\infty} e^{t_n r\cos(\theta)}r^{-1}dr$$

$$\le C\|G_0\|_{M(\Omega)}\int_{\frac{\pi t_n}{\tau\sin(\theta)}}^{\infty} e^{s\cos(\theta)}s^{-1}ds \qquad (4.230)$$

$$\le C\|G_0\|_{M(\Omega)}\frac{\tau\sin(\theta)}{\pi t_n}\int_{\frac{\pi t_n}{\tau\sin(\theta)}}^{\infty} e^{s\cos(\theta)}ds$$

$$\le C\|G_0\|_{M(\Omega)}t_n^{-1}\tau.$$

To estimate $\|J_2\|_{M(\Omega)}$, we need to use the following lemma, whose proof is presented in the following.

Lemma 4.6.

$$\left\| \beta(\lambda)^{\alpha-1}(\eta(\lambda)-\Delta)^{-1}\left(\beta(\lambda)^{1-\alpha}\frac{G_0\eta(\lambda)}{\lambda+\rho U(x)}\right) \right.$$

$$\left. - \beta_\tau(\lambda)^{\alpha-1}(\eta_\tau(\lambda)-\Delta)^{-1}\left(\beta_\tau(\lambda)^{1-\alpha}\frac{G_0\eta_\tau(\lambda)\tau e^{-\tau(\lambda+\rho U(x))}}{1-e^{-\tau(\lambda+\rho U(x))}}\right) \right\|_{M(\Omega)}$$

$$\leq C\|G_0\|_{M(\Omega)}\tau \qquad \forall \lambda \in \Gamma_{\theta,\kappa}^\tau.$$

$$(4.231)$$

Lemma 4.6 leads to

$$\|J_2\|_{M(\Omega)} \leq C\|G_0\|_{M(\Omega)}\,\tau \int_{\Gamma_{\theta,k}^\tau} e^{t_n|\lambda|\cos(\arg(\lambda))}|d\lambda|$$

$$\leq C\|G_0\|_{M(\Omega)}\,\tau \int_\kappa^{\frac{\pi}{\tau\sin(\theta)}} e^{t_n r\cos(\theta)}\,dr$$

$$+ C\|G_0\|_{M(\Omega)}\,\tau \int_{-\theta}^\theta e^{t_n\kappa\cos(\varphi)}\kappa d\varphi$$

$$\leq C\|G_0\|_{M(\Omega)}\,t_n^{-1}\tau \int_{\kappa t_n}^{\frac{\pi t_n}{\tau\sin(\theta)}} e^{s\cos(\theta)}\,ds$$

$$+ C\|G_0\|_{M(\Omega)}\,\tau\kappa \int_{-\theta}^\theta e^{T\kappa}d\varphi$$

$$\leq C\|G_0\|_{M(\Omega)}\left(t_n^{-1}+\kappa e^{\kappa T}\right)\tau$$

$$\leq C_T\|G_0\|_{M(\Omega)}\,t_n^{-1}\tau.$$

$$(4.232)$$

Thus we can get the desired results. $\qquad\qquad\Box$

Then we provide the proof of Lemma 4.6.

Proof of Lemma 4.6. Simple calculations yield

$$
\left\| \beta(\lambda)^{\alpha-1}(\eta(\lambda) - \Delta)^{-1} \left(\beta(\lambda)^{1-\alpha} \frac{G_0\eta(\lambda)}{\lambda + \rho U(x)} \right) \right.
$$

$$
\left. - \beta_\tau(\lambda)^{\alpha-1} (\eta_\tau(\lambda) - \Delta)^{-1} \left(\beta_\tau(\lambda)^{1-\alpha} \frac{G_0\eta_\tau(\lambda)\tau e^{-\tau(\lambda+\rho U(x))}}{1 - e^{-\tau(\lambda+\rho U(x))}} \right) \right\|_{M(\Omega)}
$$

$$
\leq \left\| \left(\beta(\lambda)^{\alpha-1} - \beta_\tau(\lambda)^{\alpha-1} \right) (\eta(\lambda) - \Delta)^{-1} \left(\beta(\lambda)^{1-\alpha} \frac{G_0\eta(\lambda)}{\lambda + \rho U(x)} \right) \right\|_{M(\Omega)}
$$

$$
+ \left\| \beta_\tau(\lambda)^{\alpha-1} \left((\eta(\lambda) - \Delta)^{-1} - (\eta_\tau(\lambda) - \Delta)^{-1} \right) \right.
$$

$$
\left. \times \left(\beta(\lambda)^{1-\alpha} \frac{G_0\eta(\lambda)}{\lambda + \rho U(x)} \right) \right\|_{M(\Omega)}
$$

$$
+ \left\| \beta_\tau(\lambda)^{\alpha-1} (\eta_\tau(\lambda) - \Delta)^{-1} \left(\left(\beta(\lambda)^{1-\alpha} - \beta_\tau(\lambda)^{1-\alpha} \right) \frac{G_0\eta(\lambda)}{\lambda + \rho U(x)} \right) \right\|_{M(\Omega)}
$$

$$
+ \left\| \beta_\tau(\lambda)^{\alpha-1} (\eta_\tau(\lambda) - \Delta)^{-1} \left(\beta_\tau(\lambda)^{1-\alpha} \frac{G_0 (\eta(\lambda) - \eta_\tau(\lambda))}{\lambda + \rho U(x)} \right) \right\|_{M(\Omega)}
$$

$$
+ \left\| \beta_\tau(\lambda)^{\alpha-1} (\eta_\tau(\lambda) - \Delta)^{-1} \right.
$$

$$
\left. \times \left(\beta_\tau(\lambda)^{1-\alpha} G_0\eta_\tau(\lambda) \left(\frac{1}{\lambda + \rho U(x)} - \frac{\tau e^{-\tau(\lambda+\rho U(x))}}{1 - e^{-\tau(\lambda+\rho U(x))}} \right) \right) \right\|_{M(\Omega)}
$$

$$
=: I_1 + I_2 + I_3 + I_4 + I_5.
$$

$$(4.233)$$

To get the estimates of $\beta(\lambda)^{\alpha-1} - \beta_\tau(\lambda)^{\alpha-1}$ in I_1 and $\beta(\lambda)^{1-\alpha} - \beta_\tau(\lambda)^{1-\alpha}$ in I_3, we denote $\omega = \lambda + \mu + \rho U(x)$ and use Taylor's expansion

$$
e^{-\tau\omega} = 1 - \tau\omega + \frac{1}{2}\tau^2\omega^2 \int_0^1 e^{-\theta\tau\omega}(1 - \theta)d\theta. \qquad (4.234)
$$

Thus

$$
|\beta(\lambda)^\gamma - \beta_\tau(\lambda)^\gamma| = \left| \beta(\lambda)^\gamma - \left(\frac{1 - e^{-\tau\beta(\lambda)}}{\tau} \right)^\gamma \right|
$$

$$
= \left| \beta(\lambda)^\gamma - \left(\beta(\lambda) - \tau\beta(\lambda)^2 \int_0^1 e^{-\theta\tau\beta(\lambda)}(1 - \theta)d\theta \right)^\gamma \right|
$$

$$
= |\beta(\lambda)|^\gamma \left| 1 - \left(1 - \tau\beta(\lambda) \int_0^1 e^{-\theta\tau\beta(\lambda)}(1 - \theta)d\theta \right)^\gamma \right|.
$$

$$(4.235)$$

If $\tau|\beta(\lambda)| < \frac{1}{2}$, then the following equation can be got

$$\left(1 - \frac{1}{2}\tau\beta(\lambda) \int_0^1 e^{-\theta\tau\omega}(1-\theta)d\theta\right)^\gamma = 1 + \mathcal{O}\left(\tau\beta(\lambda) \int_0^1 e^{-\theta\tau\omega}(1-\theta)d\theta\right)$$
$$= 1 + \mathcal{O}(\tau|\beta(\lambda)|).$$

(4.236)

In this case, the last two identities imply

$$|\beta(\lambda)^\gamma - \beta_\tau(\lambda)^\gamma| \le |\beta(\lambda)|^\gamma C\tau|\beta(\lambda)| \le C\tau|\lambda|^{1+\gamma}.$$

(4.237)

If $\tau|\beta(\lambda)| \ge \frac{1}{2}$, then we obtain

$$\tau|\lambda| \ge C\tau|\beta(\lambda)| \ge C \quad \forall \lambda \in \Gamma_{\theta,\kappa}^\tau,$$
$$|\beta(\lambda)^\gamma - \beta_\tau(\lambda)^\gamma| \le C|\lambda|^\gamma \le C\tau|\lambda|^{\gamma+1} \quad \forall \lambda \in \Gamma_{\theta,\kappa}^\tau.$$

(4.238)

In either case, there holds

$$|\beta(\lambda)^\gamma - \beta_\tau(\lambda)^\gamma| \le C\tau|\lambda|^{\gamma+1} \quad \forall \lambda \in \Gamma_{\theta,\kappa}^\tau,$$

(4.239)

which further implies

$$|\beta(\lambda)^{\alpha-1} - \beta_\tau(\lambda)^{\alpha-1}| \le C|\lambda|^{\alpha-1} \le C\tau|\lambda|^\alpha \quad \forall \lambda \in \Gamma_{\theta,\kappa}^\tau,$$
$$|\beta(\lambda)^{1-\alpha} - \beta_\tau(\lambda)^{1-\alpha}| \le C|\lambda|^{1-\alpha} \le C\tau|\lambda|^{2-\alpha} \quad \forall \lambda \in \Gamma_{\theta,\kappa}^\tau,$$
$$|\eta(\lambda) - \eta_\tau(\lambda)| = |\beta(\lambda)^\alpha - \beta_\tau(\lambda)^\alpha| \le C|\lambda|^\alpha \le C\tau|\lambda|^{1+\alpha} \quad \forall \lambda \in \Gamma_{\theta,\kappa}^\tau,$$

(4.240)

and

$$\|(\eta(\lambda) - \Delta)^{-1} - (\eta_\tau(\lambda) - \Delta)^{-1}\|_{M(\Omega)\to M(\Omega)}$$
$$= \|(\eta(\lambda) - \Delta)^{-1}(\eta(\lambda) - \eta_\tau(\lambda))(\eta_\tau(\lambda) - \Delta)^{-1}\|_{M(\Omega)\to M(\Omega)}$$
$$\le C\|(\eta(\lambda) - \Delta)^{-1}\|_{M(\Omega)\to M(\Omega)}$$
$$\quad \times \|(\eta(\lambda) - \eta_\tau(\lambda))\|_{C(\bar\Omega)}\|(\eta_\tau(\lambda) - \Delta)^{-1}\|_{M(\Omega)\to M(\Omega)}$$
$$\le C\tau|\lambda|^{-\alpha}|\lambda|^{1+\alpha}|\lambda|^{-\alpha}$$
$$\le C\tau|\lambda|^{1-\alpha}.$$

(4.241)

Thus

$$I_1 = \left\|(\beta(\lambda)^{\alpha-1} - \beta_\tau(\lambda)^{\alpha-1})(\eta(\lambda) - \Delta)^{-1}\left(\beta(\lambda)^{1-\alpha}\frac{G_0\eta(\lambda)}{\lambda + \rho U(x)}\right)\right\|_{M(\Omega)}$$
$$\le C\tau|\lambda|^\alpha|\lambda|^{-\alpha}\left(|\lambda|^{1-\alpha}\frac{|\lambda|^\alpha}{|\lambda|}\right)\|G_0\|_{M(\Omega)}$$
$$\le C\|G_0\|_{M(\Omega)}\tau,$$

(4.242)

$$I_2 = \left\| \beta_\tau(\lambda)^{\alpha-1} \left((\eta(\lambda) - \Delta)^{-1} - (\eta_\tau(\lambda) - \Delta)^{-1} \right) \right.$$
$$\left. \times \left(\beta(\lambda)^{1-\alpha} \frac{G_0 \eta(\lambda)}{\lambda + \rho U(x)} \right) \right\|_{M(\Omega)}$$

$$\le C\tau |\lambda|^{\alpha-1} |\lambda|^{1-\alpha} \left(|\lambda|^{1-\alpha} \frac{|\lambda|^\alpha}{|\lambda|} \right) \|G_0\|_{M(\Omega)}$$

$$\le C\|G_0\|_{M(\Omega)}\tau,$$

(4.243)

$$I_3 = \left\| \beta_\tau(\lambda)^{\alpha-1} (\eta_\tau(\lambda) - \Delta)^{-1} \left((\beta(\lambda)^{1-\alpha} - \beta_\tau(\lambda)^{1-\alpha}) \frac{G_0 \eta(\lambda)}{\lambda + \rho U(x)} \right) \right\|_{M(\Omega)}$$

$$\le C\tau |\lambda|^{\alpha-1} |\lambda|^{-\alpha} \left(|\lambda|^{2-\alpha} \frac{|\lambda|^\alpha}{|\lambda|} \right) \|G_0\|_{M(\Omega)}$$

$$\le C\|G_0\|_{M(\Omega)}\tau,$$

(4.244)

and

$$I_4 = \left\| \beta_\tau(\lambda)^{\alpha-1} (\eta_\tau(\lambda) - \Delta)^{-1} \left(\beta_\tau(\lambda)^{1-\alpha} \frac{G_0 (\eta(\lambda) - \eta_\tau(\lambda))}{\lambda + \rho U(x)} \right) \right\|_{M(\Omega)}$$

$$\le C\tau |\lambda|^{\alpha-1} |\lambda|^{-\alpha} \left(|\lambda|^{1-\alpha} \frac{|\lambda|^{1+\alpha}}{|\lambda|} \right) \|G_0\|_{M(\Omega)}$$

$$\le C\|G_0\|_{M(\Omega)}\tau.$$

(4.245)

At last, to estimate I_5, denoting $\xi = \lambda + \rho U(x)$ and using Taylor's expansion give

$$1 - e^{-\tau\xi} = \tau\xi - \tau^2\xi^2 \int_0^1 e^{-\theta\tau\xi}(1-\theta)d\theta,$$

$$\tau\xi e^{-\tau\xi} = \tau\xi - \tau^2\xi^2 \int_0^1 e^{-\theta\tau\xi}d\theta.$$

(4.246)

When $\tau|\xi| < \frac{1}{2}$, we have

$$
\begin{aligned}
&\left\| \frac{1}{\lambda + \rho U(x)} - \frac{\tau e^{-\tau(\lambda + \rho U(x))}}{1 - e^{-\tau(\lambda + \rho U(x))}} \right\|_{L^2(\Omega) \to L^2(\Omega)} \\
&= \left\| \frac{1}{\xi} - \frac{\tau e^{-\tau \xi}}{1 - e^{-\tau \xi}} \right\|_{L^2(\Omega) \to L^2(\Omega)} \\
&= \left\| \frac{1 - e^{-\tau \xi} - \tau \xi e^{-\tau \xi}}{\xi \left(1 - e^{-\tau \xi}\right)} \right\|_{L^2(\Omega) \to L^2(\Omega)} \\
&= \left\| \frac{\tau^2 \xi^2 \int_0^1 e^{-\theta \tau \xi} \theta \mathrm{d}\theta}{\tau \xi^2 \left(1 - \tau \xi \int_0^1 e^{-\theta \tau \xi}(1 - \theta) \mathrm{d}\theta\right)} \right\|_{L^2(\Omega) \to L^2(\Omega)} \\
&\leq C\tau.
\end{aligned}
\tag{4.247}
$$

When $\tau|\xi| > \frac{1}{2}$, there hold

$$
\tau|\lambda| \geq \tau|\xi - \rho U(x)| \geq \frac{1}{2} - \tau|\rho| \|U(x)\|_{C(\bar{\Omega})} \geq \frac{1}{4} \quad \text{when } \tau < \frac{1}{4|\rho| \|U(x)\|_{C(\bar{\Omega})}},
$$
$$
C\tau|\lambda| \leq |1 - e^{-\tau(\lambda + \rho U(x))}| \leq C\tau|\lambda|,
$$
$$
\tau|\lambda + \rho U(x)| \leq C \quad \text{for } \lambda \in \Gamma_{\theta,\kappa}^{\tau},
\tag{4.248}
$$

which implies

$$
\begin{aligned}
&\left\| \frac{1}{\lambda + \rho U(x)} - \frac{\tau e^{-\tau(\lambda + \rho U(x))}}{1 - e^{-\tau(\lambda + \rho U(x))}} \right\|_{L^2(\Omega) \to L^2(\Omega)} \\
&\leq \left\| \frac{1}{\lambda + \rho U(x)} \right\|_{L^2(\Omega) \to L^2(\Omega)} + \left\| \frac{\tau e^{-\tau(\lambda + \rho U(x))}}{1 - e^{-\tau(\lambda + \rho U(x))}} \right\|_{L^2(\Omega) \to L^2(\Omega)} \\
&\leq \frac{C}{|\lambda|} + \frac{C}{|\lambda|} \\
&\leq \frac{C\tau}{\tau|\lambda|} \leq C\tau.
\end{aligned}
\tag{4.249}
$$

In either case, one has

$$
\left\| \frac{1}{\lambda + \rho U(x)} - \frac{\tau e^{-\tau(\lambda + \rho U(x))}}{1 - e^{-\tau(\lambda + \rho U(x))}} \right\|_{L^2(\Omega) \to L^2(\Omega)} \leq C\tau.
\tag{4.250}
$$

Then it holds

$$I_5 = \left\| \beta_\tau(\lambda)^{\alpha-1} \left(\eta_\tau(\lambda) - \Delta \right)^{-1} \right.$$
$$\left. \times \left(\beta_\tau(\lambda)^{1-\alpha} G_0 \eta_\tau(\lambda) \left(\frac{1}{\lambda + \rho U(x)} - \frac{\tau e^{-\tau(\lambda + \rho U(x))}}{1 - e^{-\tau(\lambda + \rho U(x))}} \right) \right) \right\|_{\mathrm{M}(\Omega)}$$
$$\leq C\tau |\lambda|^{\alpha-1} |\lambda|^{-\alpha} (|\lambda|^{1-\alpha} |\lambda|^\alpha) \|G_0\|_{\mathrm{M}(\Omega)}$$
$$\leq C \|G_0\|_{\mathrm{M}(\Omega)} \tau.$$

$$(4.251)$$

Thus we complete the proof.   □

4.2.2.3   *Numerical experiments*

In this subsection, we offer some numerical results to validate error analysis. We use numerical scheme (4.206) to solve Eq. (4.151) in the one-dimensional domain $\Omega = (0,1)$. We take $T = 1$, $\mu = 0.01$, $\rho = -\mathbf{i}$ and $U(x) = x$, where the choice of the function $U(x) = x$ physically corresponds to the distribution of time average of the particles' trajectories. Here we denote G_τ^N as the numerical solution with time step size τ at time $t_N = 1$ and the order of convergence of the numerical solutions are computed by the formula

$$\frac{\ln(\|G_{2\tau}^N - G_\tau^N\| / \|G_\tau^N - G_{\tau/2}^N\|)}{\ln(2)}.$$

$$(4.252)$$

Example 4.3. In the first example, we choose smooth initial data $G_0(x) = 10x(1-x)$ and we use $\|\cdot\|_{L^2(\Omega)}$ to measure the error. To investigate the convergence in time and eliminate the influence from spatial discretization, we use finite element method with a sufficiently small mesh size $h = 1/500$ so that the error due to spatial discretization can be omitted. Table 4.5 presents the errors and convergence rates with $\tau = 1/8$, $1/16$, $1/32$, $1/64$ and $\alpha = 0.25$, 0.5, 0.75.

Example 4.4. In the second example, we choose measure data $G_0(x) = \delta(x - 1/4)$ where δ is Dirac function and we use $\|\cdot\|_{\mathrm{M}(\Omega)}$ to measure the error. To investigate the convergence in time and eliminate the influence from spatial discretization, we use finite element method with a sufficiently small mesh size $h = 1/500$ so that the error due to spatial discretization can be omitted. Table 4.6 presents the errors and convergence rates with $\tau = 1/8$, $1/16$, $1/32$, $1/64$ and $\alpha = 0.25$, 0.5, 0.75.

Table 4.5: Error and convergence rate for Example 4.3 when initial data are smooth: $G_0(x) = 10x(1 - x)$.

τ	$\alpha = 0.25$		$\alpha = 0.5$		$\alpha = 0.75$	
	Error	Rate	Error	Rate	Error	Rate
1/8	1.609E-03		2.733E-03		3.381E-03	
1/16	7.913E-04	1.0200	1.310E-03	1.0600	1.535E-03	1.1400
1/32	3.923E-04	1.0100	6.419E-04	1.0300	7.328E-04	1.0700
1/64	1.953E-04	1.0000	3.177E-04	1.0100	3.582E-04	1.0300

Table 4.6: Error and convergence rate for Example 4.4 when initial data is $G_0(x) = \delta(x - 1/4)$.

τ	$\alpha = 0.25$		$\alpha = 0.5$		$\alpha = 0.75$	
	Error	Rate	Error	Rate	Error	Rate
1/8	1.058E-03		1.553E-03		1.772E-03	
1/16	5.194E-04	1.0300	7.452E-04	1.0600	8.061E-04	1.1400
1/32	2.574E-04	1.0100	3.653E-04	1.0300	3.852E-04	1.0600
1/64	1.281E-04	1.0100	1.808E-04	1.0100	1.884E-04	1.0300

Appendix A

Fractional Calculus and Related Spaces

A.1 Continuous Function Spaces

Here we introduce some spaces to describe the continuity of functions [Adams (1975)].

Firstly, we introduce $C^m(\Omega)$, i.e., $C^m(\Omega)$ is the vector space consisting of all functions ϕ and their partial derivatives $D^\alpha \phi$ of orders $|\alpha| \leq m$ are continuous on Ω; here Ω is a domain in $\mathbb{R}^n$ (Ω is a non-empty open set in n-dimensional real Euclidean space $\mathbb{R}^n$) and m is any non-negative integer. For convenience, we abbreviate $C^0(\Omega) \equiv C(\Omega)$. When $m = \infty$, the space $C^\infty(\Omega)$ can be defined as $C^\infty(\Omega) = \cap_{m=0}^\infty C^m(\Omega)$. Moreover, we define all those functions in $C(\Omega)$ and $C^\infty(\Omega)$ with compact support in Ω as $C_0(\Omega)$ and $C_0^\infty(\Omega)$, respectively. Since Ω is an open set, we can't assure the boundedness of functions, and we introduce $C_B^m(\Omega)$ which consists of those functions $\phi \in C^m(\Omega)$ and $D^\alpha \phi$ is bounded on Ω for $0 \leq |\alpha| \leq m$. And the norm of $C_B^m(\Omega)$ can be given by

$$\|\phi\|_{C_B^m(\Omega)} = \max_{0 \leq |\alpha| \leq m} \sup_{x \in \Omega} |D^\alpha \phi(x)|. \tag{A.1}$$

Next we present the definition of spaces of bounded, uniformly continuous functions. If $\phi \in C(\Omega)$ is bounded and uniformly continuous on Ω, i.e., there are two constants M_1, M_2 such that

$$\sup_{x \in \Omega} |\phi(x)| \leq M_1;$$
$$\sup_{x \in \Omega} |\phi(x + \epsilon) - \phi(x)| \leq M_2 \quad \forall \epsilon > 0, \tag{A.2}$$

then it possesses a unique, bounded, continuous extension to the closure $\bar{\Omega}$ of Ω. Besides, we define $C^m(\bar{\Omega})$ as the vector space consisting of all those functions $\phi \in C^m(\Omega)$ with $D^\alpha \phi$ bounded and uniformly continuous on Ω

for $0 \leq |\alpha| \leq m$. It is easy to find that $C^m(\bar{\Omega})$ is a closed subspace of $C_B^m(\Omega)$ with the norm

$$\|\phi\|_{C^m(\bar{\Omega})} = \max_{0 \leq |\alpha| \leq m} \sup_{x \in \Omega} |D^\alpha \phi(x)|. \tag{A.3}$$

By the definition of $C^m(\bar{\Omega})$, it is easy to see that $\lim_{h \to 0} |D^\alpha \phi(x + h) - D^\alpha \phi(x)| = 0$ for $\phi \in C^m(\bar{\Omega})$, $x \in \Omega$ and $|\alpha| = m$, but $\sup_{x \in \Omega} |D^\alpha \phi(x + h) - D^\alpha \phi(x)|$ may go to zero slowly as $\mathcal{O}(h)$. So in order to describe the function $\phi \in C^m(\bar{\Omega})$ better, we define the space of Hölder continuous functions $C^{m,\gamma}(\bar{\Omega})$, $0 < \gamma \leq 1$ to be the subspace of $C^m(\bar{\Omega})$ consisting of those functions ϕ which satisfies

$$|D^\alpha \phi(x) - D^\alpha \phi(y)| \leq C|x - y|^\gamma, \quad \exists C \text{ is a constant, } \forall x, y \in \Omega, \quad (A.4)$$

where $0 \leq |\alpha| \leq m$. And the norm of $C^{m,\gamma}(\bar{\Omega})$ can be defined as

$$\|\phi\|_{C^{m,\gamma}(\bar{\Omega})} = \|\phi\|_{C^m(\bar{\Omega})} + \max_{0 \leq |\alpha| \leq m} \sup_{x,y \in \Omega, x \neq y} \frac{|D^\alpha \phi(x) - D^\alpha \phi(y)|}{|x - y|^\gamma}. \tag{A.5}$$

In particular, we call Hölder continuity with exponent $\gamma = 1$ as Lipschitz continuity.

Next we provide the imbedding theorem for the various continuous function spaces introduced above.

Theorem A.1 ([Adams (1975)]). *Let m be a non-negative integer and $0 < \mu < \gamma \leq 1$. Then the following imbeddings exist*

$$C^{m+1}(\bar{\Omega}) \hookrightarrow C^m(\bar{\Omega});$$
$$C^{m,\mu}(\bar{\Omega}) \hookrightarrow C^m(\bar{\Omega}); \tag{A.6}$$
$$C^{m,\gamma}(\bar{\Omega}) \hookrightarrow C^{m,\mu}(\bar{\Omega}).$$

Specially, for convex Ω, the following imbeddings hold

$$C^{m+1}(\bar{\Omega}) \hookrightarrow C^{m,1}(\bar{\Omega});$$
$$C^{m+1}(\bar{\Omega}) \hookrightarrow C^{m,\gamma}(\bar{\Omega}). \tag{A.7}$$

A.2 Lebesgue Spaces

In this section, we review some basic concepts of Lebesgue spaces [Adams (1975)]. Let Ω be a domain in $\mathbb{R}^n$ and p a positive real number. We denote by $L^p(\Omega)$ the class of all measurable functions u defined on Ω whose norm is bounded, i.e.,

$$\|u(x)\|_{L^p(\Omega)}^p = \int_\Omega |u(x)|^p dx < \infty. \tag{A.8}$$

When $p = 2$, $L^2(\Omega)$ is a Hilbert space with scalar product

$$(u, v) = \int_\Omega uv dx. \tag{A.9}$$

Note that the two functions in $L^p(\Omega)$ are equivalent means they are equal almost everywhere (a.e.) in Ω. Thus $u(x) = 0$ in $L^p(\Omega)$ means $u(x) = 0$ a.e. in Ω. When $p = \infty$, $L^\infty(\Omega)$ is the vector space consisting of all measurable and essential bounded functions $u(x)$ defined on Ω and its L^∞ norm can be defined as

$$\|u(x)\|_{L^\infty(\Omega)} = \operatorname{ess\,sup}_{x \in \Omega} |u(x)| \leq C. \tag{A.10}$$

Then we provide the Hölder's inequality and the converse of Hölder's inequality.

Theorem A.2 ([Adams (1975)]). *Let $1 < p < \infty$ and q denote the conjugate exponent defined by*

$$q = \frac{p}{p-1}, \quad \text{that is } \frac{1}{p} + \frac{1}{q} = 1, \tag{A.11}$$

which also satisfies $1 < q < \infty$. If $u \in L^p(\Omega)$ and $v \in L^q(\Omega)$, then $uv \in L^1(\Omega)$, and

$$\int_\Omega |u(x)v(x)| dx \leq \|u\|_{L^p(\Omega)} \|v\|_{L^q(\Omega)}. \tag{A.12}$$

Equality holds if and only if $|u(x)|^p$ and $|v(x)|^q$ are proportional a.e. in Ω.

Theorem A.3 ([Adams (1975)]). *A measurable function u belongs to $L^p(\Omega)$ if and only if*

$$\sup \left\{ \int_\Omega |u(x)|v(x) dx \; : \; v(x) \geq 0 \text{ on } \Omega, \; \|v\|_{L^q(\Omega)} \leq 1 \right\} \tag{A.13}$$

is finite, and then that supremum equals to $\|u\|_{L^p(\Omega)}$.

According to Hölder's inequality, we have the following two corollaries.

Corollary A.1 ([Adams (1975)]). *If $p > 0$, $q > 0$ and $r > 0$ satisfying $(1/p) + (1/q) = 1/r$, and if $u \in L^p(\Omega)$ and $v \in L^q(\Omega)$, then $uv \in L^r(\Omega)$ and $\|uv\|_{L^r(\Omega)} \leq \|u\|_{L^p(\Omega)} \|v\|_{L^q(\Omega)}$.*

Corollary A.2 ([Adams (1975)]). *Hölder's inequality can be extended to products of more than two functions. Suppose $u = \prod_{j=1}^N u_j$, where $u_j \in L^{p_j}(\Omega)$, $1 \leq j \leq N$, where $p_j > 0$. If $\sum_{j=1}^N (1/p_j) = 1/q$, then $u \in L^q(\Omega)$ and $\|u\|_{L^q(\Omega)} \leq \prod_{j=1}^N \|u_j\|_{L^{p_j}(\Omega)}$.*

A.3   Sobolev Spaces

The continuous function spaces introduced in Sec. A.1 are useful in error analysis for finite difference method but not suitable in finite element methods. Here we introduce Sobolev spaces which are suitable to estimate the error of finite element scheme. Before we provide the definitions of Sobolev spaces [Adams (1975)], we introduce the definition of weak derivative first.

For $\Omega \subset \mathbb{R}^n$ and a given function $u \in C^1(\Omega)$, then if $\phi \in C_0^\infty(\Omega)$, using integration by the parts formula, we have

$$\int_\Omega u\phi_{x_i} dx = -\int_\Omega u_{x_i}\phi dx, \quad i = 1, 2, \cdots, n, \qquad (A.14)$$

where we abbreviate $u_{x_i} = \frac{\partial u}{\partial x_i}$. More generally, for $u \in C^m(\Omega)$ and a multi-index $\alpha = (\alpha_1, \cdots, \alpha_n)$, $(|\alpha| = m)$, there holds

$$\int_\Omega uD^\alpha\phi dx = (-1)^{|\alpha|}\int_\Omega D^\alpha u\phi dx, \qquad (A.15)$$

where $D^\alpha u = \frac{\partial^{\alpha_1}}{\partial x_1^{\alpha_1}} \cdots \frac{\partial^{\alpha_n}}{\partial x_n^{\alpha_n}} u$.

Motivated by these facts, we introduce the following definitions.

Definition A.1 ([Adams (1975)]). Let α be a multi-index. Suppose $u, v \in L_{loc}^1(\Omega)$, that is, for any open set $\mathbb{D} \subset \Omega$, $u, v \in L^1(\mathbb{D})$. Denote v as the αth weak derivative of u, i.e.,

$$D^\alpha u = v, \qquad (A.16)$$

which satisfies

$$\int_\Omega uD^\alpha\phi dx = (-1)^{|\alpha|}\int_\Omega v\phi dx \quad \forall \phi \in C_0^\infty(\Omega). \qquad (A.17)$$

Based on the definition of weak derivative, we can introduce the definition of Sobolev space.

Definition A.2 ([Adams (1975)]). Let m be an integer and $1 \leq p < \infty$. The Sobolev space $W^{m,p}(\Omega)$ can be defined as

$$W^{m,p}(\Omega) = \{u \in L^p(\Omega) : D^\alpha u \in L^p(\Omega), 0 \leq |\alpha| \leq m\}. \qquad (A.18)$$

And $W^{m,p}(\Omega)$ is a Banach space with norm

$$\|u\|_{W^{m,p}(\Omega)} = \sum_{|\alpha| \leq m} \|D^\alpha u\|_{L^p(\Omega)}. \qquad (A.19)$$

Obviously, for some integers m_1, m_2, we have

$$W^{m_1,p}(\Omega) \hookrightarrow W^{m_2,p}(\Omega), \quad m_1 > m_2. \tag{A.20}$$

Specially, we have $W^{0,p}(\Omega) = L^p(\Omega)$. Moreover, we denote $H^m(\Omega) = W^{m,2}(\Omega)$ and $H^m(\Omega)$ is a Hilbert space with a scalar product

$$(u,v)_{H^m(\Omega)} = \sum_{0 \le |\alpha| \le m} (D^\alpha u, D^\alpha v). \tag{A.21}$$

Furthermore, we denote $W_0^{m,p}(\Omega)$ as the closure of $C_0^\infty(\Omega)$ in $W^{m,p}(\Omega)$ and its dual space $W^{-m,q}(\Omega) = (W_0^{m,p}(\Omega))'$ with the norm

$$\|u\|_{W^{-m,q}(\Omega)} = \sup_{v \in W_0^{m,p}(\Omega),\ v \ne 0} \frac{(u,v)}{\|v\|_{W_0^{m,p}(\Omega)}} \tag{A.22}$$

and $1/p + 1/q = 1$. Meanwhile, for $u \in W_0^{1,p}(\Omega)$, we have the Poincáre inequality

$$\|u\|_{L^p(\Omega)} \le \|\nabla u\|_{L^p(\Omega)}. \tag{A.23}$$

A.4 Definitions and Properties of Fractional Calculus

In this section, we introduce the definitions and properties of fractional calculus [Podlubny (1999)].

Here we first provide the definition of Riemann-Liouville fractional integral on $\mathbb{R}^+$,

$$_0I_t^\alpha u(t) = \frac{1}{\Gamma(\alpha)} \int_0^t (t-s)^{\alpha-1} u(s) ds, \tag{A.24}$$

where $\alpha > 0$ and $\Gamma(\alpha)$ is Gamma function.

Then we can define Riemann-Liouville fractional derivative,

$$D_t^\alpha u(t) = \frac{1}{\Gamma(n-\alpha)} \frac{d^n}{dt^n} \int_0^t \frac{u(s)}{(t-s)^{\alpha-n+1}} ds, \tag{A.25}$$

where $n - 1 < \alpha < n$, $n \in \mathbb{N}^+$. Similarly, we can define Caputo fractional derivative,

$$^C D_t^\alpha u(t) = \frac{1}{\Gamma(n-\alpha)} \int_0^t \frac{1}{(t-s)^{\alpha-n+1}} \frac{d^n}{ds^n} u(s) ds, \tag{A.26}$$

where $n - 1 < \alpha < n$, $n \in \mathbb{N}^+$.

The Laplace transforms of Riemann-Liouville fractional integral, Riemann-Liouville fractional derivative and Caputo fractional derivative can be written as [Podlubny (1999)]

$$\widehat{{}_0I_t^\alpha u}(\lambda) = \lambda^{-\alpha}\hat{u}(\lambda);$$

$$\widehat{D_t^\alpha u}(\lambda) = \lambda^\alpha \hat{u}(\lambda) - \sum_{k=0}^{n-1} \lambda^k (D_t^{\alpha-k-1}u(t))|_{t=0};$$

$$\widehat{{}^CD_t^\alpha u}(\lambda) = \lambda^\alpha \hat{u}(\lambda) - \sum_{k=0}^{n-1} \lambda^{\alpha-k-1}\frac{\partial^k}{\partial t^k}u(t)|_{t=0}.$$

(A.27)

About ${}_0I_t^\alpha$, D_t^α and ${}^CD_t^\alpha$, we have the following properties.

Proposition A.1 ([Podlubny (1999)]). *If $0 < \alpha_1, \alpha_2 < 1$ and $u(0) = 0$, then we have*

$$D_t^{\alpha_1}(D_t^{\alpha_2}u(t)) = D_t^{\alpha_2}(D_t^{\alpha_1}u(t)) = D_t^{\alpha_1+\alpha_2}u(t).$$

(A.28)

Proposition A.2 ([Podlubny (1999)]). *Let $\alpha > 0$ and n be the smallest integer larger than α.*

- *If $u(x) \in L^2(0, T)$, then*

$$D_t^\alpha({}_0I_t^\alpha u(t)) = u(t).$$

(A.29)

- *If $0 < \alpha < 1$ and $u(t) \in C^1[0, T]$, then*

$${}_0I_t^\alpha(D_t^\alpha u(t)) = u(t).$$

(A.30)

Proposition A.3 ([Podlubny (1999)]). *If $0 < \alpha < 1$, then*

$${}^CD_t^\alpha u(t) = D_t^\alpha(u(t) - u(0)).$$

(A.31)

In the following, we provide the definitions of tempered Riemann-Liouville fractional integral on $\mathbb{R}^+$,

$${}_0I_t^{\alpha,\mu}u(t) = \frac{1}{\Gamma(\alpha)}e^{-\mu t}\int_0^t (t-s)^{\alpha-1}e^{\mu s}u(s)ds,$$

(A.32)

and tempered Riemann-Liouville fractional derivative is defined as

$$D_t^{\alpha,\mu}u(t) = \frac{e^{-\mu t}}{\Gamma(n-\alpha)}\frac{d^n}{dt^n}\int_0^t \frac{e^{\mu s}u(s)}{(t-s)^{\alpha-n+1}}ds,$$

(A.33)

where $n - 1 < \alpha < n$, $n \in \mathbb{N}^+$. Similarly, we can define tempered Caputo fractional derivative

$${}^CD_t^{\alpha,\mu}u(t) = \frac{e^{-\mu t}}{\Gamma(n-\alpha)}\int_0^t \frac{1}{(t-s)^{\alpha-n+1}}\frac{d^n}{ds^n}(e^{\mu s}u(s))ds,$$

(A.34)

where $n - 1 < \alpha < n$, $n \in \mathbb{N}^+$.

The Laplace transforms of tempered Riemann-Liouville fractional integral, tempered Riemann-Liouville fractional derivative and tempered Caputo fractional derivative can be written as [Podlubny (1999)]

$$\widehat{{}_0I_t^{\alpha,\mu}u}(\lambda) = (\lambda + \mu)^{-\alpha}\hat{u}(\lambda);$$

$$\widehat{D_t^{\alpha,\mu}u}(\lambda) = (\lambda + \mu)^{\alpha}\hat{u}(\lambda) - \sum_{k=0}^{n-1}(\lambda + \mu)^k(D_t^{\alpha-k-1}(e^{\mu t}u(t)))|_{t=0};$$

$$\widehat{{}^CD_t^{\alpha,\mu}u}(\lambda) = (\lambda + \mu)^{\alpha}\hat{u}(\lambda) - \sum_{k=0}^{n-1}(\lambda + \mu)^{\alpha-k-1}\frac{\partial^k}{\partial t^k}(e^{\mu t}u(t))|_{t=0}.$$

(A.35)

Proposition A.4 ([Podlubny (1999)]). *If $0 < \alpha_1, \alpha_2 < 1$, $\mu > 0$ and $u(0) = 0$, then we have*

$$D_t^{\alpha_1,\mu}(D_t^{\alpha_2,\mu}u(t)) = D_t^{\alpha_2,\mu}(D_t^{\alpha_1,\mu}u(t)) = D_t^{\alpha_1+\alpha_2,\mu}u(t). \quad (A.36)$$

Proposition A.5 ([Podlubny (1999)]). *Let $\alpha, \mu > 0$ and n be the smallest integer larger than α.*

- *If $u(x) \in L^2(0, T)$, then*

$$D_t^{\alpha,\mu}({}_0I_t^{\alpha,\mu}u(t)) = u(t). \quad (A.37)$$

- *If $0 < \alpha < 1$ and $u(t) \in C^1[0, T]$, then*

$${}_0I_t^{\alpha,\mu}(D_t^{\alpha,\mu}u(t)) = u(t). \quad (A.38)$$

Proposition A.6 ([Podlubny (1999)]). *If $0 < \alpha < 1$ and $\mu > 0$ then*

$${}^CD_t^{\alpha,\mu}u(t) = D_t^{\alpha,\mu}(u(t) - e^{-\mu t}u(0)) = D_t^{\alpha,\mu}u(t) - e^{-\mu t}(D_t^{\alpha}u(t))|_{t=0}.$$

(A.39)

A.5 Discretization of (Tempered) Fractional Laplacian

In this section, we provide the definition of the fractional Laplacian $(-\Delta)^s$ in singular integral form [Bertoin (1996); Deng *et al.* (2018b)] firstly,

$$(-\Delta)^s u(\mathbf{x}) = -c_{n,s}\text{P.V.}\int_{\mathbb{R}^n}\frac{u(\mathbf{x}) - u(\mathbf{y})}{|\mathbf{x} - \mathbf{y}|^{n+2s}}d\mathbf{y} \quad \text{for } s \in (0, 1), \quad (A.40)$$

where $c_{n,s} = \frac{s\Gamma(\frac{n+2s}{2})}{2^{-2s}\pi^{n/2}\Gamma(1-s)}$. To describe the physical processes more accurately, the tempered fractional Laplacian has been derived in [Deng *et al.* (2018a)], i.e.,

$$-(\Delta + \mu)^s u(\mathbf{x}) = c_{n,s,\mu}\text{P.V.}\int_{\mathbb{R}^n}\frac{u(\mathbf{x}) - u(\mathbf{y})}{e^{\mu|\mathbf{x}-\mathbf{y}|}|\mathbf{x} - \mathbf{y}|^{n+2s}}d\mathbf{y} \quad \text{for } s \in (0, 1),$$

(A.41)

where

$$c_{n,s,\mu} = \begin{cases} \dfrac{\Gamma(\frac{n}{2})}{2\pi^{n/2}|\Gamma(-2s)|} & \text{for } \mu > 0 \text{ and } s \neq 1/2, \\[4mm] \dfrac{s\Gamma(\frac{n+2s}{2})}{2^{-2s}\pi^{n/2}\Gamma(1-s)} & \text{for } \mu = 0 \text{ or } s = 1/2. \end{cases} \quad \text{(A.42)}$$

There have been some works on discretizing (tempered) fractional Laplacian [Acosta and Borthagaray (2017); Duo *et al.* (2018); Duo and Zhang (2019a,b); Nie *et al.* (2019); Zhang *et al.* (2018, 2019)]. In the following, we consider the discretization of the two-dimensional tempered fractional Laplacian on a bounded domain $\Omega = (-l, l) \times (-l, l)$ with extended homogeneous Dirichlet boundary conditions: $G(x, y) \equiv 0$ for $(x, y) \in \Omega^c$ [Nie *et al.* (2019)]. We set the mesh sizes $h_1 = l/N_1$ and $h_2 = l/N_2$; denote grid points $x_i = ih_1$ and $y_j = jh_2$, $i, j \in \mathbb{Z}$, for $-N_1 \leq i \leq N_1$ and $-N_2 \leq j \leq N_2$; for convenience, let $N_1 = N_2 = N$, then we can set $h_1 = h_2 = h$. According to Eq. (A.41), we have

$$-(\Delta + \mu)^s G(x, y) = -c_{2,s,\mu} \text{P.V.} \int \int_{\mathbb{R}^2} \frac{G(\xi, \eta) - G(x, y)}{\vartheta(x, y, \xi, \eta)} d\xi d\eta, \quad \text{(A.43)}$$

where

$$\vartheta(x, y, \xi, \eta) = e^{\mu \sqrt{(\xi-x)^2 + (\eta-y)^2}} \left(\sqrt{(\xi - x)^2 + (\eta - y)^2} \right)^{2+2s}. \quad \text{(A.44)}$$

To discretize $(\Delta + \mu)^s G(x_p, y_q)$ for any $-N \leq p, q \leq N$, we first divide the integral domain into two parts for Eq. (A.43), i.e.,

$$\int \int_{\mathbb{R}^2} \frac{G(\xi, \eta) - G(x_p, y_q)}{\vartheta(x_p, y_q, \xi, \eta)} d\xi d\eta = \int \int_{\Omega} \frac{G(\xi, \eta) - G(x_p, y_q)}{\vartheta(x_p, y_q, \xi, \eta)} d\xi d\eta$$
$$+ \int \int_{\mathbb{R}^2 \backslash \Omega} \frac{G(\xi, \eta) - G(x_p, y_q)}{\vartheta(x_p, y_q, \xi, \eta)} d\xi d\eta. \quad \text{(A.45)}$$

It is easy to see that

$$\int \int_{\mathbb{R}^2 \backslash \Omega} \frac{G(\xi, \eta) - G(x_p, y_q)}{\vartheta(x_p, y_q, \xi, \eta)} d\xi d\eta = -W_{p,q}^{\infty} G(x_p, y_q), \quad \text{(A.46)}$$

where the fact $G(\xi, \eta) \equiv 0$ for $(\xi, \eta) \in \mathbb{R}^2 \backslash \Omega$ is used and

$$W_{p,q}^{\infty} = \int \int_{\mathbb{R}^2 \backslash \Omega} \frac{1}{\vartheta(x_p, y_q, \xi, \eta)} d\xi d\eta. \quad \text{(A.47)}$$

Here, $W_{p,q}^{\infty}$ can be calculated by the built-in function 'integral2.m' in MATLAB.

Next, we formulate the first integral in Eq. (A.45) as

$$\int\int_\Omega \frac{G(\xi,\eta) - G(x_p, y_q)}{\vartheta(x_p, y_q, \xi, \eta)} d\eta d\xi$$
$$= \sum_{i=-N}^{N-1} \sum_{j=-N}^{N-1} \int_{\xi_i}^{\xi_{i+1}} \int_{\eta_j}^{\eta_{j+1}} \frac{G(\xi,\eta) - G(x_p, y_q)}{\vartheta(x_p, y_q, \xi, \eta)} d\eta d\xi, \tag{A.48}$$

where $\xi_i = ih$ and $\eta_j = jh$. Denote $\mathcal{I}_{p,q} = \{(p,q), (p-1, q), (p, q-1), (p-1, q-1)\}$ and

$$\psi(x_p, y_q, \xi, \eta) = G(x_p + \xi, y_q + \eta) + G(x_p - \xi, y_q + \eta)$$
$$+ G(x_p - \xi, y_q - \eta) + G(x_p + \xi, y_q - \eta) - 4G(x_p, y_q). \tag{A.49}$$

For Eq. (A.48), when $(i,j) \in \mathcal{I}_{p,q}$, we rewrite them as

$$\int_{\xi_{p-1}}^{\xi_p} \int_{\eta_{q-1}}^{\eta_q} \frac{G(\xi,\eta) - G(x_p, y_q)}{\vartheta(x_p, y_q, \xi, \eta)} d\eta d\xi$$
$$+ \int_{\xi_p}^{\xi_{p+1}} \int_{\eta_{q-1}}^{\eta_q} \frac{G(\xi,\eta) - G(x_p, y_q)}{\vartheta(x_p, y_q, \xi, \eta)} d\eta d\xi$$
$$+ \int_{\xi_{p-1}}^{\xi_p} \int_{\eta_q}^{\eta_{q+1}} \frac{G(\xi,\eta) - G(x_p, y_q)}{\vartheta(x_p, y_q, \xi, \eta)} d\eta d\xi$$
$$+ \int_{\xi_p}^{\xi_{p+1}} \int_{\eta_q}^{\eta_{q+1}} \frac{G(\xi,\eta) - G(x_p, y_q)}{\vartheta(x_p, y_q, \xi, \eta)} d\eta d\xi$$
$$= \int_{\xi_{p-1}}^{\xi_{p+1}} \int_{\eta_{q-1}}^{\eta_{q+1}} \frac{G(\xi,\eta) - G(x_p, y_q)}{\vartheta(x_p, y_q, \xi, \eta)} d\eta d\xi$$
$$= \int_0^h \int_0^h \frac{\psi(x_p, y_q, \xi, \eta)}{e^{\mu\sqrt{\xi^2 + \eta^2}} \left(\sqrt{\xi^2 + \eta^2}\right)^{2+2s}} d\eta d\xi, \tag{A.50}$$

where we use

$$\vartheta(x, y, x - \xi, y - \eta) = \vartheta(x, y, x + \xi, y - \eta)$$
$$= \vartheta(x, y, x - \xi, y + \eta) = \vartheta(x, y, x + \xi, y + \eta). \tag{A.51}$$

Furthermore, denoting

$$\phi_\sigma(\xi, \eta) = \frac{\psi(x_p, y_q, \xi, \eta)}{e^{\mu\sqrt{\xi^2 + \eta^2}} \left(\sqrt{\xi^2 + \eta^2}\right)^\sigma}, \quad \sigma \in (2s, 2], \tag{A.52}$$

and using the weighted trapezoidal rule, we have

$$
\int_0^h \int_0^h \phi_\sigma(\xi,\eta)(\xi^2+\eta^2)^{\frac{\sigma-2-2s}{2}} d\eta d\xi
$$

$$
\approx
\begin{cases}
\dfrac{1}{4}\left(\lim_{(\xi,\eta)\to(0,0)}\phi_\sigma(\xi,\eta)+\phi_\sigma(\xi_0,\eta_1)+\phi_\sigma(\xi_1,\eta_1)+\phi_\sigma(\xi_1,\eta_0)\right)W_{0,0}, \\
\qquad\qquad\qquad\qquad\qquad\qquad\qquad\qquad\qquad \sigma\in(2s,2); \\[2mm]
\dfrac{1}{3}\left(\phi_\sigma(\xi_0,\eta_1)+\phi_\sigma(\xi_1,\eta_1)+\phi_\sigma(\xi_1,\eta_0)\right)W_{0,0}, \quad \sigma=2,
\end{cases}
$$

$$\text{(A.53)}$$

where

$$
W_{0,0}=\int_0^h \int_0^h (\xi^2+\eta^2)^{\frac{\sigma-2-2s}{2}} d\eta d\xi. \tag{A.54}
$$

Assuming that u is smooth enough, for $\sigma\in(2s,2)$, one has

$$
\lim_{(\xi,\eta)\to(0,0)}\phi_\sigma(\xi,\eta)=0; \tag{A.55}
$$

and further introduce a parameter

$$
k_\sigma=
\begin{cases}
1 & \sigma\in(2s,2), \\
\dfrac{4}{3} & \sigma=2.
\end{cases}
\tag{A.56}
$$

So, Eq. (A.53) can be rewritten as

$$
\int_0^h \int_0^h \phi_\sigma(\xi,\eta)(\xi^2+\eta^2)^{\frac{\sigma-2-2s}{2}} d\eta d\xi
$$

$$
\approx \frac{k_\sigma}{4}\left(\phi_\sigma(\xi_0,\eta_1)+\phi_\sigma(\xi_1,\eta_1)+\phi_\sigma(\xi_1,\eta_0)\right)W_{0,0}.
$$

$$\text{(A.57)}$$

For Eq. (A.48), when $(i,j)\notin\mathcal{I}_{p,q}$, denote $I_{p,q,p+i,q+j}$ as the approximation of

$$
\int_{\xi_{p+i}}^{\xi_{p+i+1}} \int_{\eta_{q+j}}^{\eta_{q+j+1}} \frac{G(\xi,\eta)-G(x_p,y_q)}{\vartheta(x_p,y_q,\xi,\eta)} d\eta d\xi; \tag{A.58}
$$

we use the bilinear interpolation to approximate $\dfrac{G(\xi,\eta)-G(x_p,y_q)}{e^{\mu\sqrt{(\xi-x)^2+(\eta-y)^2}}}$ in $[\xi_{p+i},\xi_{p+i+1}]\times[\eta_{q+j},\eta_{q+j+1}]$ and get

$$
I_{p,q,p+i,q+j}=\left(\frac{G(\xi_{p+i},\eta_{q+j})-G(x_p,y_q)}{e^{\mu h\sqrt{i^2+j^2}}}\right)W^1_{i,j}
$$

$$
+\left(\frac{G(\xi_{p+i+1},\eta_{q+j})-G(x_p,y_q)}{e^{\mu h\sqrt{(i+1)^2+j^2}}}\right)W^2_{i+1,j}
$$

$$
+\left(\frac{G(\xi_{p+i},\eta_{q+j+1})-G(x_p,y_q)}{e^{\mu h\sqrt{i^2+(j+1)^2}}}\right)W^3_{i,j+1}
$$

$$
+\left(\frac{G(\xi_{p+i+1},\eta_{q+j+1})-G(x_p,y_q)}{e^{\mu h\sqrt{(i+1)^2+(j+1)^2}}}\right)W^4_{i+1,j+1},
$$

$$\text{(A.59)}$$

where

$$W_{i,j}^1 = H_{i,j}^{\xi\eta} - \xi_{i+1}H_{i,j}^{\eta} - \eta_{j+1}H_{i,j}^{\xi} + \xi_{i+1}\eta_{j+1}H_{i,j},$$

$$W_{i,j}^2 = -\left(H_{i-1,j}^{\xi\eta} - \xi_{i-1}H_{i-1,j}^{\eta} - \eta_{j+1}H_{i-1,j}^{\xi} + \xi_{i-1}\eta_{j+1}H_{i-1,j}\right),$$

$$W_{i,j}^3 = -\left(H_{i,j-1}^{\xi\eta} - \xi_{i+1}H_{i,j-1}^{\eta} - \eta_{j-1}H_{i,j-1}^{\xi} + \xi_{i+1}\eta_{j-1}H_{i,j-1}\right),$$

$$W_{i,j}^4 = H_{i-1,j-1}^{\xi\eta} - \xi_{i-1}H_{i-1,j-1}^{\eta} - \eta_{j-1}H_{i-1,j-1}^{\xi} + \xi_{i-1}\eta_{j-1}H_{i-1,j-1},$$

$$(A.60)$$

and

$$H_{i,j} = \frac{1}{h^2}\int_{\xi_i}^{\xi_{i+1}}\int_{\eta_j}^{\eta_{j+1}}(\xi^2 + \eta^2)^{\frac{-2-2s}{2}}\,d\eta d\xi,$$

$$H_{i,j}^{\xi} = \frac{1}{h^2}\int_{\xi_i}^{\xi_{i+1}}\int_{\eta_j}^{\eta_{j+1}}\xi(\xi^2 + \eta^2)^{\frac{-2-2s}{2}}\,d\eta d\xi,$$

$$H_{i,j}^{\eta} = \frac{1}{h^2}\int_{\xi_i}^{\xi_{i+1}}\int_{\eta_j}^{\eta_{j+1}}\eta(\xi^2 + \eta^2)^{\frac{-2-2s}{2}}\,d\eta d\xi,$$

$$H_{i,j}^{\xi\eta} = \frac{1}{h^2}\int_{\xi_i}^{\xi_{i+1}}\int_{\eta_j}^{\eta_{j+1}}\xi\eta(\xi^2 + \eta^2)^{\frac{-2-2s}{2}}\,d\eta d\xi.$$

$$(A.61)$$

Then Eq. (A.48) becomes

$$\sum_{i=-N}^{N-1}\sum_{j=-N}^{N-1}\int_{\xi_i}^{\xi_{i+1}}\int_{\eta_j}^{\eta_{j+1}}\frac{G(\xi,\eta) - G(x_p, y_q)}{\vartheta(x_p, y_q, \xi, \eta)}\,d\eta d\xi$$

$$\approx \frac{k_\sigma}{4}\left(\phi_\sigma(\xi_0, \eta_1) + \phi_\sigma(\xi_1, \eta_1) + \phi_\sigma(\xi_1, \eta_0)\right)W_{0,0} + \sum_{\substack{i=-N, j=-N;\\(i,j)\notin\mathcal{I}_{p,q}}}^{N-1,N-1}I_{p,q,i,j}.$$

$$(A.62)$$

To make the form of weight $w_{p,q,i,j}^{s,\mu}$ unified, according to Eq. (A.57), we denote

$$W_{-1,-1}^1 = W_{1,-1}^2 = W_{-1,1}^3 = W_{1,1}^4 = \frac{k_\sigma}{4}\frac{W_{0,0}}{(\sqrt{2}h)^\sigma},$$

$$W_{-1,0}^1 = W_{-1,0}^3 = W_{1,0}^2 = W_{1,0}^4 = \quad\quad\quad (A.63)$$

$$W_{0,-1}^1 = W_{0,-1}^2 = W_{0,1}^3 = W_{0,1}^4 = \frac{k_\sigma}{4}\frac{W_{0,0}}{(h)^\sigma}.$$

Let $G_{p,q} = G(x_p, x_q)$. Then we have the discretization scheme

$$-(\Delta + \mu)_h^s G_{p,q} = \sum_{i=-N}^{N}\sum_{j=-N}^{N}w_{p,q,i,j}^{s,\mu}G_{i,j},\quad\quad (A.64)$$

where

$$w_{p,q,i,j}^{s,\mu} = -c_{2,s,\mu}$$

$$\begin{cases} -\left(\displaystyle\sum_{(i,j)\neq(p,q)} w_{p,q,i,j}^{s,\mu} + W_{p,q}^{\infty} \right), & i = p, j = q; \\[3.2em] \dfrac{W_{i-p,j-q}^{1} + W_{i-p,j-q}^{2}}{e^{\mu h \sqrt{(i-p)^2+(j-q)^2}}}, & -N < i < N, j = -N; \\[2.2em] \dfrac{W_{i-p,j-q}^{1} + W_{i-p,j-q}^{3}}{e^{\mu h \sqrt{(i-p)^2+(j-q)^2}}}, & i = -N, -N < j < N; \\[2.2em] \dfrac{W_{i-p,j-q}^{3} + W_{i-p,j-q}^{4}}{e^{\mu h \sqrt{(i-p)^2+(j-q)^2}}}, & -N < i < N, j = N; \\[2.2em] \dfrac{W_{i-p,j-q}^{2} + W_{i-p,j-q}^{4}}{e^{\mu h \sqrt{(i-p)^2+(j-q)^2}}}, & i = N, -N < j < N; \\[2.2em] \dfrac{W_{i-p,j-q}^{1}}{e^{\mu h \sqrt{(i-p)^2+(j-q)^2}}}, & i = -N, j = -N; \\[2.2em] \dfrac{W_{i-p,j-q}^{2}}{e^{\mu h \sqrt{(i-p)^2+(j-q)^2}}}, & i = N, j = -N; \\[2.2em] \dfrac{W_{i-p,j-q}^{3}}{e^{\mu h \sqrt{(i-p)^2+(j-q)^2}}}, & i = -N, j = N; \\[2.2em] \dfrac{W_{i-p,j-q}^{4}}{e^{\mu h \sqrt{(i-p)^2+(j-q)^2}}}, & i = N, j = N; \\[2.2em] \dfrac{W_{i-p,j-q}^{1} + W_{i-p,j-q}^{2} + W_{i-p,j-q}^{3} + W_{i-p,j-q}^{4}}{e^{\mu h \sqrt{(i-p)^2+(j-q)^2}}}, & \text{otherwise.} \end{cases}$$

$$(A.65)$$

Remark A.1. Here, we discretize the tempered fractional Laplacian satisfying homogeneous Dirichlet boundary conditions, so Eq. (A.64) can be rewritten as

$$-(\Delta+\mu)_h^s G_{p,q} = \sum_{i=-N+1}^{N-1} \sum_{j=-N+1}^{N-1} w_{p,q,i,j}^{s,\mu} G_{i,j}. \qquad (A.66)$$

Theorem A.4. *Denote $(\Delta+\mu)_h^s$ as a finite difference approximation of the tempered fractional Laplacian $(\Delta+\mu)^s$. Suppose that $G(\mathbf{x}) \in C^2(\bar{\Omega})$ is supported in an open set $\Omega \subset \mathbb{R}^2$. Then, there are*

$$\|(\Delta+\mu)^s G - (\Delta+\mu)_h^s G\|_{\infty} \le Ch^{2-2s},$$
$$\|(\Delta+\mu)^s G - (\Delta+\mu)_h^s G\| \le Ch^{2-2s} \qquad \text{for } s \in (0,1) \tag{A.67}$$

with C being a positive constant depending on s and μ. Here $\| \cdot \|_{\infty}$ and $\| \cdot \|$ mean discrete l^{∞} and l^2 norms, respectively.

Bibliography

Aaronson, J. (1997). *An Introduction to Infinite Ergodic Theory* (American Mathematical Society, Providence).

Abramowitz, M. and Stegun, I. A. (1972). *Handbook of Mathematical Functions* (Dover, New York).

Acosta, G. and Borthagaray, J. P. (2017). A fractional Laplace equation: regularity of solutions and finite element approximations, *SIAM J. Numer. Anal.* **55**, pp. 472–495.

Adams, R. A. (1975). *Sobolev Spaces* (Academic Press, New York).

Aghion, E., Kessler, D. A., and Barkai, E. (2019). From non-normalizable Boltzmann-Gibbs statistics to infinite-ergodic theory, *Phys. Rev. Lett.* **122**, p. 010601.

Agmon, N. (1984). Residence times in diffusion processes, *J. Chem. Phys.* **81**, pp. 3644–3647.

Akimoto, T. (2012). Distributional response to biases in deterministic superdiffusion, *Phys. Rev. Lett.* **108**, p. 164101.

Akimoto, T., Cherstvy, A. G., and Metzler, R. (2018). Ergodicity, rejuvenation, enhancement, and slow relaxation of diffusion in biased continuous-time random walks, *Phys. Rev. E* **98**, p. 022105.

Applebaum, D. (2009). *Lévy Processes and Stochastic Calculus* (Cambridge University Press, Cambridge).

Arendt, W., Batty, C. J., Hieber, M., and Neubrander, F. (2011). *Vector-valued Laplace Transforms and Cauchy Problems*, 2nd edn. (Birkhäuser, Basel).

Artuso, R. and Cristadoro, G. (2003). Anomalous transport: a deterministic approach, *Phys. Rev. Lett.* **90**, p. 244101.

Barabasi, A. L. and Stanley, H. E. (1995). *Fractal Concepts in Surface Growth* (Cambridge University Press, Cambridge).

Barkai, E. (2001). Fractional Fokker-Planck equation, solution, and application, *Phys. Rev. E* **63**, p. 046118.

Barkai, E. (2006). Residence time statistics for normal and fractional diffusion in a force field, *J. Stat. Phys.* **123**, pp. 883–907.

Barkai, E., Aghion, E., and Kessler, D. A. (2014). From the area under the Bessel excursion to anomalous diffusion of cold atoms, *Phys. Rev. X* **4**, p. 021036.

Barkai, E. and Burov, S. (2020). Packets of diffusing particles exhibit universal exponential tails, *Phys. Rev. Lett.* **124**, p. 060603.

Barkai, E. and Cheng, Y.-C. (2003). Aging continuous time random walks, *J. Chem. Phys.* **118**, pp. 6167–6178.

Barkai, E., Metzler, R., and Klafter, J. (2000). From continuous time random walks to the fractional Fokker-Planck equation, *Phys. Rev. E* **61**, pp. 132–138.

Barkai, E. and Silbey, R. J. (2000). Fractional Kramers equation, *J. Phys. Chem. B* **104**, pp. 3866–3874.

Bartumeus, F., Catalan, J., Fulco, U. L., Lyra, M. L., and Viswanathan, G. M. (2002). Optimizing the encounter rate in biological interactions: Lévy versus Brownian strategies, *Phys. Rev. Lett.* **88**, p. 097901.

Baule, A. and Friedrich, R. (2005). Joint probability distributions for a class of non-Markovian processes, *Phys. Rev. E* **71**, p. 026101.

Baule, A. and Friedrich, R. (2006). Investigation of a generalized Obukhov model for turbulence, *Phys. Lett. A* **350**, pp. 167–173.

Beck, C. (2001). Dynamical foundations of nonextensive statistical mechanics, *Phys. Rev. Lett.* **87**, p. 180601.

Beck, C. (2006). Superstatistical Brownian motion, *Prog. Theor. Phys. Suppl.* **162**, pp. 29–36.

Beck, C. and Cohen, E. G. D. (2003). Superstatistics, *Physica A* **322**, pp. 267–275.

Bell, J. W. (1991). *Searching Behaviour, the Behavioural Ecology of Finding Resources* (Chapman and Hall, London).

Bénichou, O., Coppey, M., Moreau, M., Suet, P.-H., and Voituriez, R. (2005a). Optimal search strategies for hidden targets, *Phys. Rev. Lett.* **94**, p. 198101.

Bénichou, O., Coppey, M., Moreau, M., Suet, P.-H., and Voituriez, R. (2005b). A stochastic model for intermittent search strategies, *J. Phys.: Condens. Matter* **17**, pp. S4275–S4286.

Bénichou, O., Loverdo, C., Moreau, M., and Voituriez, R. (2011). Intermittent search strategies, *Rev. Mod. Phys.* **83**, pp. 81–129.

Benjacob, E., Bergman, D. J., Matkowsky, B. J., and Schuss, Z. (1982). Lifetime of oscillatory steady states, *Phys. Rev. A* **26**, pp. 2805–2816.

Berg, O. G. and von Hippel, P. H. (1985). Diffusion-controlled macromolecular interactions, *Annu. Rev. Biophys. Biophys. Chem.* **14**, pp. 131–160.

Berg, O. G., Winter, R. B., and von Hippel, P. H. (1981). Diffusion-driven mechanisms of protein translocation on nucleic acids. 1. models and theory, *Biochemistry* **20**, pp. 6929–6948.

Bertoin, J. (1996). *Lévy Processes* (Cambridge University Press, Cambridge, UK).

Bhattacharya, S., Sharma, D. K., Saurabh, S., De, S., Sain, A., Nandi, A., and Chowdhury, A. (2013). Plasticization of Poly(vinylpyrrolidone) thin films under ambient humidity: Insight from single-molecule tracer diffusion dynamics, *J. Phys. Chem. B* **117**, pp. 7771–7782.

Blumen, A., Klafter, J., and Zumofen, G. (1986). *Optical Spectroscopy of Glasses* (Reidel, Dordrecht).

Bobrovsky, B. Z. and Schuss, Z. (1982). A singular perturbation method for the computation of the mean first passage time in a nonlinear filter, *SIAM J.*

Appl. Math. **42**, pp. 174–187.

Bochner, S. (1949). Diffusion equation and stochastic processes, *Proc. Natl. Acad. Sci. USA* **35**, pp. 368–370.

Bouchaud, J.-P. and Georges, A. (1990). Comment on "Stochastic pathway to anomalous diffusion", *Phys. Rev. A* **41**, pp. 1156–1157.

Bronstein, I., Israel, Y., Kepten, E., Mai, S., Shav-Tal, Y., Barkai, E., and Garini, Y. (2009). Transient anomalous diffusion of telomeres in the nucleus of mammalian cells, *Phys. Rev. Lett.* **103**, p. 018102.

Burov, S. and Barkai, E. (2008a). Critical exponent of the fractional Langevin equation, *Phys. Rev. Lett.* **100**, p. 070601.

Burov, S. and Barkai, E. (2008b). Fractional Langevin equation: overdamped, underdamped, and critical behaviors, *Phys. Rev. E* **78**, p. 031112.

Burov, S., Jeon, J.-H., Metzler, R., and Barkai, E. (2011). Single particle tracking in systems showing anomalous diffusion: the role of weak ergodicity breaking, *Phys. Chem. Chem. Phys.* **13**, pp. 1800–1812.

Burov, S., Metzler, R., and Barkai, E. (2010). Aging and nonergodicity beyond the Khinchin theorem, *Proc. Natl. Acad. Sci. USA* **107**, pp. 13228–13233.

Cairoli, A. and Baule, A. (2015a). Anomalous processes with general waiting times: functionals and multipoint structure, *Phys. Rev. Lett.* **115**, p. 110601.

Cairoli, A. and Baule, A. (2015b). Langevin formulation of a subdiffusive continuous-time random walk in physical time, *Phys. Rev. E* **92**, p. 012102.

Cairoli, A. and Baule, A. (2017). Feynman-Kac equation for anomalous processes with space- and time-dependent forces, *J. Phys. A* **50**, p. 164002.

Cairoli, A., Klages, R., and Baule, A. (2018). Weak Galilean invariance as a selection principle for coarse-grained diffusive models, *Proc. Natl. Acad. Sci. USA* **115**, pp. 5714–5719.

Carmeli, B. and Nitzan, A. (1983). Theory of activated rate processes: bridging between the Kramers limits, *Phys. Rev. Lett.* **51**, pp. 233–236.

Carmi, S. and Barkai, E. (2011). Fractional Feynman-Kac equation for weak ergodicity breaking, *Phys. Rev. E* **84**, p. 061104.

Carmi, S., Turgeman, L., and Barkai, E. (2010). On distributions of functionals of anomalous diffusion paths, *J. Stat. Phys.* **141**, pp. 1071–1092.

Chandrasekhar, S. (1943). Stochastic problems in physics and astronomy, *Rev. Mod. Phys.* **15**, pp. 1–89.

Chechkin, A. V., Seno, F., Metzler, R., and Sokolov, I. M. (2017). Brownian yet non-Gaussian diffusion: from superstatistics to subordination of diffusing diffusivities, *Phys. Rev. X* **7**, p. 021002.

Chen, M. H. and Deng, W. H. (2014). Discretized fractional substantial calculus, *M2AN Math. Model. Numer. Anal.* **49**, pp. 373–394.

Chen, M. H. and Deng, W. H. (2015). High order algorithms for the fractional substantial diffusion equation with truncated Lévy flights, *SIAM J. Sci. Comput.* **37**, pp. A890–A917.

Chen, M. H. and Deng, W. H. (2018). High order algorithm for the time-tempered fractional Feynman-Kac equation, *J. Sci. Comput.* **76**, pp. 1–21.

Chen, S., Liu, F., Zhuang, P., and Anh, V. (2009). Finite difference approximations for the fractional Fokker-Planck equation, *Appl. Math. Model.* **33**, pp. 256–273.

Chen, Y., Wang, X. D., and Deng, W. H. (2017). Localization and ballistic diffusion for the tempered fractional Brownian-Langevin motion, *J. Stat. Phys.* **169**, pp. 18–37.

Chen, Y., Wang, X. D., and Deng, W. H. (2018). Tempered fractional Langevin-Brownian motion with inverse β-stable subordinator, *J. Phys. A* **51**, p. 495001.

Chen, Y., Wang, X. D., and Deng, W. H. (2019). Subdiffusion in an external force field, *Phys. Rev. E* **99**, p. 042125.

Cherstvy, A. G., Chechkin, A. V., and Metzler, R. (2013). Anomalous diffusion and ergodicity breaking in heterogeneous diffusion processes, *New J. Phys.* **15**, p. 083039.

Cherstvy, A. G. and Metzler, R. (2013). Population splitting, trapping, and nonergodicity in heterogeneous diffusion processes, *Phys. Chem. Chem. Phys.* **15**, pp. 20220–20235.

Cherstvy, A. G. and Metzler, R. (2014). Nonergodicity, fluctuations, and criticality in heterogeneous diffusion processes, *Phys. Rev. E* **90**, p. 012134.

Cherstvy, A. G. and Metzler, R. (2015a). Ergodicity breaking, ageing, and confinement in generalized diffusion processes with position and time dependent diffusivity, *J. Stat. Mech.* **2015**, p. P05010.

Cherstvy, A. G. and Metzler, R. (2015b). Ergodicity breaking and particle spreading in noisy heterogeneous diffusion processes, *J. Chem. Phys.* **142**, p. 144105.

Cherstvy, A. G. and Metzler, R. (2016). Anomalous diffusion in time-fluctuating non-stationary diffusivity landscapes, *Phys. Chem. Chem. Phys.* **18**, pp. 23840–23852.

Chubynsky, M. V. and Slater, G. W. (2014). Diffusing diffusivity: A model for anomalous, yet Brownian, diffusion, *Phys. Rev. Lett.* **113**, p. 098302.

Coffey, W. T., Kalmykov, Y. P., and Waldron, J. T. (2004). *The Langevin Equation* (World Scientific, Singapore).

Compte, A., Metzler, R., and Camacho, J. (1997). Biased continuous time random walks between parallel plates, *Phys. Rev. E* **56**, pp. 1445–1454.

Comtet, A. and Majumdar, S. N. (2005). Precise asymptotics for a random walker's maximum, *J. Stat. Mech.*, p. P06013.

Comtet, A., Monthus, C., and Yor, M. (1998). Exponential functionals of Brownian motion and disordered system, *J. Appl. Probab.* **35**, pp. 255–271.

Coppey, M., Bénichou, O., Voituriez, R., and Moreau, M. (2004). Kinetics of target site localization of a protein on DNA: a stochastic approach, *Biophys. J.* **87**, pp. 1640–1649.

Darling, D. A. (1983). On the supremum of certain Gaussian processes, *Ann. Probab.* **11**, pp. 803–806.

Day, M. V. (1990). Large deviations results for the exit problem with characteristic boundary, *J. Math. Anal. Appl.* **147**, pp. 134–153.

de Anna, P., Le Borgne, T., Dentz, M., Tartakovsky, A. M., Bolster, D., and Davy, P. (2013). Flow intermittency, dispersion, and correlated continuous time random walks in porous media, *Phys. Rev. Lett.* **110**, p. 184502.

de Groot, S. R. and Mazur, P. (1969). *Non-Equilibrium Thermodynamics* (North-Holland, Amsterdam).

Debye, P. (1945). *Polar Molecules* (Dover, New York).

Dechant, A., Lutz, E., Kessler, D. A., and Barkai, E. (2014). Scaling Green-Kubo relation and application to three aging systems, *Phys. Rev. X* **4**, p. 011022.

Deng, W. H. and Barkai, E. (2009). Ergodic properties of fractional Brownian-Langevin motion, *Phys. Rev. E* **79**, p. 011112.

Deng, W. H., Chen, M. H., and Barkai, E. (2015). Numerical algorithms for the forward and backward fractional Feynman-Kac equations, *J. Sci. Comput.* **62**, pp. 718–746.

Deng, W. H., Li, B. Y., Qian, Z., and Wang, H. (2018a). Time discretization of a tempered fractional Feynman-Kac equation with measure data, *SIAM J. Numer. Anal.* **56**, pp. 3249–3275.

Deng, W. H., Li, B. Y., Tian, W. Y., and Zhang, P. W. (2018b). Boundary problems for the fractional and tempered fractional operators, *Multiscale Model. Simul.* **16**, pp. 125–149.

Deng, W. H., Wang, X. D., and Zhang, P. W. (2020). Anisotropic nonlocal diffusion operators for normal and anomalous dynamics, *Multiscale Model. Simul.* **18**, pp. 415–443.

Deng, W. H., Wu, X. C., and Wang, W. L. (2017). Mean exit time and escape probability for the anomalous processes with the tempered power-law waiting times, *Europhys. Lett.* **117**, p. 10009.

Deng, W. H. and Zhang, Z. J. (2017). Numerical schemes of the time tempered fractional Feynman-Kac equation, *Comput. Math. Appl.* **73**, pp. 1063–1076.

Denisov, S. I., Horsthemke, W., and Hänggi, P. (2009). Generalized Fokker-Planck equation: derivation and exact solutions, *Eur. Phys. J. B* **68**, pp. 567–575.

Dhar, A. and Majumdar, S. N. (1999). Residence time distribution for a class of Gaussian Markov processes, *Phys. Rev. E* **59**, pp. 6413–6418.

Dieterich, P., Klages, R., and Chechkin, A. V. (2015). Fluctuation relations for anomalous dynamics generated by time-fractional Fokker-Planck equations, *New J. Phys.* **17**, p. 075004.

Duo, S. W., van Wyk, H. W., and Zhang, Y. Z. (2018). A novel and accurate finite difference method for the fractional Laplacian and the fractional Poisson problem, *J. Comput. Phys.* **355**, pp. 233–252.

Duo, S. W. and Zhang, Y. Z. (2019a). Accurate numerical methods for two and three dimensional integral fractional Laplacian with applications, *Comput. Methods Appl. Mech. Engrg.* **355**, pp. 639–662.

Duo, S. W. and Zhang, Y. Z. (2019b). Numerical approximations for the tempered fractional Laplacian: error analysis and applications, *J. Sci. Comput.* **81**, pp. 569–593.

e Silva, M. S., Stuhrmann, B., Betz, T., and Koenderink, G. H. (2014). Time-resolved microrheology of actively remodeling actomyos in networks, *New J. Phys.* **16**, p. 075010.

Edwards, S. F. and Wilkinson, D. R. (1982). The surface statstics of a granular aggregate, *Proc. R. Soc. London* **381**, pp. 17–31.

Einstein, A. (1956). *Investigation on the Theory of the Brownian Movement* (Dover, New York).

Embrechts, P. and Maejima, M. (2002). *Selfsimilar Processes* (Princeton University Press, Princeton).

Erdelyi, A. (1981). *Higher Transcendental Functions* (Krieger Publishing Company, Malabar).

Eule, S. and Friedrich, R. (2009). Subordinated Langevin equations for anomalous diffusion in external potentials–biasing and decoupled external forces, *Europhys. Lett.* **86**, pp. 30008–30013.

Eule, S., Friedrich, R., Jenko, F., and Kleinhans, D. (2007). Langevin approach to fractional diffusion equations including inertial effects, *J. Phys. Chem. B* **111**, pp. 11474–11477.

Eule, S., Zaburdaev, V., Friedrich, R., and Geisel, T. (2012). Langevin description of superdiffusive Lévy processes, *Phys. Rev. E* **86**, p. 041134.

Feller, W. (1971). *An Introduction to Probability Theory and its Applications* (John Wiley & Sons, New York).

Flajolet, P. and Louchard, G. (2001). Analytic variations on the airy distribution, *Algorithmica* **31**, pp. 361–377.

Flajolet, P., Poblete, P., and Viola, A. (1998). On the analysis of linear probing hashing, *Algorithmica* **22**, pp. 490–515.

Fogedby, H. C. (1994). Langevin equations for continuous time Lévy flights, *Phys. Rev. E* **50**, pp. 1657–1660.

Friedrich, R., Jenko, F., Baule, A., and Eule, S. (2006a). Anomalous diffusion of inertial, weakly damped particles, *Phys. Rev. Lett.* **96**, p. 230601.

Friedrich, R., Jenko, F., Baule, A., and Eule, S. (2006b). Exact solution of a generalized Kramers-Fokker-Planck equation retaining retardation effects, *Phys. Rev. E* **74**, p. 041103.

Froemberg, D. and Barkai, E. (2013a). No-go theorem for ergodicity and an Einstein relation, *Phys. Rev. E* **88**, p. 024101.

Froemberg, D. and Barkai, E. (2013b). Random time averaged diffusivities for Lévy walks, *Eur. Phys. J. B* **86**, p. 331.

Froemberg, D. and Barkai, E. (2013c). Time-averaged Einstein relation and fluctuating diffusivities for the Lévy walk, *Phys. Rev. E* **87**, p. 030104(R).

Fujiwara, M., Sengupta, P., and McIntire, S. L. (2002). Regulation of body size and behavioral state of C. elegans by sensory perception and the EGL-4 cGMP-dependent protein kinase, *Neuron* **36**, pp. 1091–1102.

Gajda, J. and Magdziarz, M. (2010). Fractional Fokker-Planck equation with tempered α-stable waiting times: Langevin picture and computer simulation, *Phys. Rev. E* **82**, p. 011117.

Gajda, J. and Magdziarz, M. (2011). Kramers' escape problem for fractional Klein-Kramers equation with tempered α-stable waiting times, *Phys. Rev. E* **84**, p. 021137.

Gall, J. L. (1991). Brownian excursions, trees and measure-valued branching processes, *Ann. Probab.* **19**, pp. 1399–1439.

Gardiner, C. W. (1983). *Handbook of Stochastic Methods for Physics, Chemistry and the Natural Sciences* (Springer-Verlag, Berlin).

Geman, H. and Yor, M. (1993). Bessel processes, Asian options, and perpetuities, *Math. Finance* **3**, pp. 349–375.

Godec, A. and Metzler, R. (2013). Finite-time effects and ultraweak ergodicity breaking in superdiffusive dynamics, *Phys. Rev. Lett.* **110**, p. 020603.

Godec, A. and Metzler, R. (2017). First passage time statistics for two-channel diffusion, *J. Phys. A* **50**, p. 084001.

Godrèche, C. and Luck, J. M. (2001). Statistics of the occupation time of renewal processes, *J. Stat. Phys.* **104**, pp. 489–524.

Golding, I. and Cox, E. C. (2006). Physical nature of bacterial cytoplasm, *Phys. Rev. Lett.* **96**, p. 098102.

Gradshteyn, I. S., Ryzhik, I. M., Geraniums, Y. V., and Tseytlin, M. Y. (1980). *Table of Integrals, Series, and Products* (Academic Press, USA).

Grebenkov, D. S. (2007). NMR survey of reflected Brownian motion, *Rev. Mod. Phys.* **79**, pp. 1077–1137.

Hapca, S., Crawford, J. W., and Young, I. M. (2009). Anomalous diffusion of heterogeneous populations characterized by normal diffusion at the individual level, *J. R. Soc. Interface* **6**, pp. 111–122.

He, Y., Burov, S., Metzler, R., and Barkai, E. (2008). Random time-scale invariant diffusion and transport coefficients, *Phys. Rev. Lett.* **101**, p. 058101.

Heinsalu, E., Patriarca, M., Goychuk, I., and Hänggi, P. (2007). Use and abuse of a fractional Fokker-Planck dynamics for time-dependent driving, *Phys. Rev. Lett.* **99**, p. 120602.

Hill, S., Burrows, M. T., and Hughes, R. N. (2000). Increased turning per unit distance as an area-restricted search mechanism in a pause-travel predator, juvenile plaice, foraging for buried bivalves, *J. Fish Biol.* **56**, pp. 1497–1508.

Hou, R. and Deng, W. H. (2018). Feynman-Kac equations for reaction and diffusion processes, *J. Phys. A* **51**, p. 155001.

Isaacson, E. and Keller, H. B. (1996). *Analysis of Numerical Methods* (Wiley, New York).

Itô, K. (1950). Stochastic differential equations in a differentiable manifold, *Nagoya Math. J.* **1**, pp. 35–47.

Jain, R. and Sebastian, K. L. (2018). Diffusing diffusivity: Fractional Brownian oscillator model for subdiffusion and its solution, *Phys. Rev. E* **98**, p. 052138.

Jeon, J.-H., Chechkin, A. V., and Metzler, R. (2014). Scaled Brownian motion: a paradoxical process with a time dependent diffusivity for the description of anomalous diffusion, *Phys. Chem. Chem. Phys.* **16**, pp. 15811–15817.

Jeon, J.-H. and Metzler, R. (2010). Fractional Brownian motion and motion governed by the fractional Langevin equation in confined geometries, *Phys. Rev. E* **81**, p. 021103.

Jeon, J.-H. and Metzler, R. (2012). Inequivalence of time and ensemble averages in ergodic systems: exponential versus power-law relaxation in confinement, *Phys. Rev. E* **85**, p. 021147.

Jeon, J.-H., Tejedor, V., Burov, S., Barkai, E., Selhuber-Unkel, C., Berg-Sørensen, K., Oddershede, L., and Metzler, R. (2011). In vivo anomalous diffusion and weak ergodicity breaking of lipid granules, *Phys. Rev. Lett.* **106**, p. 048103.

Jespersen, S., Metzler, R., and Fogedby, H. C. (1999). Lévy flights in external force fields: Langevin and fractional Fokker-Planck equations and their solutions, *Phys. Rev. E* **59**, pp. 2736–2745.

Jin, B. T., Lazarov, R., and Zhou, Z. (2016). Two fully discrete schemes for fractional diffusion and diffusion-wave equations with nonsmooth data, *SIAM J. Sci. Comput.* **38**, pp. A146–A170.

Kac, M. (1949). On distributions of certain Wiener functionals, *Trans. Amer. Math. Soc.* **65**, pp. 1–13.

Kac, M. (1951). *On Some Connections between Probability Theory and Differential and Integral Equations* (University of California Press, Berkeley).

Karatsas, I. and Shreve, S. (1977). *Brownian Motion and Stochastic Calculus* (Springer, New York).

Kardar, M., Parisi, G., and Zhang, Y.-C. (1986). Dynamical scaling of growing interfaces, *Phys. Rev. Lett.* **56**, pp. 889–892.

Karlin, S. and Taylor, H. M. (1981). *A Second Course in Stochastic Processes* (Academic Press, New York).

Kepten, E., Bronshtein, I., and Garini, Y. (2011). Ergodicity convergence test suggests telomere motion obeys fractional dynamics, *Phys. Rev. E* **83**, p. 041919.

Klafter, J., Blumen, A., and Shlesinger, M. F. (1987). Stochastic pathway to anomalous diffusion, *Phys. Rev. A* **35**, pp. 3081–3085.

Klafter, J., Blumen, A., Shlesinger, M. F., and Zumofen, G. (1990). Reply to "Comment on 'Stochastic pathway to anomalous diffusion' ", *Phys. Rev. A* **41**, pp. 1158–1159.

Klafter, J. and Shlesinger, M. F. (1986). On the relationship among three theories of relaxation in disordered systems, *Proc. Natl. Acad. Sci. USA* **83**, pp. 848–851.

Klafter, J. and Silbey, R. (1980). Derivation of the continuous-time random-walk equation, *Phys. Rev. Lett.* **44**, pp. 55–58.

Klafter, J. and Sokolov, I. M. (2011). *First Steps in Random Walks From Tools to Applications* (Oxford University Press, New York).

Klafter, J. and Zumofen, G. (1994). Lévy statistics in a Hamiltonian system, *Phys. Rev. E* **49**, pp. 4873–4877.

Kleinhans, D. and Friedrich, R. (2007). Continuous-time random walks: simulation of continuous trajectories, *Phys. Rev. E* **76**, p. 061102.

Klimontovich, Y. L. (1990). Ito, stratonovich and kinetic forms of stochastic equations, *Physica A* **163**, pp. 515–532.

Kou, S. C. and Xie, X. S. (2004). Generalized Langevin equation with fractional Gaussian noise: subdiffusion within a single protein molecule, *Phys. Rev. Lett.* **93**, p. 180603.

Kramer, D. L. and McLaughlin, R. L. (2015). The behavioral ecology of intermittent locomotion, *American Zoologist* **41**, pp. 137–153.

Kramers, H. A. (1940). Brownian motion in a field of force and the diffusion model of chemical reactions, *Physica* **7**, pp. 284–304.

Kubo, R. (1966). The fluctuation-dissipation theorem, *Rep. Prog. Phys.* **29**, pp.

255–284.

Kühn, T., Ihalainen, T. O., Hyväluoma, J., Dross, N., Willman, S. F., Langowski, J., Vihinen-Ranta, M., and Timonen, J. (2011). Protein diffusion in mammalian cell cytoplasm, *PLoS ONE* **6**, p. e22962.

Kumar, A. and Vellaisamy, P. (2015). Inverse tempered stable subordinators, *Statist. Probab. Lett.* **103**, pp. 134–141.

Langevin, P. (1908). On the theory of Brownian motion, *C. R. Acad. Sci.* **146**, pp. 530–533.

Le Vot, F., Abad, E., and Yuste, S. B. (2017). Continuous-time random-walk model for anomalous diffusion in expanding media, *Phys. Rev. E* **96**, p. 032117.

Leibovich, N. and Barkai, E. (2019). Infinite ergodic theory for heterogeneous diffusion processes, *Phys. Rev. E* **99**, p. 042138.

Lenk, R. and Gellert, W. (1974). *Fachlexikon Physik* (Deutsch, Zürich).

Lévy, P. (1939). Sur certains processus stochastiques homogènes, *Compos. Math.* **7**, pp. 283–339.

Li, C., Deng, W. H., and Zhao, L. J. (2019). Well-posedness and numerical algorithm for the tempered fractional differential equations, *Discrete Contin. Dyn. Syst. Ser. B* **24**, pp. 1989–2015.

Li, L., Nørrelykke, S. F., and Cox, E. C. (2008). Persistent cell motion in the absence of external signals: a search strategy for eukaryotic cells, *PLoS ONE* **3**, p. e2093.

Liemert, A., Sandev, T., and Kantz, H. (2017). Generalized Langevin equation with tempered memory kernel, *Physica A* **466**, pp. 356–369.

Lim, S. C. and Muniandy, S. V. (2002). Self-similar Gaussian processes for modeling anomalous diffusion, *Phys. Rev. E* **66**, p. 021114.

Lomholt, M. A., Koren, T., Metzler, R., and Klafter, J. (2008). Lévy strategies in intermittent search processes are advantageous, *Proc. Natl. Acad. Sci. USA* **105**, pp. 11055–11059.

Louchard, G. (1984). Kac's formula, Lévy's local time and Brownian excursion, *J. Appl. Prob.* **21**, pp. 479–499.

Lubelski, A., Sokolov, I. M., and Klafter, J. (2008). Nonergodicity mimics inhomogeneity in single particle tracking, *Phys. Rev. Lett.* **100**, p. 250602.

Lubich, C. (1988a). Convolution quadrature and discretized operational calculus. I, *Numer. Math.* **52**, pp. 129–145.

Lubich, C. (1988b). Convolution quadrature and discretized operational calculus. II, *Numer. Math.* **52**, pp. 413–425.

Lubich, C. (2004). Convolution quadrature revisited, *BIT* **44**, pp. 503–514.

Lubich, C., Sloan, I. H., and Thomée, V. (1996). Nonsmooth data error estimates for approximations of an evolution equation with a positive-type memory term, *Math. Comp.* **65**, pp. 1–17.

Lutz, E. (2001). Fractional Langevin equation, *Phys. Rev. E* **64**, p. 051106.

Maćkała, A. and Magdziarz, M. (2019). Statistical analysis of superstatistical fractional Brownian motion and applications, *Phys. Rev. E* **99**, p. 012143.

Maeda, Y. T., Tlusty, T., and Libchaber, A. (2012). Effects of long DNA folding and small RNA stem-loop in thermophoresis, *Proc. Natl. Acad. Sci. USA*

109, pp. 17972–17977.

Magdziarz, M. (2009). Langevin picture of subdiffusion with infinitely divisible waiting times, *J. Stat. Phys.* **135**, pp. 763–772.

Magdziarz, M. and Weron, A. (2007). Competition between subdiffusion and Lévy flights: a Monte Carlo approach, *Phys. Rev. E* **75**, p. 056702.

Magdziarz, M., Weron, A., and Klafter, J. (2008). Equivalence of the fractional Fokker-Planck and subordinated Langevin equations: the case of a time-dependent force, *Phys. Rev. Lett.* **101**, p. 210601.

Magdziarz, M., Weron, A., and Weron, K. (2007). Fractional Fokker-Planck dynamics: stochastic representation and computer simulation, *Phys. Rev. E* **75**, p. 016708.

Magdziarz, M. and Zorawik, T. (2017). Aging ballistic Lévy walks, *Phys. Rev. E* **95**, p. 022126.

Majumdar, S. N. (2005). Brownian functionals in physics and computer science, *Curr. Sci.* **89**, pp. 2076–2092.

Majumdar, S. N. and Comtet, A. (2002). Local and occupation time of a particle diffusing in a random medium, *Phys. Rev. Lett.* **89**, p. 060601.

Majumdar, S. N. and Comtet, A. (2005). Airy distribution function: from the area under a Brownian excursion to the maximal height of fluctuating interfaces, *J. Stat. Phys.* **119**, pp. 777–826.

Majumdar, S. N., Randon-Furling, J., Kearney, M. J., and Yor, M. (2008). On the time to reach maximum for a variety of constrained Brownian motions, *J. Phys. A* **41**, p. 365005.

Mandelbrot, B. B. and Ness, J. W. V. (1968). Fractional Brownian motions, fractional noises and applications, *SIAM Rev.* **10**, pp. 422–437.

Mast, C. B., Schink, S., Gerland, U., and Braun, D. (2013). Escalation of polymerization in a thermal gradient, *Proc. Natl. Acad. Sci. USA* **110**, pp. 8030–8035.

Mathai, A. M. and Saxena, R. K. (1978). *The H-Function with Applications in Statistics and Other Disciplines* (Wiley Eastern, New Delhi).

Mathai, A. M., Saxena, R. K., and Haubold, H. J. (2009). *The H-Function, Theory and Applications* (Springer, Berlin).

McConnell, J. R. (1980). *Rotational Brownian Motion and Dielectric Theory* (Academic Press, New York).

Meerschaert, M. M., Benson, D. A., and Bäumer, B. (1999). Multidimensional advection and fractional dispersion, *Phys. Rev. E* **59**, pp. 5026–5028.

Meerschaert, M. M. and Sabzikar, F. (2013). Tempered fractional Brownian motion, *Stat. Probab. Lett.* **83**, pp. 2269–2275.

Meerschaert, M. M. and Sikorskii, A. (2011). *Stochastic Models for Fractional Calculus* (De Gruyter, Germany).

Metzler, R. (2000). Generalized Chapman-Kolmogorov equation: a unifying approach to the description of anomalous transport in external fields, *Phys. Rev. E* **62**, pp. 6233–6245.

Metzler, R. (2001). Non-homogeneous random walks, generalised master equations, fractional Fokker-Planck equations, and the generalised Kramers-Moyal expansion, *Eur. Phys. J. B* **19**, pp. 249–258.

Metzler, R., Barkai, E., and Klafter, J. (1999a). Anomalous diffusion and relaxation close to thermal equilibrium: a fractional Fokker-Planck equation approach, *Phys. Rev. Lett.* **82**, pp. 3563–3567.

Metzler, R., Barkai, E., and Klafter, J. (1999b). Deriving fractional Fokker-Planck equations from a generalised master equation, *Europhys. Lett.* **46**, pp. 431–436.

Metzler, R., Jeon, J.-H., Cherstvy, A. G., and Barkai, E. (2014). Anomalous diffusion models and their properties: non-stationarity, non-ergodicity, and ageing at the centenary of single particle tracking, *Phys. Chem. Chem. Phys.* **16**, pp. 24128–24164.

Metzler, R. and Klafter, J. (2000a). From a generalized Chapman-Kolmogorov equation to the fractional Klein-Kramers equation, *J. Phys. Chem. B* **104**, pp. 3851–3857.

Metzler, R. and Klafter, J. (2000b). The random walk's guide to anomalous diffusion: a fractional dynamics approach, *Phys. Rep.* **339**, pp. 1–77.

Metzler, R. and Klafter, J. (2000c). Subdiffusive transport close to thermal equilibrium: from the Langevin equation to fractional diffusion, *Phys. Rev. E* **61**, pp. 6308–6311.

Meyer, P., Barkai, E., and Kantz, H. (2017). Scale-invariant Green-Kubo relation for time-averaged diffusivity, *Phys. Rev. E* **96**, p. 062122.

Min, W., Luo, G., Cherayil, B. J., Kou, S. C., and Xie, X. S. (2005). Observation of a power-law memory kernel for fluctuations within a single protein molecule, *Phys. Rev. Lett.* **94**, p. 198302.

Molina-Garcia, D., Sandev, T., Safdari, H., Pagnini, G., Chechkin, A., and Metzler, R. (2018). Crossover from anomalous to normal diffusion: truncated power-law noise correlations and applications to dynamics in lipid bilayers, *New J. Phys.* **20**, p. 103027.

Montroll, E. W. and Weiss, G. H. (1965). Random walks on lattices. II, *J. Math. Phys.* **6**, pp. 167–181.

Naeh, T., Klosek, M. M., Matkowsky, B. J., and Schuss, Z. (1990). A direct approach to the exit problem, *SIAM J. Appl. Math.* **50**, pp. 595–627.

Nelson, J. (1999). Continuous-time random-walk model of electron transport in nanocrystalline TiO2 electrodes, *Phys. Rev. B* **59**, pp. 15374–15380.

Neusius, T., Sokolov, I. M., and Smith, J. C. (2009). Subdiffusion in time-averaged, confined random walks, *Phys. Rev. E* **80**, p. 011109.

Nie, D. X., Sun, J., and Deng, W. H. (2019). Numerical algorithms of the two-dimensional Feynman-Kac equation for reaction and diffusion processes, *J. Sci. Comput.* **81**, pp. 537–568.

Niemann, M., Barkai, E., and Kantz, H. (2006). Renewal theory for a system with internal states, *Math. Model. Nat. Phenom.* **11**, pp. 191–293.

O'Brien, W. J., Browman, H. I., and Evans, B. I. (1990). Search strategies of foraging animals, *Am. Sci.* **78**, pp. 152–160.

Øksendal, B. (2005). *Stochastic Differential Equations* (Springer-Verlag, Berlin).

Paradisi, P., Kaniadakis, G., and Scarfone, A. M. (2015). The emergence of self-organization in complex systems-Preface, *Chaos Soliton. Fract.* **81**, pp. 407–411.

Pfister, G. and Scher, H. (1978). Dispersive (non-Gaussian) transient transport in disordered solids, *Adv. Phys.* **27**, pp. 747–798.

Pierce-Shimomura, J. T., Morse, T. M., and Lockery, S. R. (1999). The fundamental role of pirouettes in Caenorhabditis elegans chemotaxis, *J. Neurosci.* **19**, pp. 9557–9569.

Piryatinska, A., Saichev, A. I., and Woyczynski, W. A. (2005). Models of anomalous diffusion: the subdiffusive case, *Physica A* **349**, pp. 375–420.

Podlubny, I. (1999). *Fractional Differential Equations* (Academic, San Diego).

Pollak, E. and Talkner, P. (1993). Activated rate processes: Finite-barrier expansion for the rate in the spatial-diffusion limit, *Phys. Rev. E* **47**, pp. 922–933.

Polyanin, A. D., Zaitsev, V. F., and Moussiaux, A. (2002). *Handbook of First-order Partial Differential Equations* (Taylor & Francis, London).

Porrà, J. M., Wang, K.-G., and Masoliver, J. (1996). Generalized Langevin equations: anomalous diffusion and probability distributions, *Phys. Rev. E* **53**, pp. 5872–5881.

Prabhakar, T. R. (1971). A singular integral equation with a generalized Mittag Leffler function in the kernel, *Yokohama Math. J.* **19**, pp. 7–15.

Rebenshtok, A., Denisov, S., Hänggi, P., and Barkai, E. (2014a). Infinite densities for Lévy walks, *Phys. Rev. E* **90**, p. 062135.

Rebenshtok, A., Denisov, S., Hänggi, P., and Barkai, E. (2014b). Non-normalizable densities in strong anomalous diffusion: beyond the central limit theorem, *Phys. Rev. Lett.* **112**, p. 110601.

Rebenshtok, A., Denisov, S., Hänggi, P., and Barkai, E. (2016). Complementary densities of Lévy walks: typical and rare fluctuations, *Math. Model. Nat. Phenom.* **11**, pp. 76–106.

Redner, S. (2001). *A Guide to First-Passage Processes* (Cambridge University Press, Cambridge).

Revuz, D. and Yor, M. (1990). *Continuous Martingales and Brownian Motion* (Springer-Verlag, Berlin).

Risken, H. (1989). *The Fokker-Planck Equation* (Springer-Verlag, Berlin).

Sadhu, T., Delorme, M., and Wiese, K. J. (2018). Generalized arcsine laws for fractional Brownian motion, *Phys. Rev. Lett.* **120**, p. 040603.

Safdari, H., Cherstvy, A. G., Chechkin, A. V., Thiel, F., Sokolov, I. M., and Metzler, R. (2015). Quantifying the non-ergodicity of scaled Brownian motion, *J. Phys. A* **48**, p. 375002.

Samanta, N. and Chakrabarti, R. (2016). Tracer diffusion in a sea of polymers with binding zones: Mobile vs. frozen traps, *Soft Matter* **12**, pp. 8554–8563.

Samko, S. G., Kilbas, A. A., and Marichev, O. I. (1993). *Fractional Integrals and Derivatives: Theory and Applications* (Gordon and Breach, New York).

Samorodnitsky, G. and Taqqu, M. (1994). *Stable Non-Gaussian Random Processes: Stochastic Models with Infinite Variance* (Chapman and Hall, New York).

Sanders, D. P. and Larralde, H. (2006). Occurrence of normal and anomalous diffusion in polygonal billiard channels, *Phys. Rev. E* **73**, p. 026205.

Schehr, G. and Le-Doussal, P. (2010). Extreme value statistics from the real space renormalization group: Brownian motion, Bessel processes and continuous time random walks, *J. Stat. Mech.*, p. P01009.

Scher, H. and Lax, M. (1972). Continuous time random walk model of hopping transport: application to impurity conduction, *J. Non-Crystalline Solids* **8-10**, pp. 497–504.

Scher, H. and Lax, M. (1973a). Stochastic transport in a disordered solid. I. Theory, *Phys. Rev. B* **7**, pp. 4491–4502.

Scher, H. and Lax, M. (1973b). Stochastic transport in a disordered solid. II. Impurity conduction, *Phys. Rev. B* **7**, pp. 4502–4519.

Scher, H. and Montroll, E. W. (1975). Anomalous transit-time dispersion in amorphous solids, *Phys. Rev. B* **12**, pp. 2455–2477.

Schertzer, D., Larchevêque, M., Duan, J., Yanovsky, V. V., and Lovejoy, S. (2001). Fractional Fokker-Planck equation for nonlinear stochastic differential equations driven by non-Gaussian Lévy stable noises, *J. Math. Phys.* **42**, pp. 200–212.

Schulz, J. H. P., Barkai, E., and Metzler, R. (2014). Aging renewal theory and application to random walks, *Phys. Rev. X* **4**, p. 011028.

Shlesinger, M. F., Zaslavsky, G. M., and Frisch, U. (1995). *Lévy Flights and Related Topics* (Springer-Verlag, Berlin).

Sinai, Y. G. (1983). The limiting behavior of a one-dimensional random walk in a random medium, *Theor. Probab. Appl.* **27**, pp. 247–258.

Ślęzak, J., Metzler, R., and Magdziarz, M. (2018). Superstatistical generalised langevin equation: non-gaussian viscoelastic anomalous diffusion, *New J. Phys.* **20**, p. 023026.

Sokolov, I. M., Heinsalu, E., Hänggi, P., and Goychuk, I. (2009). Universal fluctuations in subdiffusive transport, *Europhys. Lett.* **86**, p. 30009.

Sokolov, I. M. and Klafter, J. (2006). Field-induced dispersion in subdiffusion, *Phys. Rev. Lett.* **97**, p. 140602.

Solomon, T. H., Weeks, E. R., and Swinney, H. L. (1993). Observation of anomalous diffusion and Lévy flights in a two-dimensional rotating flow, *Phys. Rev. Lett.* **71**, pp. 3975–3978.

Song, M. S., Moon, H. C., Jeon, J.-H., and Park, H. Y. (2018). Neuronal messenger ribonucleoprotein transport follows an aging Lévy walk, *Nat. Commun.* **9**, pp. 344–351.

Sposini, V., Chechkin, A. V., Seno, F., Pagnini, G., and Metzler, R. (2018). Random diffusivity from stochastic equations: comparison of two models for Brownian yet non-Gaussian diffusion, *New J. Phys.* **20**, p. 043044.

Srivastava, H. M., Gupta, K. C., and Goyal, S. P. (1982). *The H-Functions of One and Two Variables with Applications* (South Asian Publishers, New Delhi).

Srivastava, H. M. and Kashyap, B. R. K. (1982). *Special Functions in Queuing Theory and Related Stochastic Processes* (Academic Press, New York).

Stone, L. D. (1975). *Theory of Optimal Search* (Academic Press, New York).

Stratonovich, R. (1966). A new representation for stochastic integrals and equations, *J. SIAM Control* **4**, pp. 362–371.

Takacs, L. (1995). Limit distributions for the Bernoulli meander, *J. Appl. Prob.*

32, pp. 375–395.

Tejedor, V., Bénichou, O., Voituriez, R., Jungmann, R., Simmel, F., Selhuber-Unkel, C., Oddershede, L. B., and Metzler, R. (2010). Quantitative analysis of single particle trajectories: mean maximal excursion method, *Biophys. J.* **98**, pp. 1364–1372.

Thaler, M. and Zweimüller, R. (2006). Distributional limit theorems in infinite ergodic theory, *Probab. Theory Rel.* **135**, pp. 15–52.

Thiel, F. and Sokolov, I. M. (2014). Scaled Brownian motion as a mean-field model for continuous-time random walks, *Phys. Rev. E* **89**, p. 012115.

Toyota, T., Head, D. A., Schmidt, C. F., and Mizuno, D. (2011). Non-Gaussian athermal fluctuations in active gels, *Soft Matter* **7**, pp. 3234–3239.

Tschoegl, N. W. (1989). *The Phenomenological Theory of Linear Viscoelastic Behavior* (Springer, Heidelberg).

Turgeman, L., Carmi, S., and Barkai, E. (2009). Fractional Feynman-Kac equation for non-Brownian functionals, *Phys. Rev. Lett.* **103**, p. 190201.

Vahabi, M., Schulz, J. H. P., Shokri, B., and Metzler, R. (2013). Area coverage of radial Lévy flights with periodic boundary conditions, *Phys. Rev. E* **87**, p. 042136.

van Kampen, N. G. (1992). *Stochastic Processes in Physics and Chemistry* (North-Holland, Amsterdam).

von Smoluchowski, M. (1906). Zur kinetischen theorie der brownschen molekularbewegung und der suspensionen, *Ann. Phys.* **326**, pp. 756–780.

Wang, B., Anthony, S. M., Bae, S. C., and Granick, S. (2009). Anomalous yet Brownian, *Proc. Natl. Acad. Sci. USA* **106**, pp. 15160–15164.

Wang, B., Kuo, J., Bae, S. C., and Granick, S. (2012). When Brownian diffusion is not Gaussian, *Nat. Mater.* **11**, pp. 481–485.

Wang, W., Cherstvy, A. G., Chechkin, A. V., Thapa, S., Seno, F., Liu, X., and Metzler, R. (2020a). Fractional Brownian motion with random diffusivity: Emerging residual nonergodicity below the correlation time, *J. Phys. A* **53**, p. 474001.

Wang, W., Cherstvy, A. G., Liu, X., and Metzler, R. (2020b). Anomalous diffusion and nonergodicity for heterogeneous diffusion processes with fractional Gaussian noise, *Phys. Rev. E* **102**, p. 474001.

Wang, W. L., Barkai, E., and Burov, S. (2020c). Large deviations for continuous time random walks, *Entropy* **22**, p. 697.

Wang, W. L. and Deng, W. H. (2018). Aging Feynman-Kac equation, *J. Phys. A* **51**, p. 015001.

Wang, X. D. and Chen, Y. (2021). Ergodic property of Langevin systems with superstatistical, uncorrelated or correlated diffusivity, *Physica A* **577**, p. 126090.

Wang, X. D., Chen, Y., and Deng, W. H. (2018). Feynman-Kac equation revisited, *Phys. Rev. E* **98**, p. 052114.

Wang, X. D., Chen, Y., and Deng, W. H. (2019a). Aging two-state process with Lévy walk and Brownian motion, *Phys. Rev. E* **100**, p. 012136.

Wang, X. D., Chen, Y., and Deng, W. H. (2019b). Lévy-walk-like Langevin dynamics, *New J. Phys.* **21**, p. 013024.

Wang, X. D., Deng, W. H., and Chen, Y. (2019c). Ergodic properties of heterogeneous diffusion processes in a potential well, *J. Chem. Phys.* **150**, p. 164121.

Weber, S. C., Spakowitz, A. J., and Theriot, J. A. (2010). Bacterial chromosomal loci move subdiffusively through a viscoelastic cytoplasm, *Phys. Rev. Lett.* **104**, p. 238102.

Weron, A., Magdziarz, M., and Weron, K. (2008). Modeling of subdiffusion in space-time-dependent force fields beyond the fractional Fokker-Planck equation, *Phys. Rev. E* **77**, p. 036704.

West, B. J., Bulsara, A. R., Lindenberg, K., Seshadri, V., and Shuler, K. E. (1979). Stochastic processes with non-additive fluctuations, *Physica A* **97**, pp. 211–233.

Winter, R. B. and von Hippel, P. H. (1981). Diffusion-driven mechanisms of protein translocation on nucleic acids. 2. The Escherichia coli lac repressor-operator interaction: equilibrium measurements, *Biochemistry* **20**, pp. 6948–6960.

Wolf, K. B. (1979). *Integral Transforms in Science and Engineering* (Plenum Press, New York).

Wu, X. C., Deng, W. H., and Barkai, E. (2016). Tempered fractional Feynman-Kac equation: theory and examples, *Phys. Rev. E* **93**, p. 032151.

Xu, P. B. and Deng, W. H. (2018a). Fractional compound Poisson processes with multiple internal states, *Math. Model. Nat. Phenom* **13**, p. 10.

Xu, P. B. and Deng, W. H. (2018b). Lévy walk with multiple internal states, *J. Stat. Phys.* **173**, pp. 1598–1613.

Yor, M. (2000). *Exponential Functionals of Brownian Motion and Related Processes* (Springer, Berlin).

Yuste, S. B., Abad, E., and Escudero, C. (2016). Diffusion in an expanding medium: Fokker-Planck equation, Green's function, and first-passage properties, *Phys. Rev. E* **94**, p. 032118.

Zaburdaev, V., Denisov, S., and Klafter, J. (2015). Lévy walks, *Rev. Mod. Phys.* **87**, pp. 483–530.

Zaburdaev, V., Schmiedeberg, M., and Stark, H. (2008). Random walks with random velocities, *Phys. Rev. E* **78**, p. 011119.

Zhang, Z. J. and Deng, W. H. (2017). Numerical approaches to the functional distribution of anomalous diffusion with both traps and flights, *Adv. Comput. Math.* **43**, pp. 699–732.

Zhang, Z. J., Deng, W. H., and Karniadakis, G. E. (2018). A Riesz basis Galerkin method for the tempered fractional Laplacian, *SIAM J. Numer. Anal.* **56**, pp. 3010–3039.

Zhang, Z. J., Deng, W. H., and Tao, F. H. (2019). Finite difference schemes for the tempered fractional Laplacian, *Numer. Math. Theor. Meth. Appl.* **12**, pp. 492–516.

Zhou, H. X., Rivas, G., and Minton, A. P. (2008). Macromolecular crowding and confinement: biochemical, biophysical, and potential physiological consequences, *Annual Rev. Biophys.* **37**, pp. 375–397.

Zimmerman, S. B. and Minton, A. P. (1993). Macromolecular crowding: biochemical, biophysical, and physiological consequences, *Annu. Rev. Biophys. Biomol. Struct.* **22**, pp. 27–65.

Zumofen, G. and Klafter, J. (1993). Scale-invariant motion in intermittent chaotic systems, *Phys. Rev. E* **47**, pp. 851–863.

Index